Gagan Bansal, Yashvir Singh (Eds.)
Biotribology

Green Tribology and Tribochemistry

Edited by
Yashvir Singh, Nishant Kumar Singh

Volume 2

Biotribology

Friction, Wear and Lubrication in Biological Systems

Edited by
Gagan Bansal and Yashvir Singh

DE GRUYTER

Editors
Dr Gagan Bansal
Department of Mechanical Engineering
Graphic Era (Deemed to be University)
Dehradun 248002
Uttarakhand
India
gaganbansal12345@gmail.com

Dr Yashvir Singh
Department of Mechanical Engineering
Harcourt Butler Technical University
Kanpur 208002
Uttar Pradesh
India
yashvirsingh21@gmail.com

ISBN 978-3-11-156294-0
e-ISBN (PDF) 978-3-11-156317-6
e-ISBN (EPUB) 978-3-11-156322-0

Library of Congress Control Number: 2025945634

Bibliographic information published by the Deutsche Nationalbibliothek
The Deutsche Nationalbibliothek lists this publication in the Deutsche Nationalbibliografie; detailed bibliographic data are available on the Internet at http://dnb.dnb.de.

Cover image: Azulillo/iStock/Getty Images Plus
Typesetting: Integra Software Services Pvt. Ltd.
Printing and binding: CPI books GmbH, Leck

www.degruyterbrill.com
Questions about General Product Safety Regulation:
productsafety@degruyterbrill.com

Preface

The field of **biotribology** represents an essential bridge between engineering and biology, providing a scientific framework to understand how friction, wear, and lubrication phenomena occur in living systems. From the smooth motion of human joints to the durability of implants in biomedical applications, tribological behavior plays a fundamental role in health, mobility, and the design of life-saving technologies. The natural world, through millions of years of evolution, has refined complex biological surfaces and mechanisms that manage mechanical interactions with extraordinary efficiency. Studying and emulating these systems can lead to significant advancements in medical device engineering, regenerative medicine, prosthetics, orthopedics, and many other areas.

This book is the outcome of years of academic pursuit, interdisciplinary exploration, and curiosity-driven research. It aims to provide an in-depth and holistic perspective on the principles, mechanisms, and applications of biotribology. The core motivation behind writing this book is to offer students, researchers, and professionals a unified resource that combines foundational tribological theory with its biological implications and real-world biomedical applications.

The initial chapters are devoted to establishing the fundamental principles of friction, wear, and lubrication in the context of biological interfaces. These concepts, though rooted in mechanical engineering, gain new dimensions when applied to complex biological environments, where the materials are often soft, hydrated, viscoelastic, and responsive. Biological tissues such as cartilage, skin, and mucous membranes exhibit intricate surface morphologies and biochemical compositions that make their tribological behavior uniquely adaptive and efficient. Understanding such systems requires a blend of mechanical analysis, material science, cell biology, and physiology – an interdisciplinary challenge that this book attempts to address systematically.

Subsequent chapters delve into specialized topics such as the tribology of synovial joints, the biomechanics of the human skin, tribological characteristics of dental and oral structures, and the influence of biolubricants such as synovial fluid, saliva, and mucus. A particular emphasis is placed on **wear mechanisms** in biological systems and their impact on cellular response and inflammation, especially in the context of orthopedic and dental implants. The interaction of implants with surrounding biological tissues, and the role of surface coatings, nano-engineering, and hydrogels in enhancing tribological performance, are thoroughly discussed.

Additionally, the book explores modern experimental techniques and computational tools used to investigate biotribological phenomena. From in vitro simulations to finite element modeling and advanced surface characterization, the integration of technology in understanding biological interfaces is emphasized to equip researchers with practical insights for innovation and development.

https://doi.org/10.1515/9783111563176-202

One of the most dynamic sections of the book discusses **biomimicry in tribology**, highlighting how nature-inspired designs can lead to smarter, more efficient materials and devices. The study of gecko feet, fish scales, insect wings, and other natural surfaces has inspired remarkable engineering breakthroughs in adhesion, drag reduction, and wear resistance – ideas that are increasingly being adapted in biomedical device fabrication.

This book would not have been possible without the support and encouragement of numerous mentors, colleagues, students, and institutions. Their invaluable insights, critical questions, and collaborative spirit provided the necessary motivation to transform scattered ideas and research into a structured academic resource. I also extend my gratitude to the researchers across the world whose pioneering work has shaped the field of biotribology and continues to expand its horizons.

In an age where medical technology is advancing rapidly, the need for biocompatible, durable, and high-performance materials is greater than ever. It is my hope that this book serves as both an educational foundation and a source of inspiration for future exploration in biotribology. Whether you are a student just beginning your journey, a researcher looking for multidisciplinary insights, or a clinician exploring better prosthetic solutions, I believe this book will offer valuable knowledge and direction.

Dr Gagan Bansal
Graphic Era (Deemed to be University), Dehradun

Contents

List of authors

Lazreg Hadji
University of Tiaret
BP 78 Zaaroura, Tiaret 14000, Algeria
lazreg.hadji@univ-tiaret.dz
Chapter 1

Vartika Agarwal
Department of Computer Applications
Graphic Era (Deemed to be University)
Dehradun, Uttarakhand, India
vartikaagarwal2015@gmail.com
Chapters 1, 11, 14, and 15

Pavan Gangwar
Department of Electrical Engineering
United College of Engineering and Research
Prayagraj, Uttar Pradesh, India
pavan.bmas@gmail.com
Chapter 11 and 15

Kartikey Raina
Graphic Era (Deemed to be University)
Dehradun, Uttarakhand, India
Kartikeyaraina@geu.ac.in
Chapter 1

Smriti Tandon Gupta
Graphic Era (Deemed to be University)
Dehradun, Uttarakhand, India
smriti.tandon.mba@geu.ac.in
Chapter 1

Royal Madan
Graphic Era (Deemed to be University)
Dehradun, Uttarakhand, India
royalmadan6293@gmail.com
Chapter 1

Adeeba Mirza
Department of Chemistry
Graphic Era (Deemed to be University)
Dehradun 248002, Uttarakhand, India
adeebamirza4jan@gmail.com
Chapter 2

Ishika Sharma
Department of Chemistry
Graphic Era (Deemed to be University)
Dehradun 248002, Uttarakhand, India
ishikasharm125@gmail.com
Chapter 2

Abhilasha Mishra
Department of Chemistry
Graphic Era (Deemed to be University)
Dehradun 248002, Uttarakhand, India
abhi1680@gmail.com
Chapter 2

Hasan Koten
Department of Mechanical Engineering
Engineering Faculty
University of Medeniyet
Kadıköy 34000, Istanbul, Turkey
hasan.koten@medeniyet.edu.tr
Chapter 2

Annu Rani
G. D. Goenka University
Gurugram, Haryana, India
annuchaudhary39@gmail.com
Chapter 3

T. P. Singh
Department of Farm Machinery and Power Engineering
G. B. Pant University of Agriculture and Technology
Pantnagar, Uttarakhand, India
tpsingh_62@yahoo.co.in
Chapter 3

Raushan Singh
Utah State University
Logan, UT, USA
raushan.singh@usu.edu
Chapter 3

https://doi.org/10.1515/9783111563176-204

Sabra Banu
Department of Pharmacy
JBIT College of Pharmacy
Dehradun, Uttarakhand, India
sabrabanu5969@gmail.com
Chapter 4

Sachin Kumar
School of Pharmacy
Graphic Era Hill University
Dehradun, Uttarakhand, India
srai24304@gmail.com
Chapter 4

Sarla Yadav
Department of Applied Chemistry
Delhi Technological University (formerly Delhi College of Engineering)
Delhi 110042, India
sarlayadav_2k21phdac505@dtu.ac.in
Chapter 5

Gunjan Varshney
Department of Applied Chemistry
Delhi Technological University (formerly Delhi College of Engineering)
Delhi 110042, India
varshneygunjan111@gmail.com
Chapter 5

Pooja Yadav
Department of Applied Chemistry
Delhi Technological University (formerly Delhi College of Engineering)
Delhi 110042, India
pyadav6651@gmail.com
Chapter 5

Riya Malik
Department of Applied Chemistry
Delhi Technological University (formerly Delhi College of Engineering)
Delhi 110042, India
riyamalik3098@gmail.com
Chapter 5

Raminder Kaur
Department of Applied Chemistry
Delhi Technological University (formerly Delhi College of Engineering)
Delhi 110042, India
rkaurdtu@gmail.com
Chapter 5

Neetu Sharama
Department of Chemistry
Graphic Era (Deemed to be University)
Dehradun, Uttarakhand, India
neetu.cherry@gmail.com
Chapter 6

Ved Prakash
Department of Chemistry
Graphic Era (Deemed to be University)
Dehradun, Uttarakhand, India
vedprakashdhayani28@gmail.com
Chapter 6

Bhawana
Department of Chemistry
Graphic Era (Deemed to be University)
Dehradun, Uttarakhand, India
bishtbhawana29@gmail.com
Chapter 6

Diksha
Scientist-B
Department of Microbiology
AIIMS Rishikesh, Uttarakhand, India
rani.diksha06@gmail.com
Chapter 6

Tanuj Kumar Bisht
Shree Siddhi Vinayak Group of Institutes
Bareilly, Uttar Pradesh, India
tanuj.k.bisht@gmail.com
Chapter 6

Lazreg Hadji
Department of Civil Engineering
University of Tiaret
BP 78 Zaaroura, Tiaret 14000, Algeria
lazreg.hadji@univ-tiaret.dz
Chapter 6

Pranjal Pratap Singh
Siddhartha Institute of Pharmacy
Dehradun 248013, Uttarakhand, India
Pranjalsingh3555@gmail.com
Chapter 7

Vishal Chhabra
Graphic Era Hill University
Dehradun 248001, Uttarakhand, India
vishalchhabra25@gmail.com
Chapter 7

Priyadarsini Morampudi
Department of Mechanical Engineering
VNR Vignana Jyothi Institute of Engineering and Technology
Hyderabad 500090, Telangana, India
satyaprasad.sv@gmail.com
Chapters 8 and 12

Venkata Satya Prasad Somayajula
Department of Mechanical Engineering
VNR Vignana Jyothi Institute of Engineering and Technology
Hyderabad 500090, Telangana, India
satyaprasad.sv@gmail.com
Chapters 8 and 12

P. Rajasekar
Department of Biotechnology
Rajalakshmi Engineering College
Chennai, Tamil Nadu, India
rajasekar.panchamurthy@gmail.com
Chapter 9

Swetha P.
Department of Biotechnology
Rajalakshmi Engineering College
Chennai, Tamil Nadu, India
swepandiaraj@gmail.com
Chapter 9

Suwathi S.
Department of Biotechnology
Rajalakshmi Engineering College
Chennai, Tamil Nadu, India
suwathisuresh31@gmail.com
Chapter 9

S. Rishikesavan
Department of Biotechnology
Rajalakshmi Engineering College
Chennai, Tamil Nadu, India
sarishikesavan@gmail.com
Chapter 9

Rekha Goswami
Department of Environmental Sciences
Graphic Era Hill University
Dehradun, Uttarakhand, India
rgoswami.2324@gmail.com
Chapter 10

Sargam Bahl
School of Management Studies
Graphic Era Hill University
Dehradun, Uttarakhand, India
bahl.sargam@gmail.com
Chapter 10

Naveen Negi
School of Management Studies
Graphic Era Hill University
Dehradun, Uttarakhand, India
nnegi@gehu.ac.in
Chapter 10

Kamal Kant Joshi
Department of Environmental Sciences
Graphic Era Hill University
Dehradun, Uttarakhand, India
kkjoshi@gehu.ac.in
Chapter 10

Abdul Jabbar Ansari
Department of Mechanical Engineering
Bansal Institute of Engineering and Technology
Lucknow 226201, Uttar Pradesh, India
techno.abdul@gmail.com
Chapters 11 and 15

Gagan Bansal
Department of Mechanical Engineering
Graphic Era (Deemed to be University)
Dehradun 24800, Uttarakhand, India
gaganbansal12345@gmail.com
Chapters 11, 14, and 15

Emad Kadum Njim
Ministry of Industry and Minerals
State Company for Rubber and Tires Industries
Najaf, Iraq
emad.njim@gmail.com
Chapter 11

Dr M. Karthik
Department of Electrical and Electronics Engineering
SRM Madurai College for Engineering and Technology
Sivaganga, Tamil Nadu, India
kartik.m1992@gmail.com
Chapter 13

Ashwani Kharola
Department of Mechanical Engineering
Graphic Era (Deemed to be University)
Dehradun 248001, Uttarakhand, India
ashwanikharola.me@geu.ac.in
Chapter 14

Lazreg Hadji, Vartika Agarwal Pavan Gangwar, Kartikey Raina, Smriti Tandon Gupta, and Royal Madan

1 Introduction to biotribology

Abstract: Biotribology, the study of friction, wear, and lubrication in biological systems, plays a pivotal role in understanding the mechanical interactions within the human body and between biological tissues and artificial implants. Biotribology, an interdisciplinary field at the intersection of biology, materials science, and mechanical engineering, investigates the principles of friction, wear, and lubrication within biological systems. This book presents a comprehensive exploration of biotribology across 15 chapters, covering foundational concepts, specialized applications, and emerging challenges. The introductory chapter sets the stage by defining biotribology and establishing its significance in the development of medical devices and prosthetics. Following this, the fundamentals of biological surfaces are examined to highlight their unique structural and chemical features that influence tribological behavior.

Chapters on friction, wear, and lubrication in biological systems delve into the core mechanisms that govern joint mobility, cellular interactions, and tissue preservation under physiological loads. The tribological behavior of critical anatomical structures – such as articular cartilage, dental tissues, and soft tissues – is analyzed to demonstrate how nature has evolved highly efficient systems for managing motion and minimizing damage. Special emphasis is placed on the biotribology of implantable medical devices, where understanding surface interactions and wear mechanisms is crucial for improving device longevity and biocompatibility.

Subsequent chapters explore the role of biomaterials and surface modifications in optimizing tribological performance, while the integration of nanotribology offers insights into molecular-level interactions at biological interfaces. The book also presents advances in computational modeling and experimental techniques, enabling better simulation and measurement of complex biotribological phenomena. Clinical applications are reviewed to underscore the direct impact of biotribology in orthopedic, dental, and surgical outcomes. Finally, the text concludes with future perspectives and challenges, highlighting the need for multidisciplinary research to address gaps in simulating real-life biological environments and designing next-generation biomedical devices. This comprehensive work serves as a valuable resource for researchers, engineers, and clinicians dedicated to improving healthcare through tribological innovations.

Keywords: Biotribology, characterization, material analysis, surface property, nanotechnology, artificial intelligence

https://doi.org/10.1515/9783111563176-001

1.1 Introduction

The origins of tribology as a scientific field can be traced back to the seminal work of Leonardo da Vinci, who made early observations about frictional forces. However, the term "tribology" itself was coined much later in 1966 following the Jost Report in the UK, which emphasized the industrial importance of studying friction, lubrication, and wear. As traditional tribology matured with applications in mechanical systems and manufacturing processes, researchers began to recognize its relevance in biological systems.

Biotribology formally emerged in the 1970s, particularly with advancements in biomedical engineering and orthopedic implant development. The need to understand the wear of artificial joints, such as hip and knee prostheses, catalyzed a deeper inquiry into the mechanics of biological tissues. Pioneering studies focused on articular cartilage and the remarkable low-friction properties of synovial joints, leading to a specialized area of research known today as biotribology. The 1980s and 1990s saw a surge in studies related to joint replacements, biocompatibility of materials, and simulation of in vivo conditions using tribometers and joint simulators.

The term "biotribology" was first introduced by Professor Duncan Dowson in the 1970s, emphasizing the importance of tribological principles in biological contexts. Biotribology encompasses the study of friction, wear, and lubrication phenomena in biological systems, including natural joints, skin, eyes, and interactions between medical devices and biological tissues. Understanding these interactions is crucial for the design of prosthetics, implants, and other biomedical devices that mimic or interact with biological tissues.

The current book will highlight various technical, experimental, hypothetical, computational as well as social trends related to biotribology. Also, the chapter-wise description will enhance the research ability of a researcher working on biotribology (Fig. 1.1). The emerging trend, research gaps, AI, and future trends are added in the subsequent chapters.

Technological innovation has significantly influenced the growth of biotribology. The development of joint simulators enabled researchers to replicate physiological motion and loading conditions in laboratory settings, allowing for the assessment of wear rates and lubrication behavior in artificial joints. Pin-on-disk and reciprocating tribometers were adapted to mimic biological conditions, while specialized tribological testing platforms were created for dental, ocular, and dermal applications.

The advent of atomic force microscopy (AFM) and scanning electron microscopy has revolutionized the study of biological surfaces at micro and nanoscale. These tools allow for surface topography mapping, detection of wear scars, and evaluation of adhesive forces, especially in studies involving cartilage, skin, and soft tissues.

Figure 1.2 cohesively presents how natural biological systems (joints, teeth, and soft tissue) maintain efficient motion through specialized surfaces and lubrication. It also depicts how implants and medical devices must be tribologically optimized to

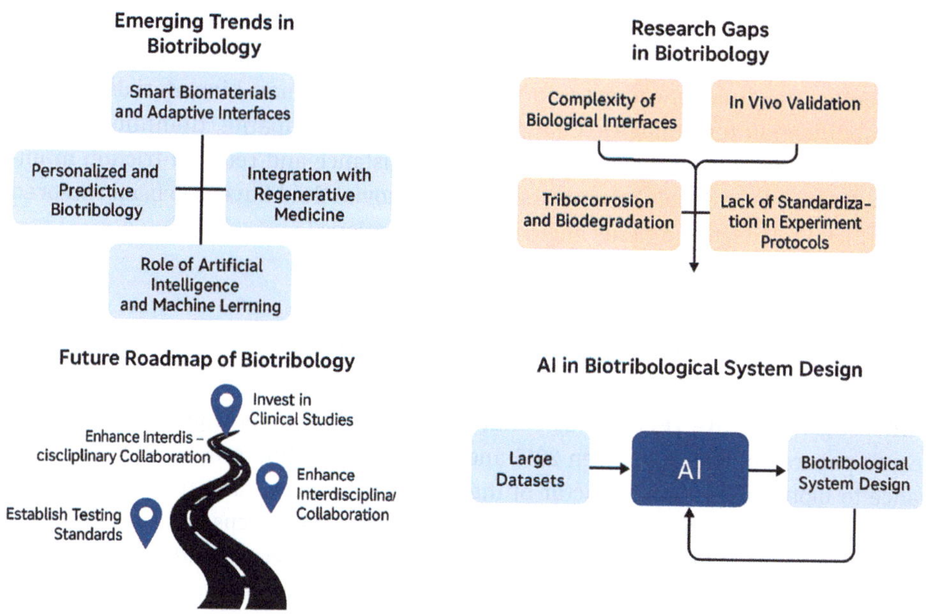

Fig. 1.1: Schematic diagram to highlight emerging trends, research gaps, future map, and AI in biotribology.

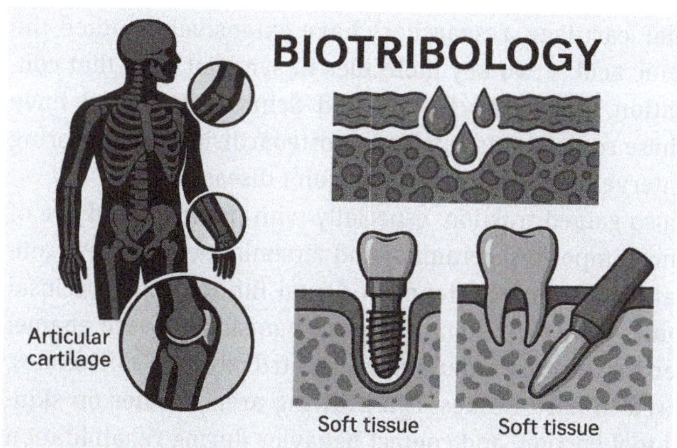

Fig. 1.2: System in the human body.

mimic or withstand these environments. Computational modeling has become a cornerstone in modern biotribology. Finite element analysis (FEA), multibody dynamics simulations, and molecular dynamics (MD) are widely used to simulate complex tribological interactions in biological joints and tissue-material interfaces. These computational tools not only enhance understanding of biomechanical environments but also reduce the dependency on costly and time-consuming experimental procedures.

Another major advancement is in the development of smart biomaterials and surface modification technologies. Coatings such as hydroxyapatite, titanium nitrides, and diamond-like carbon have improved wear resistance and reduced friction in implants. Hydrogels and biolubricants that mimic synovial fluid have also been explored for artificial cartilage and tissue engineering applications.

1.2 Literature review

A thorough literature review was conducted, analyzing several peer-reviewed research papers published between 2015 and 2023. The selection criteria included relevance to biotribology, impact factor of the publishing journal, and the novelty of research findings. The studies were categorized based on their focus areas such as joint tribology, dental tribology, skin tribology, implantable devices, and computational modeling. The literature on biotribology has expanded tremendously over the past few decades. Early works by Mow and Ratcliffe in the 1980s contributed significantly to our understanding of cartilage biomechanics. Their studies detailed the biphasic theory of cartilage mechanics, which accounts for both the solid matrix and fluid flow, influencing later tribological investigations.

In the domain of articular cartilage, researchers have extensively studied the roles of lubricin and hyaluronic acid – two key molecules in synovial fluid that contribute to boundary lubrication. Works by Jay [3] and Schmidt et al. [6] have highlighted how damage to these components can lead to osteoarthritis, underscoring the need for biotribological interventions in degenerative joint diseases.

Dental biotribology has also gained traction, especially with the increased use of restorative materials like resin composites, ceramics, and zirconia. Research by Lambrechts and Vanherle [4] evaluated wear resistance in dental fillings under occlusal forces, while more recent studies have focused on wear mechanisms between enamel and artificial materials under mastication simulations. The tribology of soft tissues, though less explored compared to hard tissues, is a growing area. Studies on skin-implant interactions, surgical tool friction, and contact behavior during rehabilitation have opened new applications of biotribology. In ocular tribology, tear film stability and blinking mechanics are examined to improve contact lens design and treat dry eye syndrome [5].

Significant research has also been conducted on implantable medical devices. Investigations into the wear particles generated by joint prostheses, particularly polyethylene and cobalt-chromium alloy debris, have shown their association with inflammatory responses and implant loosening. This has prompted the development of ultra-high-molecular-weight polyethylene (UHMWPE) with cross-linking treatments and ceramic-on-ceramic bearings.

The field of nanotribology has added a new dimension to biological interface analysis. Molecular interactions between cells and substrates are being studied using advanced AFM techniques. Literature from the past two decades – such as work by Bhushan [1] – details nanoscale friction, adhesion, and surface energy measurements on biological samples.

In terms of modeling and simulation, a plethora of studies have simulated joint mechanics, cartilage stress-strain distribution, and wear prediction using FEA. Authors like Bergmann et al. [2] have contributed significant datasets of joint contact forces during daily activities, which are now used in validating tribological models.

Lastly, clinical literature has increasingly incorporated biotribological considerations. Outcomes of total knee arthroplasty and total hip arthroplasty are now evaluated not just for alignment and fixation but also for wear performance and tribological compatibility. Evidence-based studies help refine materials and surgical approaches.

Biotribology is inherently multidisciplinary, requiring collaboration among engineers, clinicians, material scientists, biologists, and physicists. Its principles are applied in the design and testing of orthopedic implants, dental prostheses, ocular devices, dermal adhesives, and even robotic surgery instruments. As personalized medicine and regenerative therapies evolve, the relevance of biotribology continues to expand [7–9].

The field also interfaces with regulatory and safety standards such as those set by ISO and ASTM for implant testing. Biotribological data are crucial for gaining approvals from regulatory bodies like the U.S. FDA and the European Medicines Agency.

Advancements in biotribology have led to significant improvements in the longevity and performance of medical implants, development of biomimetic materials, and enhanced understanding of disease mechanisms related to wear and lubrication deficiencies. The detailed summary of the literature based on biotribology is outlined in Tab. 1.1.

Tab. 1.1: Summary of literature review in biotribology.

Authors	Title	Findings and limitations	Current findings in biotribology	Area of research
Meng et al. [10]	A review of advances in tribology in 2020–2021	Comprehensive overview of tribology advancements; limited focus on biotribology specifics	Highlighted emerging trends in biotribology including nanotribology and computational modeling	General tribology

Tab. 1.1 (continued)

Authors	Title	Findings and limitations	Current findings in biotribology	Area of research
Zhang et al. [11]	A review of the biotribology of medical devices	Identified tribological issues in medical devices; lacked in-depth analysis of wear mechanisms	Emphasized the need for improved materials and surface treatments in implants	Medical devices
Lin and Klein [12]	Recent progress in cartilage lubrication	Discussed mechanisms of cartilage lubrication; limited experimental validation	Proposed hydration lubrication as a key mechanism in joint health	Joint tribology
Cao et al. [13]	Interactions between bilayers of phospholipids mixture extracted from human osteoarthritic synovial fluid	Investigated lipid interactions in synovial fluid; study limited to in vitro conditions	Provided insights into boundary lubrication in osteoarthritic joints	Joint tribology
Clegg et al. [14]	Tribomechanical properties and bioactivity of additively manufactured PAEK materials for load-bearing medical applications	Evaluated tribological properties of PAEK materials; limited clinical data	Demonstrated potential of PAEK materials in load-bearing implants	Implant materials
Wagner et al. [15]	Biotribology of vascular devices: a review of tissue/device friction	Reviewed frictional interactions in vascular devices; limited in vivo data	Highlighted the importance of understanding tissue-device interactions	Vascular devices
Bhambure and Rao [16]	A review on biotribology in engineering and medical applications	Covered various applications of biotribology; lacked quantitative analysis	Emphasized the role of surface morphology in biotribological performance	General biotribology
Bowland et al. [17]	Review of the biomechanics and biotribology of osteochondral grafts used for surgical interventions in the knee	Evaluated biomechanical stability and biotribological behavior of osteochondral grafts; limited to knee joint	Provided insights into graft performance and areas for improvement	Osteochondral grafts

Tab. 1.1 (continued)

Authors	Title	Findings and limitations	Current findings in biotribology	Area of research
Ungar and Zhou [18]	Dental biotribology: wearing away the boundary between biology and engineering	Discussed tribological aspects of dental materials; lacked experimental data	Highlighted the importance of understanding wear mechanisms in dental applications	Dental tribology
Akdogan et al. [19]	Biotribology of cartilage wear in knee and hip joints review of recent developments	Reviewed recent developments in cartilage wear; limited to in vitro studies	Emphasized the need for improved materials and testing methods	Joint tribology
Ruggiero and Zhang [20]	Editorial: Biotribology and biotribocorrosion properties of implantable biomaterials	Introduced the concept of biotribocorrosion; lacked detailed analysis	Highlighted the significance of corrosion in implant performance	Implant materials
Popoola et al. [21]	High cycle in vitro hip wear of and in vivo biological response to vitamin E blended highly cross-linked polyethylene	Evaluated wear performance of polyethylene; limited in vivo data	Demonstrated improved wear resistance with vitamin E blending	Implant materials
Sava et al. [22]	Normal load scaling of friction in Gemini hydrogels	Investigated friction behavior in hydrogels; limited to specific materials	Provided insights into load-dependent friction in soft materials	Soft tissue tribology
Ji and Zhang [23]	Current strategies for the treatment of early stage osteoarthritis	Reviewed treatment strategies; limited focus on tribology	Emphasized the role of lubrication in joint health	Joint tribology
Poliakov et al. [24]	Current trends in improving of artificial joints design and technologies for their arthroplasty	Discussed design improvements in artificial joints; lacked tribological analysis	Highlighted the importance of tribology in joint design	Joint tribology
Dini et al. [25]	Progression of biotribocorrosion in implant dentistry	Reviewed tribocorrosion in dental implants; limited experimental data	Emphasized the need for corrosion-resistant materials	Dental tribology
Hua et al. [26]	Wear test apparatus for friction and wear evaluation hip prostheses	Developed a wear test apparatus; limited to specific prosthesis types	Provided a method for evaluating wear in hip prostheses	Implant testing

1.3 Current findings in biotribology

- **Joint tribology**: Studies have elucidated the role of synovial fluid components, such as lubricin and hyaluronic acid, in maintaining low friction in joints. Hydration lubrication has emerged as a significant mechanism in cartilage function.
- **Dental tribology**: Research has focused on the wear resistance of dental materials and the impact of saliva composition on oral lubrication.
- **Skin tribology**: Investigations into skin friction have implications for prosthetic design, cosmetic formulations, and understanding skin disorders.
- **Implantable devices**: Advancements in surface engineering and material science have led to the development of implants with improved wear resistance and biocompatibility.
- **Computational modeling**: FEA and MD simulations are increasingly used to predict tribological behavior in biological systems.

1.4 Areas of research

- **Nanotribology**: Exploring friction and wear at the nanoscale to understand molecular interactions in biological systems.
- **Biomimetic materials**: Designing materials that replicate the tribological properties of natural tissues.
- **Surface modification techniques**: Developing coatings and treatments to enhance the tribological performance of implants.
- **In vivo and in vitro testing**: Establishing standardized testing protocols to evaluate biotribological properties.

1.5 Outline of the current book

Biotribology is a multidisciplinary field that bridges biology, engineering, and materials science. The brief literature review highlights significant progress in understanding the tribological behavior of biological systems and the development of advanced materials and devices. However, challenges remain in replicating complex biological environments in experimental setups and translating laboratory findings into clinical applications. Future research should focus on integrating computational models with experimental data, developing standardized testing methods and exploring the role of biotribology in emerging biomedical technologies. The current book is outlined in 15 chapters as outlined below.

Chapter 1: Introduction to biotribology

Biotribology is an interdisciplinary field that explores the principles of friction, wear, and lubrication within biological systems. Bridging biomechanics, materials science, and medicine, it aims to understand how biological tissues and artificial implants interact under motion and load. The growing demand for advanced biomedical devices, aging populations, and chronic joint disorders has elevated the importance of biotribological research. This chapter lays the foundational understanding of biotribology, highlighting its history, scope, and relevance to healthcare innovations such as joint replacements, dental implants, and cardiovascular devices.

Chapter 2: Fundamentals of biological surfaces

Biological surfaces are uniquely adapted to their functions, often exhibiting hierarchical structures, hydration layers, and complex biochemical compositions. These surfaces play a critical role in regulating tribological behavior within the body, influencing factors such as adhesion, deformation, and wear. Understanding their structural, chemical, and mechanical properties is essential for designing biocompatible materials and interpreting biological interactions. This chapter delves into the morphology and functional adaptations of natural biological surfaces that serve as benchmarks for synthetic materials.

Chapter 3: Friction and wear mechanisms in biological systems

Friction and wear in biological systems are highly dynamic and governed by a combination of mechanical, chemical, and biological factors. These processes occur at various scales – from the nano- to the macro-level – and impact critical physiological functions such as joint mobility, cardiovascular flow, and cellular integrity. This chapter explores the diverse wear mechanisms (abrasive, adhesive, fatigue, and corrosive) and frictional behaviors observed in living organisms, with an emphasis on how these affect long-term tissue health and implant performance.

Chapter 4: Lubrication in biological systems

Lubrication in the human body is a highly evolved phenomenon involving fluid films, synovial fluids, mucosal secretions, and biolubricants like glycoproteins. These natural lubricants reduce friction and wear in joints, eyes, oral cavities, and internal organs, facilitating smooth biological function. This chapter examines the types and mechanisms of lubrication in vivo – boundary, mixed, and hydrodynamic lubrication – highlighting the molecular constituents responsible for their remarkable performance and stability under complex physiological conditions.

Chapter 5: Biotribology of articular cartilage

Articular cartilage is one of the most tribologically efficient tissues in the human body, enabling near-frictionless motion over a lifetime. Its unique structure – composed of chondrocytes embedded in a dense extracellular matrix – plays a vital role in load distribution and lubrication. This chapter discusses the hierarchical structure of cartilage, its wear resistance, and the pathomechanics of degenerative diseases like osteoarthritis. Advances in cartilage repair and synthetic replacements are also reviewed from a tribological perspective.

Chapter 6: Dental biotribology

The oral cavity presents a complex tribological environment characterized by intermittent contact, saliva-based lubrication, and diverse material interfaces. Dental biotribology studies the wear of natural enamel, dentin, and restorative materials during mastication and abrasion. This chapter explores the tribological implications of chewing forces, food properties, and dental prosthetics, emphasizing the importance of wear resistance and biocompatibility in the longevity of dental materials and devices.

Chapter 7: Soft tissue biotribology

Soft tissues such as skin, tendons, and mucosa experience various forms of frictional contact, often under shear, tension, or compression. Understanding their tribological behavior is critical in areas like wound healing, prosthetic interface design, and surgical procedures. This chapter highlights the structural and functional complexities of soft tissues, their interaction with both natural and synthetic materials, and the strategies employed to minimize frictional damage and promote healing.

Chapter 8: Biotribology of implantable medical devices

Implantable devices such as joint prostheses, heart valves, and spinal implants must operate seamlessly within the tribologically challenging environment of the human body. This chapter focuses on the wear mechanisms, material selection, and lubrication challenges associated with implants. It also discusses the impact of implant wear debris on biological tissues and immune responses as well as recent innovations in surface coatings and design improvements aimed at enhancing implant longevity.

Chapter 9: Biomaterials and surface modifications

Biomaterials used in tribological applications must balance mechanical strength, corrosion resistance, and biocompatibility. Surface modifications – such as coatings, texturing, and chemical treatments – play a crucial role in enhancing tribological performance. This chapter reviews the properties of commonly used biomaterials (e.g., titanium, UHMWPE, and ceramics), their failure modes, and the role of advanced surface engineering techniques in improving wear resistance and biological integration.

Chapter 10: Nanotribology at biological interfaces

Nanotribology investigates friction and wear phenomena at the nanometer scale, providing insight into interactions between biomolecules, cells, and nanostructured surfaces. At biological interfaces, nanotribology explains how atomic-scale features influence lubrication, adhesion, and wear. This chapter discusses experimental tools such as AFM and MD simulations that enable the study of nanoscale interactions, offering new perspectives on cell-material and protein-surface interactions in biomedical contexts.

Chapter 11: Computational modeling of biotribological systems

Computational modeling has emerged as a powerful tool to predict and analyze tribological phenomena in biological systems, helping overcome experimental limitations. FEA, multibody dynamics, and molecular simulations allow the virtual exploration of stress distributions, deformation behavior, and wear patterns. This chapter introduces key modeling techniques, their applications in implant design and tissue mechanics, and the challenges in replicating complex biological environments computationally.

Chapter 12: Experimental techniques in biotribology

The study of biotribology requires specialized experimental setups that mimic physiological conditions and measure tribological parameters with precision. From pin-on-disk tribometers to custom joint simulators and AFM, a variety of instruments are employed to characterize wear, friction, and lubrication behavior. This chapter presents an overview of these experimental methods, including sample preparation, testing protocols, and interpretation of results, offering best practices for reliable data generation.

Chapter 13: Clinical applications of biotribology

Biotribology has significant clinical implications, influencing the design, performance, and longevity of orthopedic implants, dental restorations, cardiovascular devices, and surgical tools. This chapter illustrates how tribological insights have translated into improved patient outcomes, reduced revision surgeries, and enhanced device safety. Case studies and clinical trials are reviewed to demonstrate the role of biotribology in evidence-based medical device development.

Chapter 14: Future perspectives and challenges

Despite substantial advancements, biotribology faces several challenges, including simulating real biological conditions, integrating multiscale data, and addressing the variability of human anatomy and pathology. Future research directions point toward smart materials, bioinspired lubricants, personalized implants, and machine learning in predictive modeling. This chapter outlines the future landscape of biotribology, emphasizing interdisciplinary collaboration and innovation to address current limitations and meet emerging healthcare demands.

Chapter 15: Conclusion and summary

Biotribology stands at the intersection of science, engineering, and medicine, offering vital solutions to challenges in healthcare and biomedical innovation. This concluding chapter synthesizes the key themes and discoveries discussed throughout the book, highlighting the integration of fundamental research with applied technologies. It reiterates the importance of a biotribological perspective in improving human health, extending the life of implants, and guiding the next generation of biomedical device design.

References

[1] Bhushan B. (2011). *Nanotribology and Nanomechanics: An Introduction*. Springer.

[2] Bergmann G, Graichen F, Rohlmann A. (2001). Hip joint loading during walking and running, measured in two patients. *Journal of Biomechanics*, 34(7), 859–869. https://doi.org/10.1016/S0021-9290(01)00040-3.

[3] Jay GD. (2000). Lubricin and surfactant protein A as lubricating molecules in articular cartilage. *Nature*, 410(6828), 660–663. https://doi.org/10.1038/35070585.

[4] Lambrechts P, Vanherle G. (1994). Clinical relevance of physical and mechanical properties of composite resins. *Operative Dentistry*, 19(3), 130–140.

[5] Mow VC, Ratcliffe A. (1997). Structure and function of articular cartilage and meniscus. In Nordin M, Frankel VH. (eds.), *Basic Biomechanics of the Musculoskeletal System* (3rd ed., pp. 113–177). Lippincott Williams & Wilkins.

[6] Schmidt TA, Gastelum NS, Nguyen QT, Schumacher BL, Sah RL. (2007). Boundary lubrication of articular cartilage: Role of synovial fluid constituents. *Arthritis & Rheumatism*, 56(3), 882–891. https://doi.org/10.1002/art.22446.

[7] Vanherle G, Lambrechts P. (1989). Wear of dental composites. *Journal of Dental Research*, 68(12), 1855–1861. https://doi.org/10.1177/00220345890680120501.

[8] ISO 14242-1:2014. (2014). Implants for surgery – Wear of total hip-joint prostheses – Part 1: Loading and displacement parameters for wear-testing machines and corresponding environmental conditions for test. International Organization for Standardization.

[9] ASTM F732-17. (2017). *Standard Test Method for Wear Testing of Polymeric Materials Used in Total Joint Prostheses*. ASTM International.

[10] Meng Y, et al. (2022). A review of advances in tribology in 2020–2021. *Friction*, 10(10), 1443–1595.

[11] Zhang X, et al. (2021). A review of the bio-tribology of medical devices. *Friction*, 10, 4–30.

[12] Lin W, Klein J (2023). Recent progress in cartilage lubrication. arXiv preprint arXiv:2311.02681.

[13] Cao Y, et al. (2022). Interactions Between Bilayers of Phospholipids Mixture Extracted from Human Osteoarthritic Synovial Fluid. *arXiv preprint arXiv:2204.12128*.

[14] Clegg, B. A., Shrestha, D. R., & Emami, N. (2023). Tribo-mechanical properties and bioactivity of additively manufactured PAEK materials for load bearing medical applications: a systematic review. *Biotribology*, 35, 100263.

[15] Wagner, R. M., Maiti, R., Carré, M. J., Perrault, C. M., Evans, P. C., & Lewis, R. (2021). Bio-tribology of vascular devices: A review of tissue/device friction research. *Biotribology*, 25, 100169.

[16] Bhambure, Shilpa and Rao, A. S., A Review on Biotribology in Engineering and Medical Applications (December 13, 2018). Proceedings of TRIBOINDIA-2018. http://dx.doi.org/10.2139/ssrn.3321040.

[17] Bowland P, Ingham E, Jennings L, Fisher J. (2015). Review of the biomechanics and biotribology of osteochondral grafts used for surgical interventions in the knee. *Proceedings of the Institution of Mechanical Engineers, Part H: Journal of Engineering in Medicine*, 229(12), 879–888.

[18] Ungar PS, Zhou Z. (2017). Dental biotribology: Wearing away the boundary between biology and engineering. *Biosurface and Biotribology*, 3(4), 115–118.

[19] Gulsen, A., Merve, G., & Meltem, P. (2018). Biotribology of cartilage wear in knee and hip joints review of recent developments. In *IOP Conference Series: Materials Science and Engineering* (Vol. 295, No. 1, p. 012040). IOP Publishing.

[20] Ruggiero A, Zhang Y. (2020). Biotribocorrosion of biomaterials. *Biotribology*, 24, 100135.

[21] Popoola, O. O., Villasenor, D. A. O., Fryman, J. C., Mimnaugh, K., & Rufner, A. (2018). High cycle in vitro hip wear of and in vivo biological response to vitamin E blended highly crosslinked polyethylene. *Biotribology*, 16, 10–16.

[22] Sava M, et al. (2018). Friction in hydrogels. Soft Matter, 10, 34–56.

[23] Ji X, Zhang H. (2019). Current strategies for the treatment of early stage osteoarthritis. *Frontiers in Mechanical Engineering*, 5, 57.

[24] Poliakov A, et al. (2020). Trends in artificial joint design. Biomedical Engineering Letters, Vol 1, 16–35.

[25] Dini C, Costa RC, Sukotjo C, Takoudis CG, Mathew MT, Barão VA. (2020). Progression of bio-tribocorrosion in implant dentistry. *Frontiers in Mechanical Engineering*, 6, 1.

[26] Hua Z, Dou P, Jia H, Tang F, Wang X, Xiong X, . . . Jin Z. (2019). Wear test apparatus for friction and wear evaluation hip prostheses. *Frontiers in Mechanical Engineering*, 5, 12.

Adeeba Mirza, Ishika Sharma, Abhilasha Mishra, and Hasan Koten

2 Fundamentals of biological surfaces

Abstract: Biotribology can be defined as the study of friction, wear, and associated lubrication in all biological systems by using the knowledge of mechanical engineering in materials sciences and biology in general. This will help researchers study tissue structure, composition, and function by maintaining tissue integrity through cellular interactions. The details of the extracellular matrix constituents – collagen, elastin, glycoproteins, and proteoglycans – are then expounded upon, highlighting their roles in maintaining the shape and functionality of tissues. Additionally, it trades with the hierarchical organization of tissues, ranging from primary tissues, such as epithelial and connective tissues, to complex structures such as organs and organ systems. The chapter also discusses the mechanical and functional interactions that occur at interfaces between various tissues, such as those between muscle and tendon junctions, epithelial-mesenchymal, tendon-bone, and synovial joints. Lubricants are substances that reduce any unnecessary friction caused by continuous movement in biological systems. More particularly, it highlights several biological lubricants such as synovial fluid, mucus, and saliva, their composition, functions, and roles in lubrication and wear protection for tissues. The study also discusses boundary layers in biological systems, including friction reduction, and how tissue wear and tear can be influenced. The hydrodynamic, boundary, and mixed lubrication mechanisms are discussed to explain how these processes enable the smooth functioning of biological surfaces. Finally, this chapter reviews the properties and adaptations of tissues under mechanical load. The study covers bone microstructure and mechanical properties, the types and functions of cartilage in load distribution, and its composition, and the role of tendons and ligaments in the transmission of forces. Additionally, it provides information on the adaptation of load-bearing tissues to mechanical loads, stress-strain relationships, and viscoelasticity. In addition, it accounts for the pathophysiological conditions of tissues, such as osteoarthritis, tendinitis, and even fractures, to provide a deep understanding of how such incidents alter joint function and the entire health of the tissues. This chapter provides an overview, in general, of the basic principles underlying the functioning of biological surfaces, relating tissue structure and function to their implications for biotribology. Understanding these basics forms the necessary foundation for developing appropriate strategies for solving friction, wear, and lubrication problems in biological systems.

Keywords: Biotribology, friction, lubricants, mechanisms, tendon, bones

https://doi.org/10.1515/9783111563176-002

2.1 Introduction to biological surfaces and tribology

Biological surfaces are the interfaces at which living tissues interact with their environment, which could either be internal or external to the outside world. These surfaces are thus crucial in maintaining the functionality of organs, tissues, and cells, as they take on a front-line role in protecting, interacting, lubricating, and managing friction [1, 2]. Understanding their structure and functioning is important in many scientific fields, but particularly in the domain of tribology, which is considered the most outstanding. It deals with friction, wear, and lubrication. Biological surfaces are the outermost layers of tissues exposed to adjacent tissues, fluids, or external environments [3]. These range from the skin, internal linings of the respiratory and digestive tracts, interfaces between bones and muscles, organs, and joint surfaces to being frequently covered with epithelial cells, connective tissues, or extracellular matrices (ECMs), and many of these are also bathed in fluids such as mucus, synovial fluid, or saliva [4, 5]. Biological surfaces are important structural elements that promote the health and function of living organisms. Their special structures and properties provide them with various functions such as protection barriers, selective permeability, cellular communication, and signaling [6]. Surfaces such as the skin or the lining of organs act as barriers against physical damage and invasion by microbes or other harmful substances [7]. For instance, mucosal surfaces provide protection against ingested pathogens, whereas skin protects against UVR and other environmental factors [8]. Some biological surfaces, such as those lining blood vessels or the intestines, serve the purpose of controlling what comes into and out of tissues, allowing appropriate nutrients to enter and remove waste products [9, 10]. This selective permeability should be maintained within the body to maintain homeostasis. Where tissues slide relative to each other, like in joints or the gut, biological surfaces reduce friction and limit wear and tear along with specialized lubricants like synovial fluid or mucus [11].

Biological surfaces act as a platform for cell-to-environmental communication. Receptors on the cell surface may perceive chemical cues, thereby responding to changes in the external environment [12]. This process is crucial in processes such as immunity, wound healing, and cellular growth. This knowledge is applicable not only to physiology and biology but also to medical and engineering fields, especially those engaged in designing medical devices, implants, and biomaterials that operate inside the body. When the science of friction, wear, and lubrication is applied to biological systems, this is known as biotribology [13, 14]. Biotribology deals with how biological surfaces interact with one another and how such interactions can be optimized or mimicked in artificial systems. Therefore, we provide an overview of how biological surfaces contribute to the tribological behavior of living systems. The resistance caused by two surfaces gliding against each other is called friction. In biological systems, controlling friction is important for preserving tissues [15–17]. Therefore, the cartilage covering the ends of the bones within a synovial joint reduces friction during movement and provides smooth and painless movement. Excessive friction creates wear and tissue damage, like in instances such as osteoarthritis, where breakdown of the cartilage leads to a failure of friction-painless movement of the joint. As a result of mechanical action, materials being worn from a surface are known as wear [18]. In biological systems, wear can occur as the result of friction between tissues, which can persist for a long time, leading to gradual wearing of protective surfaces, such as cartilage in joints or enamel in teeth [19] because living matter biological surfaces often possess mechanisms that reduce the wear-like regeneration of skin cells or the release of lubricating fluids such as saliva and synovial fluid. In cases where such mechanisms fail or are overwhelmed, wear causes degeneration and disease [20].

Biological systems have established different types of lubrication mechanisms for friction control and wear limitation. Lubrication can be achieved through fluid lubrication. Thin films consisting of synovial fluid, saliva, and mucus keep the surfaces apart so that they slide smoothly over each other, thereby avoiding direct contact and friction [21]. Boundary lubrication, lubricin, a constituent of synovial fluid or mucins in mucus, leads to the formation of a coating as a protective shield against friction at the interface [22]. Intermittent lubrication often occurs in biological systems under fluid and boundary lubrication conditions, depending on the type of motion or load applied. For instance, the lubrication in synovial joints may require more fluid lubrication in certain rapid motions but may move toward boundary lubrication under higher loads [23]. The significance of knowledge of the tribological properties of biological surfaces has brought some critical applications in medicine and engineering. The design of artificial joints or prosthetic devices requires intimate knowledge of how natural joints handle friction and wear. Materials chosen for these devices must replicate the low friction characteristics of cartilage for long-term performance. Knowledge of the wear mechanisms of enamel and the role of saliva in oral lubrication puts in the hands of dentistry, the possibility of developing materials for fillings, crowns, and dentures resistant to chewing forces. In tissue engineering, the surfaces to be designed should allow cell growth but with reduced friction and wear [24, 25]. Therefore, for example, when designing scaffolds for skin or cartilage repair, the surface should mimic that of its natural tissue and withstand the stress of mechanical loads. Catheters and implants that interact with biological tissues greatly benefit from surface designs that reduce friction and wear. This will, in turn, promote minimal tissue irritation and allow a longer product lifespan. Understanding how biological surfaces function in relation to tribology can help develop better therapeutic approaches and design materials that integrate with the body's own functions [26].

2.2 Structure and composition of biological tissues and interfaces

The organization and function of living beings are dictated by the structure and composition of biological tissues and their interfaces. Biological tissues include specialized cells, fluids, and extracellular components that perform various functions. This chapter discusses the cellular and extracellular components that form the basis of biological tissues and their hierarchical organization [27].

2.2.1 Cellular and extracellular components

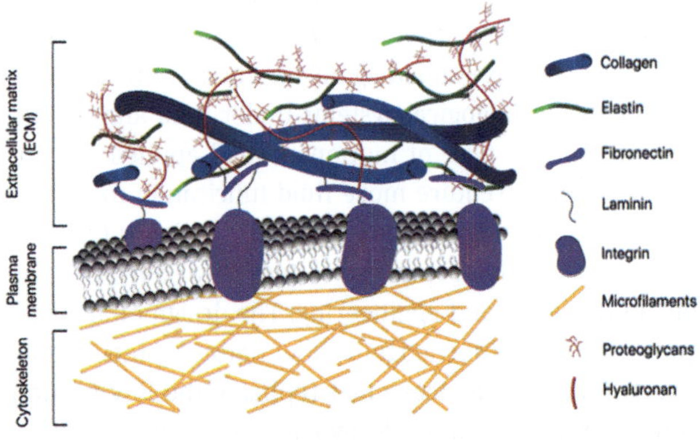

Fig. 2.1: Schematic representation of extracellular matrix and its major components. Reproduced with permission [28].

The ECM supports the structural integrity and metabolic diversity of the various cell types comprising biological tissues. The epithelial cells are very close together and form a continuous sheet. Their ECM tendon is sparse in comparison with the ECM of a normal animal, and it attaches to the basement membrane binding to the underlying connective tissue. Epithelial cells are polarized with one apical surface facing the lumen or environment and a basal surface attaching to the basement membrane. The epithelial cells form a layer that protects the tissues underneath from mechanical damage, pathogens, and water loss. Epithelial cells have specialized absorptive and secretory functions in the gastrointestinal and glandular organs to absorb nutrients, hormones, and mucus [29–31]. They often bear sensory receptors that aid in the perception of environmental changes for functions in relation to sensory organs such as taste and smell and some touch. Connective tissue cells are loosely arranged within a loose ECM containing fibers and ground substances as shown in Fig. 2.1. The relative amount of ECM differs according to the form of connective tissue. Connective tissues provide structural support to organs and tissues and maintain their shape and integrity. Adipose tissue stores energy in the form of fat, whereas other connective tissues store minerals, such as calcium, in bones. Blood is a class of connective tissue that is meant to transport nutrients, gases, and waste throughout the body [32].

Muscle cells, myocyte, are spindle-shaped and are specialized for contraction. Sarcomeres are constituents of contractile proteins found only in striated muscle; smooth muscle cells appear more spindle-shaped. Muscle cells enable the movement of the body and internal organs. Skeletal muscles are responsible for all voluntary movements [33]. Conversely, the cardiac and smooth muscles perform all involuntary functions such as beating the heart and peristalsis. Muscle contraction produces heat to regulate body temperature. Nervous tissue comprises neurons along with glial cells, also known as neuroglia. Neurons receive support, protection, and nourishment from glial cells, while they themselves are specialized for signal transduction. Neurons deliver electrical signals throughout the body to allow distant parts of the nervous system to communicate [34]. Glial cells provide support and protection to neurons through their participation in the achievement of homeostasis, myelination, and effects on the immune response. The ECM is an extensive network of proteins and carbohydrates that provides structural and metabolic support to neighboring cells. The composition varies according to the tissue type and functional need [35].

Collagen is the most abundant protein in the ECM, providing tensile strength and structural support. Various types of collagens exist, each with different functions. For example, Type I collagen is found in bones and tendons and type II collagen is in cartilage. The collagen fibers have an arrangement similar to a weave that resists stretching and deformation [36]. Elastin provides tissue elasticity and resilience. It enables stretching and recoiling of structures such as the skin, lungs, and blood vessels. Elastin in the ECM is essential for maintaining the shape and proper functioning of organs during repetitive stretching. Glycoproteins, fibronectin and laminin, are two types of glycoproteins that are crucial for cell adhesion and signaling. They act as physical bridges between cells and the ECM, thus facilitating communication and coordination of functional operations between tissues [37]. Thus, glycoproteins contribute to controlling cell behavior during development and wound healing. A protein core to which one or more chains of glycosaminoglycan (GAG) are linked defines a proteoglycan. Due to their ability to hold onto water, they provide the ECM with toughness and hydration and a gel-like consistency that offers support and cushioning. Proteoglycans are abundant in cartilage and are essential for the structural integrity of tissues [38].

2.2.2 Tissue hierarchy and organization tissues

Tissue hierarchy and organization tissues of the body are classified into levels at which complex biological functions are enabled. The levels include primary tissues, complex tissue structures, organs, and organ systems as the basic level of organization [39], as shown in Fig. 2.2.

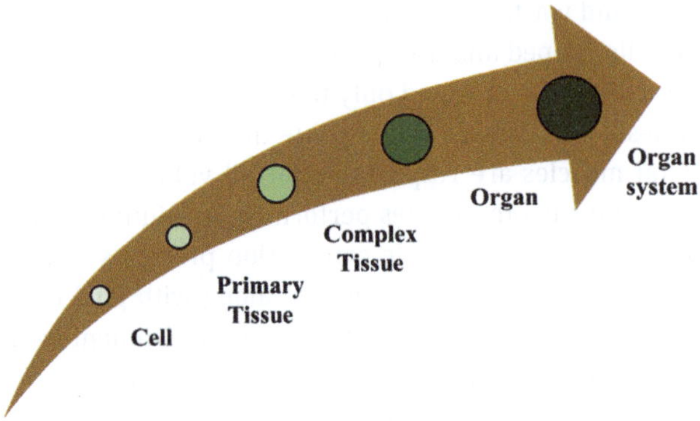

Fig. 2.2: Tissue hierarchy.

There are four primary types of tissues in the human body as shown in Fig. 2.3, which are distinguished by different structures and functions. Epithelial tissue, composed of closely packed cells arranged in protective membranes, is responsible for absorption, secretion, and sensation. Epithelial tissue is typically simple or stratified based on the number of layers and cell shape (squamous, cuboidal, or columnar). Connective tissue comprises various cell types and large amounts of ECM. Connective tissues provide support, bind, and protect tissues. They are separated into blood, cartilage, bone, adipose tissue, dense and loose connective tissue, and so on. The contractile cells comprise three types of muscle tissue: skeletal (controlled voluntarily), cardiac (controlled involuntarily by the heart), and smooth muscle (involuntary control in organs and blood vessels). The nervous tissue is composed of neurons and glial cells that facilitate communication within the body through electrical signals and neuronal support functions [40].

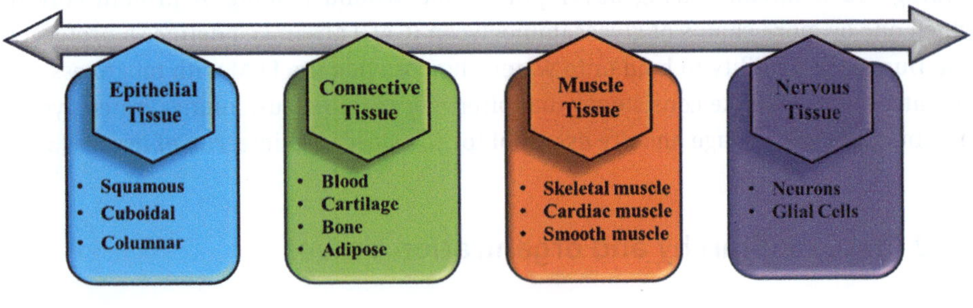

Fig. 2.3: Types of tissues in the human body and their functions.

2.2.3 Complex tissue structure

The structures of more than two types of tissues are combined to perform specialized functions. An example is the heart. It contains cardiac muscle tissue, connective tissue, and nervous tissue for blood pumping. Organ systems are a collection of organs that combine to carry out complex functions of life. One example is the digestive system, which consists of the mouth, esophagus, stomach, intestines, and related organs including the pancreas and liver. It also facilitates the digestion and absorption of nutrients [41, 42]. Tissues and organs can make biological systems complex and specialize with their hierarchical organization. Thus, diverse physiological processes can be made efficient, coordinated, and responsive to body needs.

2.3 Attributes of surface area

The operation of biological surfaces during tribological interactions is fundamentally influenced by their surface attributes. They explain how tissues respond to mechanical stresses, interact with liquids, and communicate with adjacent cells and tissues. The important surface properties of a material include its chemical composition, topography, roughness, surface energy, and wetting behavior [43].

2.3.1 Chemical composition of biological surfaces

The chemical composition of biological surfaces significantly determines their mechanical properties and potential interactions with other materials or biological entities. This depends on the number of biomolecules present, such as proteins, carbohydrates, and lipids, which have a great impact on the chemical characteristics of these surfaces. The cell membrane is rich in molecules of lipids, with a lipid bilayer acting as a physical barrier and a surface lubrication effector [44]. The specific molecules of lipids present affect whether the surface is more hydrophobic or hydrophilic, thus affecting the interaction with water and other molecules. Biological surfaces may contain receptors, enzymes, or structural proteins. Interactions with other cells or with the ECM are often aided by the outermost surface proteins of cells. Proteins such as collagen, laminin, and fibronectin are integral parts of the surfaces of connective tissues and contribute to their structure as well as influence cell adhesion. Carbohydrates are components of glycoproteins or glycolipids and are frequently located within the glycocalyx coat of cells. Physical barriers, protecting cells, and facilitating interactions are some purposes that can be served by these agents. Mineral components, the hardness and toughness imparted to the tissues of bones and teeth, are due to the presence of an inorganic material such as hydroxyapatite in their surfaces [45, 46]. These minerals serve to impart such surfaces with appreciable stiffness and exhibit a certain resistance to abrasion.

2.3.2 Topography and roughness

The topography and roughness of a biological surface significantly influence the mechanical interactions: friction, adhesion, and wear resistance. In biological tissues, the three-dimensional architecture, including ridges, valleys, and pores, is called surface topography. Such features can be extremely varied from smooth surfaces of articular cartilage to complex textures of intestinal epithelial surfaces [47]. The topographical features of a surface can control cell behavior, including migration and adhesion, and the transport of fluids across the surface, either by overlying or invading the surface. Roughness is a measure of surface texture height variation that is commonly quantified by height variations. Rough surfaces can provide more attachment area to cells, increasing contact with extracellular molecules in biological contexts. However, if the roughness is too high, friction and wear become considerable problems, particularly at joints or moving interfaces. The biological surfaces are just as important in interactions at the cellular and molecular levels by their roughness at the nanoscale level. The nano-roughness of the bone surface is used to influence the adhesion of osteoblasts, which are cells used to promote bone formation [48, 49]. Nanoscale texturing of the glycocalyx layer of endothelial cells is known to affect blood flow and reduce shear stress.

2.3.3 Wetting and surface energy

A pivotal characteristic that influences the way a liquid interacts with a biological surface is the concept of surface energy, which simultaneously dictates both the wettability and the dynamics of boundary-layer phenomena. Surface energy is the energy associated with the surface of a material, which arises from unsatisfied bonds at the interface. A surface with high energy tends to attract molecules from the liquid phase, leading to greater wetting. Many biological surfaces like cell membranes, for example, have moderate to low surface energy because of their lipid constituents and help regulate interactions with aqueous environments [50]. Wettability is the ability of a liquid to maintain contact with a solid surface; basically it is a balance between adhesive and cohesive forces. For example, hydrophilic surfaces tend to have higher surface energies, enabling water to spread over the surface completely. Hydrophobic surfaces have low surface energies, so water tends to bead up. The wetting characteristics of biological surfaces depend on how lubricants such as synovial fluid or mucus spread over them, thus influencing their lubrication and frictional behavior. For tissues like articular cartilage, a balance between surface energy and wettability is vital because the synovial fluid must form a stable, low-friction film to allow for the possibility of movement at the joint [51]. The interfacial energy of the cartilage supports this layer, thereby decreasing friction and wear.

2.4 Interaction mechanisms

Biological surfaces interact through several molecular and biomechanical mechanisms that determine how surfaces interact with or adhere to one another as well as the amount of friction generated between surfaces and how tissues react to their environments.

2.4.1 Molecular interactions

Molecular interactions are fundamental to biological surface functions and interactions. Several chemical and physical forces are given below.

2.4.1.1 Van der Waals forces

Weak attractions exist between molecules if they are held in proximity. These forces are important for the weak adhesive forces between biological molecules such as those in cell membranes.

Intermolecular interactions of hydrogen: The phenomenon of hydrogen bonding is a significant factor in the structural configuration of biological macromolecules including deoxyribonucleic acid and various proteins [52]. In addition, these interactions influence the way water molecules engage with biological surfaces.

2.4.1.2 Electrostatic interactions

Other molecules or surfaces may find electrostatic interactions appealing or repulsive due to the presence of charged groups on biological surfaces. For example, the negative charge on the glycocalyx surface can cause the repulsion of other negatively charged molecules [53]. This is relevant to cellular interactions.

2.4.1.3 Hydrophobic interactions

Hydrophobic interactions dominate the interactions of lipid-rich surfaces such as cell membranes. Hydrophobic areas in aqueous solvents tend to coaggregate and hence facilitate the self-assembly of lipid bilayers, further determining the interaction of proteins along with other molecules with membrane surfaces [54].

2.4.2 Biological response to surface characteristics

The biological response to material surface properties is essential in tissue engineering, implant design, and cell behavior. Adhesion molecules are the molecules that the cells attach with integrins, binding them to specific extracellular proteins. Roughness, stiffness, and composition are surface attributes that influence cell adhesion, proliferation, and differentiation. The body mounts an immunological defense against foreign objects or injured tissue surfaces. The degree of inflammation and final integration of the biomaterials used for implants with surrounding tissues are determined by their surface characteristics [55]. The chemical constituency and surface characteristics, such as porosity and roughness, affect the assimilation of implants into biological tissues. Each of these elements may facilitate tissue proliferation and cellular adhesion.

2.4.3 Glycocalyx and proteins

The glycocalyx layer present on the cellular surface, along with various proteins, assumes a pivotal function in governing the interactions between biological surfaces and their surrounding environment: surface-bound proteins can bridge cell-surface interactions and lubrication as well as influence the biocompatibility of materials. Lubricin is an important component of joint lubrication for reducing friction between cartilage surfaces. Many cell types, most notably epithelial and endothelial, have an external dense carbohydrate-rich layer known as the glycocalyx, which provides a physical and chemical barrier that modulates the interaction of cells with immune cells and their control of fluid flow over the cell. In the vascular system, it reduces shear stress caused by flowing blood, whereas in the digestive and respiratory tracts, it contributes to mucous production and protects against mechanical stress [56]. A well-organized glycocalyx can participate in low-friction environments and protect itself from damage under such dynamic conditions.

2.5 Interfaces between tissues

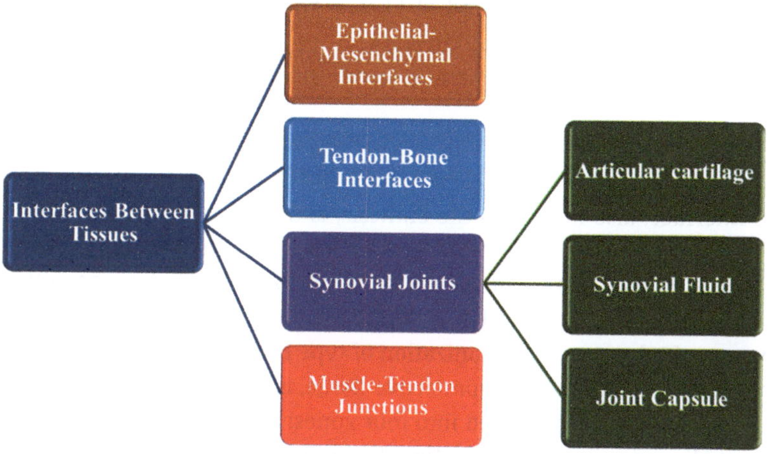

Fig. 2.4: Types of interfaces between tissues.

To preserve the body's mechanical performance, structural integrity, and intertissue communication, interfaces between its various tissues are essential. These interfaces provide seamless transitions, allowing tissues with different properties to work together efficiently [57]. The types of interfaces between tissues are schematically represented in Fig. 2.4. Understanding these interfaces is essential for exploring how forces are transmitted and absorbed, and how damage to one area can impact neighboring tissues.

2.5.1 Epithelial-mesenchymal interfaces

The epithelial-mesenchymal interface represents the boundary between the epithelial layers and the underlying connective tissue (mesenchyme). This interface is critical during development, tissue repair, and the progression of certain diseases including cancer. For EMT of tissues and organs, transition from epithelial to mesenchymal cells is required during embryogenesis. EMT allows cells to become more migratory, breaking away from epithelial sheets to form new structures such as hearts, muscles, and bones [58].

The interface of epithelial-mesenchymal is significant in wound healing, as it allows them to undergo EMT, which allows them to displace into the wound, regenerate, and then return back to mesenchymal-epithelial transition state once the repair is complete in case of skin or epithelial tissue damage. Cancer progression metastasis is the process in which carcinoma cells break away from a large main epithelial tumor, penetrate surrounding tissues, and travel to other parts of the body via blood vessel movements. Abnormal EMT is implicated in this process. In this case, disruption of the epithelial-mesenchymal interface permitted cancer cells to be spread and infiltrate further organs [59]. A unique kind of ECM, a basement membrane, is the characterization of epithelial-mesenchymal interface.

2.5.2 Tendon-bone interfaces

The tendon-bone interface, also known as the enthesis, is the point at which tendons attach to bones, allowing the transmission of forces generated by muscle contractions to the skeleton. The structure of this interface is highly specialized to manage the transition between the flexible and rigid bones, thereby preventing injury and ensuring efficient movement [60]. A specific organization of type I collagen fibers in tendon, which is a highly organized hypocellular connective tissue, is displayed in Fig. 2.5.

The tendon-bone interface is organized into four zones: the tendon zone, which is composed mainly of aligned collagen fibers, offers flexibility and tensile strength.

Fibrocartilage zone: Collagen fibers gradually blend into fibrocartilage, acting as a transition zone and reducing stress concentration.

Mineralized: The fibrocartilage becomes mineralized, providing a gradual increase in stiffness as it approaches the bone.

Bone zone: The bone zone is where the mineralized fibrocartilage merges with dense, rigid bone, ensuring strong anchorage. The graded structure at the tendon-bone interface distributes mechanical stress more evenly, thereby reducing the risk of injury. It allows the interface to withstand high forces without tearing, which is particularly important during activities such as jumping, lifting, and sprinting [62]. The tendon-bone interface has limited healing capacity. Injuries to this area, such as rotator cuff tears and Achilles tendon ruptures, often require surgical intervention because the natural regeneration process cannot fully recreate the complex, graded structure.

2.5.3 Muscle-tendon junctions

The muscle-tendon junction (MTJ) is the connection between muscle fibers and tendons, ensuring effective transfer of muscle-generated forces to tendons, which in turn transmit these forces to bones to produce movement. The MTJ has a highly folded membrane at the ends of muscle fibers that interlocks with the tendon's collagen fibers. This folded arrangement increases the surface area of contact and reduces the stress at the junction during muscle contraction. The actin cytoskeleton of the muscle is attached to the ECM of the tendon via integrins and other adhesion molecules at the MTJ [63]. This allows the force generated by muscle contraction to be more efficiently transferred to the tendon. These molecular connections are critical for main-

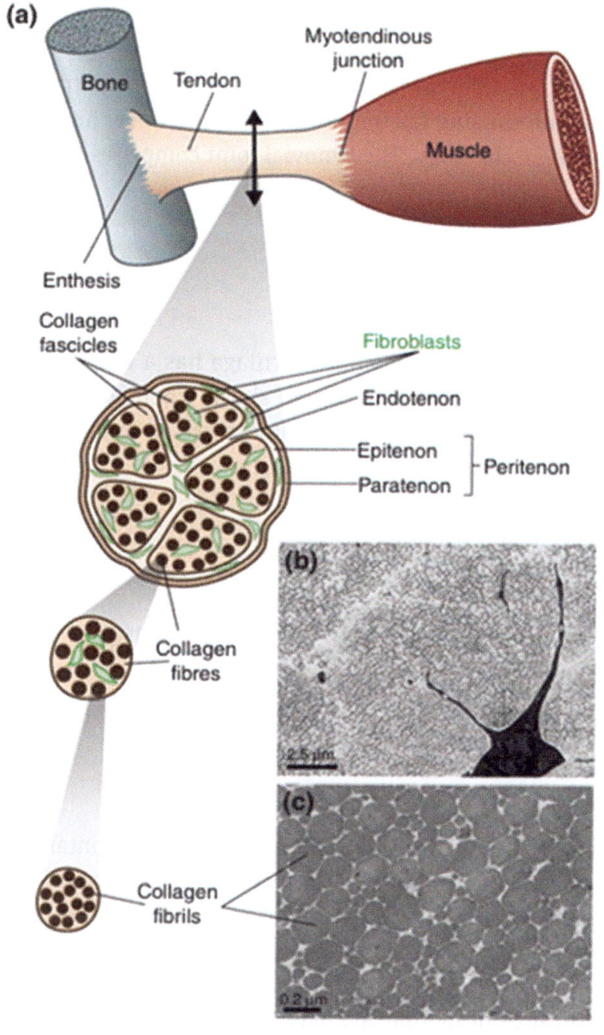

Fig. 2.5: Tendon organization: (a) tendon is linking muscle to bone (b and c) collagen fibers and fibrils at different scales with electron microscopy. Reproduced with permission [61].

taining the structural integrity of the MTJ under repeated loading. The MTJ adapts to mechanical loading through structural changes such as increased folding and reinforcement of molecular anchors. This adaptation allows the rat to become stronger and more resilient in response to repetitive physical activity such as exercise or sports [64]. Strains and tears frequently occur in the MTJ, particularly in activities involving eccentric contractions, in which muscle lengthens during contraction. Inflammation, pain, and decreased mobility may result from these injuries and frequently require physical therapy or surgical correction.

2.5.4 Synovial joints

Synovial joints are highly specialized structures that facilitate smooth, controlled movement between bones while minimizing friction and wear. The knee, hip, and shoulder joints are the most mobile in the body. The synovial joint comprises the following parts (shown in Fig. 2.6).

2.5.4.1 Articular cartilage

Covering the extremities of bones, this smooth, hyaline cartilage has a low friction surface that facilitates movement. Additionally, it absorbs shock when running or walking. Lubrication of the joint occurs through synovial fluid secreted by the synovial membrane, lining the inside of the joint capsule [65].

2.5.4.2 Synovial fluid

This viscous fluid fills the joint cavity, reducing friction between the cartilage surfaces and providing nutrients to the avascular cartilage.

2.5.4.3 Joint capsule

The joint capsule encloses the joint space, providing structural support and containing the synovial fluid. The ligaments stabilize the joint. It also serves as a medium for nutrient exchange and waste removal from cartilage, which lacks its own blood supply. Synovial fluid's primary function is to lessen friction between the surfaces of synovial joint cartilage.

According to their form and range of motion, synovial joints can be categorized as follows:

Joints with a ball and socket, like the hip and shoulder: Permit rotation in addition to other directions of movement.

Hinge joints (e.g., knee and elbow): Allow flexion and extension in one plane.

Pivot joints (e.g., radioulnar joint): Allow rotation around a single axis.

Gliding joints (e.g., wrist and ankle): These joints allow sliding movements between bones.

Wear and degeneration: The health of synovial joints is influenced by factors such as age, mechanical load, and activity levels [67].

Degenerative conditions, such as osteoarthritis, cause increased friction, pain, and limit of movement by breaking down of cartilage. Due to the decreasing ability of cartilage to regenerate various therapies are introduced such as injecting hyaluronic

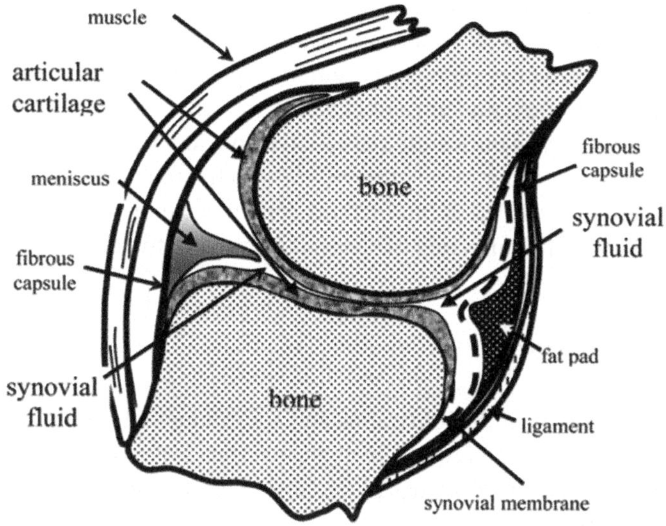

Fig. 2.6: Diagrammatic representation of a synovial joint. Reproduced with permission [66].

acid (visco-supplementation) and medicinal therapy (stem cell therapy) to restore joint function in degenerative disorders.

2.6 Friction mechanism and measurements

Friction on biological surfaces is a complex phenomenon influenced by the interactions among different tissues, fluids, and surface topographies. Depending on the biological context, it is a key fundamental element in restricting and aiding movement. The friction resistance is dependent on the presence of lubricants, surface roughness, and contact nature. Skin, cartilage, mucosal membranes, and tissues with varying degrees of elasticity are considered in biological system. The nature of friction is determined by how surfaces come into contact like in elastic contact tissues like skin and cartilage [68]. Friction is influenced by elastic deformation, where the surface adapts to the load but returns to its original shape after the load is removed. In plastic contact, permanent changes are observed in the surfaces when material exceeds its capacity of elastic deformation, leading to wear and increase in friction. Dry friction occurs when there is little to no lubricating fluid between surfaces, as observed in certain skin-to-surface contacts. This type of friction can cause irritation or damage if sustained over time. Lubricated friction is more common in internal biological systems, such as synovial joints, where synovial fluid lubricates cartilage surfaces, thereby reducing friction and preventing wear. Lubricated friction is typically lower than dry friction because of the separation of surfaces by a thin fluid layer. Boundary lubrication occurs when a thin layer of lubricant is present, but the surfaces still come into contact. It is common in slow joint movements. Mixed lubrication features partial contact between surfaces, while some areas remain separated by fluid [69]. Hydrodynamic lubrication occurs when the movement is fast enough to keep surfaces fully separated by a fluid film, which significantly reduces friction. Molecular interactions such as hydrogen bonding, van der Waals forces, and electrostatic interaction affect friction at the microscopic level. These interactions can enhance or reduce the adhesive forces between surfaces, thus affecting the overall friction coefficient.

2.6.1 Factors affecting friction

The properties of the surfaces, nature of the lubricants, and environmental conditions are key factors that influence the friction level of biological surfaces. Friction can be affected by the microscopic topography of biological surfaces including features such as valleys and ridges [70]. A smooth surface of the articular cartilage minimizes friction, whereas rougher surfaces aid in creating more resistance. The composition of tissues, including the concentration of proteins such as collagen and elastin, affects their frictional behavior. For example, cartilage, which has high water content and an organized collagen matrix, is optimized for low-friction movement. The water content of tissue influences friction, especially in structures like skin and mucosal membranes. Hydration and dehydration affect friction by creating a smooth surface, resulting in a reduction in friction and an increase in friction, drying, and cracking damage, respectively. Biological lubricants such as synovial fluid, mucus, and saliva play crucial roles in reducing friction. The composition of proteins and glycoproteins affects their ability to form effective boundary layers and reduce wear. The friction lines generated by the force are applied to a surface [71, 72]. Higher loads can increase the contact area between surfaces, which increases friction. This is particularly applicable to load-carrying joints such as the knees and hips. The velocity at which surfaces slide against one another decides the type of lubrication mechanism. It was observed that at low and high velocities, the boundary and hydrodynamic lubrication are dominant, respectively. Changes in temperature can modify the viscosity and elasticity of biological lubricants. The friction increased as the viscosity of synovial fluid decreased at higher temperatures. When the fluid is reduced to a point at which it cannot form a protective film between surfaces, friction increases. Measuring friction accurately on biological surfaces is important for understanding how changes might shift friction from normal or near-normal values and for developing therapeutic interventions in the context of diseases such as osteoarthritis and skin disorders. Several techniques have been developed to measure friction in biological systems [73].

The pin-on-disk tribometer method involves applying a pin against a rotating disk to mimic the interaction between two surfaces. The frictional forces were recorded as the interactions of two surfaces rubbing against each other. Research using this method is well-suited for frictional cartilage, skin, and other similar soft tissues. Atomic force microscopy (AFM), a high-resolution technique, measures friction at the nanoscale [74]. This involves pulling a sharp tip over the surface of a biological sample and measuring the resistance encountered. AFM is especially useful for understanding the molecular interactions that contribute to friction on soft, hydrated surfaces such as mucous membranes. Tribometers with contact force sensors: These devices can measure the frictional forces between surfaces under controlled conditions such as varying loads, speeds, and lubrication states. They are often used to study the mechanics of joints by simulating the sliding of cartilage surfaces with and without synovial fluid. Minute biological samples, such as cartilage and skin, are tested using microscale testing to measure their functional properties under conditions. Such tests can be designed to better simulate a more natural environment in which the tissue exists, such as under pressure and temperature changes, for example. Techniques such as ultrasound and MRI offer noninvasive in vivo measurement techniques that help quantify friction in living organisms. It is possible that through joint function studies and lubrication in patients with arthritis, variations in the thickness of cartilage and synovial fluid flow can be observed. Surface roughness and tissue topography are measured using surface profilometry [75]. It is particularly crucial to explain why and how changes in surface structure at the microscopic level affect friction, for instance, in skin as it ages or in articular cartilage during degenerative diseases.

2.7 Properties of biological lubricants and boundary layers

Biological lubricants play a crucial role in maintaining smooth body systems by reducing friction and wear between moving tissues. The chemical involved in making the mixture often gets involved in the composition, and such biological mixtures allow the tissues to slide over each other with ease without causing mechanical damage. This must work effectively for the composition and specific needs of each tissue environment.

2.7.1 Biological lubricants

Biological lubricants, such as synovial fluid, mucus, and saliva reduce friction, protect tissues and facilitate smooth functioning of body systems. These biological fluids can fulfill the demands presented in a specific setting by providing lubrication, padding, and safeguarding against mechanical stress and microbial invasion. In summary, these biological lubricants facilitate physiological processes that take place with slight tissue wear and tear during joint movements, digestive processes, or even breathing processes.

Synovial fluid is the other specialized biological lubricant present in the joint cavities of synovial joints. It plays an essential role in the effortless movement of articulating bones and in the well-being of joints. It comprises lubricin, proteins, plasma filtrate, and hyaluronic acid, and it is a viscous and non-Newtonian fluid. It produces a viscoelastic gel that absorbs compressive stress to cushion joint surfaces. A glycoprotein that lines the surface of articular cartilage lowers the friction between the opposing surfaces of the cartilage. Synovial fluid comprises proteins, albumin, and globulins, which preserve its biochemical characteristics. The fluid also contains a plasma filtrate that supplies it with nutrients and transports waste products from the avascular articular cartilage. The synovial fluid performs several essential functions, such as reducing friction, absorbing shock, transporting nutrients, and homeostasizing joint fluid in synovial joints [76]. The viscoelastic properties of synovial fluid allow it to leave a thin film over the surface of cartilage, thereby diminishing the friction between bones during movement. This is important for lubrication, especially in joints such as knees, hips, and shoulders. The gel-like property of synovial fluid allows it to absorb shock and thus dissipate the impact forces encountered during walking, running, or jumping. The exchange of nutrients and waste products is maintained between the synovial membrane and the cartilage to maintain healthy avascular cartilage. Synovial fluid fills the joint space and thus generates a frictionless environment within the joint that favors the structures it consists of and contributes to health and integrity while at the same time preventing wear and damage [77]. The unique properties of synovial fluid that make it so important for joint health are due to its composition. For instance, synthesis of synovial fluid composition, such as a decrease in viscosity, is prevalent in osteoarthritis and can affect joint function, causing pain and degeneration.

Mucus is another significant biological lubricant that covers the epithelial surfaces of many organs such as the respiratory, gastrointestinal, and reproductive tracts. Its main function is to provide protection and lubrication for these systems. This is a complex gel comprising mostly water, mucins (glycoproteins), enzymes, salts, and antimicrobial peptides. These large glycoproteins in this network are highly dense and give mucus a gel-like texture. The carbohydrate side chains within the mucins hold water molecules in a hydrated layer that withstands dehydration. It usually is 90–95% water to be fluid and spread easily along the epithelial surfaces. Lysozymes and lactoferrin are other types of enzymes in mucosa that break down bacterial cell walls, thereby providing antimicrobial defense [78]. These actions help maintain the pH and osmotic balance of mucus, allowing it to be functional in all tissues. These smaller proteins protect the body from pathogens by disrupting the membranes; thus they assist the immune system at the mucous membrane. The mucus plays several critical roles in various organ systems. This acts physically as a barrier that traps dust, pathogens, and other debris, and this prevents them from passing into the underlying epithelial cells. For instance, it acts as a filter to remove particles from the breathing passage, which is particularly significant here. Mucus does not cause friction between surfaces, as in the act of digesting food, due to its gel-like nature, thus enabling food to pass down the esophagus and easily into the intestines. It prevents moisture from drying up on epithelial surfaces. The mucomocytic system plays a significant role in several physiological functions and should be considered an integral part of the human organism. The respiratory tract and its mucous membranes are such exposed tissues. Trapping pathogens containing antimicrobial peptides in the mucus plays an important role in the innate immune system, preventing infections at mucosal surfaces. The composition and viscosity of mucus can vary with conditions such as dehydration, infection, or chronic diseases like cystic fibrosis, affecting its protective and lubricative functions [79, 80].

The main functions of saliva are related to digestion, lubrication, and oral health and are multifunctional fluids secreted by the salivary glands of the oral cavity. The composition of this material allows an environment to be kept moist within the oral cavity and allows it to perform various functions. It contains a mixture of electrolytes, enzymes, mucins, antimicrobial agents, and water. Water constitutes about 98–99% of saliva, which is the primary constituent responsible for its fluidity and facilitating moistening of the oral cavity. Sodium, calcium, phosphate, bicarbonate, and potassium are present in its ion content. These factors play a crucial role in maintaining pH balance and serving as the buffering capacity of saliva to protect teeth from acid erosion. Starch and fat digestion are achieved by amylase and lipase enzymes, respectively [81]. Both are more likely to be swallowed and digested in the stomach than outside in their intact forms. Glycoproteins in saliva make it a lubricant and increase the viscosity of saliva, forming a protective layer above the oral mucosa. Lysozyme, lactoferrin, and immunoglobulins are major constituents of saliva. The agent controls bacterial growth in the oral cavity and inhibits infections. Saliva performs many critical oral health maintenance functions. Mucins in saliva serve as a protective film over the tissues of the oral cavity, making it easier to chew food, form a bolus, and swallow. This lubricant also facilitates speaking by preventing the oral surfaces from sticking together. The salivary enzymes, mainly amylase, begin the breakdown of carbohydrates into simpler sugars at the mouth and then further break down easily accessible starch, a form of complex carbohydrates. Saliva cushions foods or drinks that are acidic, thus preventing their acidic attack on teeth and thereby preventing tooth damage through erosion and cavities. Saliva is also rich in growth factors that stimulate tissue repair, thereby healing within the oral cavity such as minor injuries such as cuts or abrasions. Growth of bacteria and fungi in the oral cavity is regulated by added antimicrobial agents by limiting their development, thus preventing oral candidiasis and many such infections. The importance of saliva in oral cavity health could be attributed to the multifunctionality of saliva. Conditions such as dry mouth or dry mouth can severely impair oral function and health by causing problems with swallowing and speaking and an increase in dental decay and oral infections [81].

2.7.2 Boundary layers in biological systems

These are thin layers that form at interfaces between different types of biological tissues or between tissues and biological fluids. These layers are important for movement mechanics within the body because they manage friction, reduce wear, and allow smooth interaction among moving parts. Understanding the properties and functions of such boundary layers involving friction, lubrication, and wear in biological systems is crucial for further exploration of the biotribological behavior of tissues. Boundary layers in biological systems are transitionally specialized zones providing a pathway to facilitate fluid-tissue surface interactions such as between synovial fluid or mucus and the surfaces of cartilage, epithelial linings, and cell membranes. All layers have properties appropriate to the environment in which they are exposed; however, they allow movement with minimal protection. Biological lubricants may constitute boundary layers: glycoproteins, mucins, and hyaluronic acid. The composition of boundary layers can vary with tissue and fluid [82]. Many boundary layers are gel-like or viscous in consistency, permitting them to coat surfaces but still deform enough to track movements. They are usually very well hydrated, so they have a smooth surface and a low friction interface. Hydration is especially important in joint spaces where synovial fluid functions as a barrier layer over the surface of cartilage. The thickness of the boundary layer may differ depending on the mechanical demands placed on the tissue. For instance, in synovial joints, the boundary layer is relatively thin but effective at reducing frictional resistance, whereas mucus layers within the respiratory tract should be thicker to protect epithelial surfaces from the influence of the environment. Biological boundary layers are not static constructs that passively await alterations in mechanical load, movement, and environmental conditions. This dynamic behavior is evident, particularly in synovial joints, where the thickness and viscosity of the boundary layer are sensitive to changes in joint movements and load conditions. The sensitivity of boundary layers to the physiological environment ensures that they can maintain their protective and lubricative functions despite changes in the environment. Boundary layers function as a medium for biochemical interactions between cells and their environment. They can serve as a lubrication component that guards tissues from pathogenic sites simultaneously and facilitates the passage of nutrients and gas exchange, along with the removal of waste products. For example, mucous lining in the respiratory tract guards the tissue by providing lubricant function to the airway surfaces while catching particles and microorganisms, thus preventing them from accessing other deep tissues. In many biological systems, boundary layers are constructed primarily to reduce friction between moving surfaces [83]. This function decreases friction, which is one of the most important functions in maintaining and prolonging the life of many tissues with repetitive movement.

In synovial joints like the knee and hip, the boundary layer of synovial fluid reduces friction between the cartilage surfaces articulating with one another. This lubrication allows the bones to glide freely over one another without causing damage to the cartilage. Synovial fluid is produced as a thin film over the surface of the cartilage, creating a slippery interface that minimizes resistance during joint movements. At these interfaces, several mechanisms contribute to friction reduction: This arises because molecules such as lubricin adhere to the cartilage surfaces, preventing them from coming into direct contact with each other and allowing for sliding. When this movement occurs, synovial fluid is squeezed into the space between the cartilage surfaces and a fluid film, which creates surface separation and further reduces friction. In most cases, both the boundary and hydrodynamic lubricants are combined to minimize friction, even when the loads and movements are varied. The interiors of many organs, such as the esophagus, stomach, and intestines, are guarded by the boundary layer of mucus in the gastrointestinal tract. Directly, that friction can be reduced by boundary layers and impacts the wear and tear that tissues experience over time. Because these layers are important for maintaining tissue integrity and protecting against mechanical stress-induced damage, they reduce wear and tear in high-motion environments. In synovial joints, the boundary layer of synovial fluid protects the articular cartilage from wear and erosion. This is because it protects the smooth surface of the cartilage from frictionally induced wear and erosion that may occur if the bones undergo repeated movements due to friction elimination. Lubricin in the synovial fluid also prevents adhesive wear, whereby the opposite cartilage surfaces might otherwise stick and damage occurs during movement. When the synovial boundary layer is damaged, for example, in osteoarthritis, increased friction and loss of lubrication lead to breakdown of the cartilage. The boundary layer offers less resistance to the surfaces of the bones, thus leading to more direct contact between the bone surfaces, which results in pain and inflammation and further joint damage. Mechanical and chemical damage to epithelial tissue is prevented by boundary layer of mucus. For example, the mucus layer of the digestive tract acts as a barrier against acid corrosion, thus preventing ulcers. It also acts as a lubricant. Once the integrity of the mucus layer is lost, epithelial tissues come into increased friction along with the corrosive actions of digestive enzymes, resulting in gastritis or peptic ulcers. In contact systems, for example, as tissues slide against each other, eyelids overlying the eye or tongue in the oral cavity border layers can prevent microtrauma that add up. For example, a tear film prevents repetitive blinking from removing the layers of the corneal surface, which would otherwise compromise the integrity of the eye surface.

The boundary layers also play a role in preventing acute injuries. For example, the fluid in the joints enables the absorption and spreading of the load on the surfaces of the joint, and this prevents injuries such as tearing of the cartilage and sprains in the joint when a person performs any high-impact activity. Similarly, the mucus in the respiratory system helps prevent rips in the tissues in the airways when one coughs or inhales dry or polluted air. Therefore, failure of the protective roles of the boundary layers could cause accelerated wear and tear and subsequently lead to chronic conditions, along with an ever-decreasing functionality of the tissues involved. Thus, the health and maintenance of these layers are crucial to overall body wellness, especially in systems in which mechanical interactions are frequently intense [84, 85].

2.7.3 Lubrication mechanisms

Biological systems are mainly lubricated by mechanisms that decrease friction and wear, thus leading to energy loss in interacting surfaces. Various mechanisms have been adapted for different demands on the mechanical environment of tissues, ensuring smooth movement with minimal damage. Hydrodynamic, boundary, and mixed lubrication are the fundamental lubrications occurring in biological systems. Each mechanism works under different conditions and provides specific benefits to maintain tissue functionality and integrity. Hydrodynamic lubrication is a mechanism in which a fluid film separates two moving surfaces without their surfaces coming into direct contact with each other, which causes a reduction in friction. It can be very effective in terms of reducing friction if the surfaces are in relative motion to each other at an adequate speed and the lubricant can maintain a fluid film. Hydrodynamic lubrication, like hydrodynamics, finds an important place in biological systems, for example, at the sites of joints and other tissues with sliding movements. Hydrodynamic lubrication obeys a principle that, wherever there is a relative movement of two surfaces with a fluid layer between them, the motion of the two surfaces causes pressure to develop within the fluid to produce a load-bearing film [86]. This pressure keeps surfaces apart, allowing the fluid to carry the load, which heavily reduces friction between the surfaces. The size of the fluid film depends on the fluid's viscosity, relative speed of the surfaces, and geometry of the contact area. In synovial joints, such as the knee and hip joints, synovial fluid creates a hydrodynamic film in the cartilage surfaces of the bones. When a person walks or runs, synovial fluid is drawn into the joint space, creating a cushion of fluid to support and lubricate the joint. Hydrodynamic friction can be effective for steady motions such as walking or cycling if the relative motion and pressure of the joint surfaces are maintained. Hydrodynamic friction effectively reduces friction and helps to smoothen the distribution of loads by equalizing the forces on the joint surfaces. Its effectiveness becomes much weaker in slow or static movements in which the fluid film cannot be maintained, then, other lubrication mechanisms must come into effect. A lube boundary occurs when the film thickness of a lubricant is less than the effective clearance between the opposing surfaces. In this case, the sliding operation relies mainly on friction reduction based on the direct adsorption of lubricating molecules onto the surfaces. It is worth noting that this mechanism plays an especially important role in biological systems where movement is slow or when the load is high, which makes it difficult to maintain a full fluid film. The method follows the principle of monolayer adsorption of lubricating molecules such as proteins or glycoproteins on the tissue surface. Such molecules form a slippery coating at the interface between the touching surfaces, even when they are in intimate contact. The viscosity of the liquid does not affect the boundary lubrication. Moreover, it depends on the chemical interactions between surfaces and lubricant molecules. Lubricin, a glycoprotein that is synthesized in synovial fluid, is the most important molecule associated with boundary lubrication within a joint.

This produces a coating over the articular cartilage, so the surfaces of the cartilage never touch each other during low-motion activities like standing or walking slowly. The mucins attached to the corneal epithelium create a smooth surface that the eyelids move over when they blink, forming a tear film on the surface of the eye. The mucins present in saliva form a protective layer over the oral mucosa, enabling food to move rather smoothly and reducing friction between the surfaces of the mouth [87].

The combined action of mixed lubrication helps protect the articular cartilage because it diminishes friction and provides some degree of shock absorption, thus supporting the longevity and health of joints. Because mixed lubrication allows tissues to adapt to various loading conditions, it offers flexibility in the way lubrication is supposed to be achieved based on the situation. Mixed lubrication provides the best performance by filling the gap between low friction within hydrodynamic lubrication and protection by boundary lubrication. Thus, it is well-suited for biological systems that are often subject to diverse mechanical conditions. Control of the transition between these two mechanisms is often problematic, and changes in the composition of lubricants, as observed in arthritis with reduced levels of lubricant lubricin, can compromise mixed lubrication function [88].

2.8 Biomechanics of load-bearing tissues

Mechanical loads are carried by specific tissues in the body that resist forces in different directions and maintain structural integrity. Bones, cartilage, tendons, and ligaments are some examples that also have different structures. These interactions form a fundamental platform for the biomechanics of movement, stability, and load distribution throughout the human body [89]. Such tissues are studied in terms of composition and mechanical behavior because this is crucial to understanding responses to stress and injury and how such tissues may adapt to those physical demands.

2.8.1 Structure and function of load-bearing tissues

Load-bearing tissues are built to be able to tolerate specific mechanical loads: compressive, tensile, or shear forces. Tissues have evolved to provide a mix of strength, flexibility, and resilience so that movement by the body does not injure them. Bone, cartilage, tendons, and ligaments are the key load-bearing tissues. Their detailed structural views, mechanical properties, and biomechanical roles are listed below.

2.8.1.1 Bone

The bone is a major structural component of the skeletal system and provides support, protection, and leverage for movement. It has the unique ability to withstand compressive forces and simultaneously allowing dynamic activity because of its high rigidity and flexibility. The highly organized microstructure of bone can be distinguished as trabecular or spongy bone and cortical or compact bone. Cortical bone forms a strong, dense, compact outside layer and comprises approximately 80% of the skeletal mass. It has high resistance to bending and torsion. Cortical bone contains tightly packed osteons called Haversian systems (cylindrical structures parallel to the long axis of the bone). A central canal houses blood vessels and nerves in each osteon, with concentric layers of calcified matrix surrounding it. Trabecular bone occurred at the ends of long bones and vertebrae. This sponge-like porous bone has a honeycomb-like structure. It is lighter and more pliable because its open network of thin beams absorbs shock and distributes forces. Mineral and calcium exchange occurs because of the high surface area of the trabecular bone. Mechanical properties are also essential for bone, particularly for its role as a load-bearing material [90]. Due to the presence of hydroxyapatite in its mineral content, it has compressive strength, which enables it to resist deformation under pressure. However, the same material is relatively more sensitive to tensile and shear forces, leading to increased fracture under such conditions. Bones are viscoelastic because they deform under stress and return to their original shape when the load is withdrawn. This property allows it to absorb impact forces during falls and prevents fractures. Bones are isotropic, which means that their strengths and stiffness vary in the directions in which the force is applied. It is strongest along its long axis in bone, for example, in the femur, where it typically experiences compressive forces during weight-bearing activities [91]. From birth until death, bone may maintain its structural integrity and adjust to variations in physical activity through mechanical loading-induced bone remodeling, such as bone production and resorption.

2.8.1.2 Cartilage

A specific type of connective tissue found at a joint's surface is termed cartilage. The slippery nature ensures smooth movement and helps distribute loads across joints. This arrangement can withstand both compression and shear forces and is therefore fundamental to the function of a joint. There are three major categories of cartilage, with each being specially designed to play a different mechanical role: hyaline, fibrocartilage, and elastic cartilage. Among all, hyaline cartilage is abundant and can be found on articular surfaces of bones, the nose and trachea, and ends of ribs. It has a smooth, glassy appearance and comprises chondrocytes embedded in a matrix rich in type II collagen and proteoglycans. Its main functions are to reduce friction and absorb compressive forces in the joints. Fibrocartilage is found in intervertebral discs, menisci, and pubic symphysis. It is strengthened by dense collagen fibers of type I collagen to impart tensile strength and resistance to shear forces. Therefore, it is an essential support and shock absorber in regions under heavy stress.

Elastic cartilage: It is found in structures such as the ear and epiglottis. It contains elastic fibers in addition to collagen. This structure provides a flexible support and allows structures to retain their form in the presence of bending forces [92]. Cartilage plays a crucial role in distributing mechanical stress as well as preventing the overuse of a joint. As it is full of proteoglycans, the ECM of cartilage may get squashed; however, it has a great ability to recover its shape and hence can be considered as a great shock-absorbing material. The smooth surface of hyaline cartilage with synovial fluid enables the low-friction movement of the bones in synovial joints, thereby lessening the damage when movements are repeated. Cartilage can distribute loads over much larger areas and does not place undue stress on points of the underlying bone. This treatment also prevents damage to bone surfaces and ensures joint function. Cartilage is required to provide flexibility and to resist compressive forces. Its role in the maintenance of joint health is therefore vital. However, self-repairing ability is bounded [93]. Additionally, cartilage damage and degeneration are the causes of conditions like osteoarthritis.

2.8.1.3 Tendons and ligaments

The tendons and ligaments are fibrous connective tissues that play significant roles in transmitting forces and maintaining joint stability. Their structural makeup is also similar but plays different roles: tendons connect bones to muscles enabling movement, and ligaments connecting other bones with each other stabilize. These are densely packed collagen fibers with only a population of fibroblasts and small ECM components tenocyte in the case of tendons. Here again, type I collagen constitutes the majority, leading to tendons and ligaments with considerable tensile strength. The collagen fibers are arranged in parallel bundles, so these tissues can resist tensile loads in a specific direction. Ligaments and tendons are critical for joint stability and efficient force transmission. The tendon transmits forces produced by muscle contractions to the bones, enabling movement of the limbs and other body components. They resist tension loads and have a zig-ragged collagen matrix structure that linearizes with tension to provide some elasticity and prevent rupture. Ligaments help joints by providing bony attachment and restriction of extreme movements such as hyperextension and lateral displacement. Their viscoelastic nature means that they stretch a little with load but mainly stabilize joints and prevent injuries such as dislocation. Some tendons, for example, the Achilles tendon, store and release elastic energy during running or jumping. This elastic energy storage capacity reduces the metabolic cost of movement in most cases because energy transfer is more efficient. The tendons and ligaments provide structural support and enable them to withstand tensile stress while allowing them to perform physical activities. They support the joint, enabling dynamic movements while maintaining stability in both [94–96]. Poor blood supply makes it difficult to heal from injuries such as sprains or tears, and rehabilitation programs are often specific to the injury.

2.8.2 Adaptation and mechanical properties

The mechanical properties of biological tissues are critical for understanding how the body responds to forces and stresses during everyday activities. These properties include stress-strain behavior, which determines how tissues adapt to mechanical loading, absorb energy, and deform over time. The capacity of tissues to adapt allows the body to withstand various physical demands, thus ensuring the structural integrity and function of organs and musculoskeletal structures. The stress-strain relationship is fundamental to understanding how materials, including biological tissues, respond to mechanical forces. The internal force per unit area within a material is referred to as stress, whereas the strain measures the deformation in length relative to the original length of the material. Stress is measured in units such as pascals or megapascals, whereas strain has no unit as it is a dimensionless quantity [97]. When a rising stress is applied to a material, it initially changes in a reversible manner, but as the stress is removed, its original shape is returned. This region is known as the elastic region. The yield point marks the transition between elastic and plastic deformation. The stiffness of a material in elastic regions was measured by Young's modulus. The stress-to-strain ratio provides information about how much strain results from a given stress. Biological materials such as bones have great stiffness and represent high Young modulus, whereas soft tissues such as skin and cartilage show more flexibility, which indicates low values. The relationship between stress and strain is given by a stress-strain curve graph, where Young's modulus is represented by the initial slope. The curve shows different regions including the elastic region, yield point, plastic region, and ultimate tensile strength. For biological tissues, the curve is often nonlinear, reflecting the complex arrangement of cells and ECM components. For example, tendons and ligaments have an initial "toe region" where collagen fibers straighten, followed by a linear region where the tissue becomes stiffer. Understanding stress-strain relationships is crucial for analyzing how tissues behave under different loads, such as how bones resist compression or how tendons stretch under tension during physical activities [98, 99].

2.8.3 Pathophysiology of load-bearing tissues

Load-bearing tissues in the musculoskeletal system include bones, cartilage, tendons, and ligaments, and these ensure that the entire musculoskeletal system continues to operate properly. Although they do an excellent job, numerous pathologies hamper the working of tissues, bringing with it pain and disability and impaired quality of life. In this subsection, we discuss the very common pathophysiologies of load-bearing tissue, osteoarthritis, tendonitis, fracture mechanics, and their healing [100].

2.8.3.1 Osteoarthritis

It moves beyond merely being a chronic disease of articular cartilage degeneration knotted to joint pain, stiffness, and limitation of movement. It is the most common form of arthritis, and this condition can affect a person immensely. The causes of osteoarthritis include multifactorial mechanical, biological, and biochemical components. Over time, repetitive joint loading, obesity, and joint injury increase the mechanical stress on cartilage, causing wear and tear [101]. The chondrocytes present in OA are of lesser assistance in the production and maintenance of ECM. This causes the destruction of collagen and proteoglycans, and structural integrity is lost. The exposure of the joints to such factors leads to further joint damage through the release of cytokines, for example, IL-1 and TNF-alpha which trigger a cascade of inflammation pathways. These mediators are also responsible for bone marrow edema and synovial membrane inflammation, which further exacerbates joint pain. Changes in the subchondral bone and cartilage loss occur simultaneously. Degradation of articular cartilage and changes in the bone underlying it result in functional impairments. The naked nerve endings of the underlying subchondral bone are exposed because of the loss of cartilage, and through movement and weight-bearing activity pain is produced. Morning stiffness or stiffness following periods of inactivity commonly affects most patients with OA because it results in limitations in mobility and function. The loss of range of motion results from joint deformities and pain, which makes activities of daily living challenging [102, 103].

2.8.3.2 Tendonitis

It is an inflammatory condition affecting the tendons. The etiology is often an overuse or acute injury. It is commonly observed in athletes and individuals performing repetitive movements. Repetitive movements cause microtears in the tendon fibers. The resulting inflammation causes pain. Activities that cause tendonitis include running, jumping, and throwing. It can suddenly present with an acute injury, such as falls and heavy lifting, that causes localized pain and inflammation at the site of the tendon. Tendons lose elasticity with advancing age and become more vulnerable to injury [104]. Degenerative changes may result in diminished blood flow and healing potential, thereby worsening the effects of overuse. Poor biomechanics may be caused by improper footwear, muscle imbalance, and poor technique while performing the activity. This can increase the probability of developing tendonitis because these factors tend to exert stress on the tendons. The consequences of tendonitis can be debilitating: the main symptom is localized pain that worsens with activity and improves with rest. Pain is often accompanied by tenderness and swelling. The inflammatory condition may lead to the weakening of the involved tendon, which limits strength or endurance during specific activities. Tendonitis can strongly affect both daily activity and sport performance, eventually resulting in a reduction in participation in any form of physical activity, with further potential for lifelong disability unless treated. Tendonitis treatment primarily focuses on rest, physical therapy, anti-inflammatory drugs, and sometimes corticosteroid injections aimed at tackling inflammation to facilitate healing. Fractures are the broken continuity of bone, usually caused by trauma, stress, or underlying pathologies. Healing after fracture is a complex biological process that restores the integrity and function of the bone [105]. There are basically four major types of fractures: closed fractures: the bone is broken but does not cut through the skin; open fractures: the bone is broken and cut through the skin, increasing the risk of infection; stress fractures: small cracks in the bone are caused by repeated stress or overuse and are usually found in athletes; pathological fractures: fractures occurring in weak bones due to disease, for example, osteoporosis, or tumors [106].

Healing of fractures, in their natural sequence, follows a definite pattern: immediately after fracture, blood vessel rupture results in a hematoma that develops at the fractured site. This sets off the inflammatory response, which introduces immune cells to the fracture site to clean up debris. Several days later, a soft callus of cartilage begins to form around the fracture. Chondrogenesis stabilizes the fracture. After weeks, the soft callus transforms into a hard callus of woven bone through a process known as ossification, and the fracture becomes more stable. The final stage is the remodeling of the woven bone into lamellar bone, the bone's original shape and strength are regained through months to years. Remodeling is influenced by mechanical loading, and proper stress, healing, and strength are maximally attained. Some factors may delay the healing process. Among these factors, younger people generally heal faster than older people because of better blood supply and metabolic activity [107, 108].

Healing is also dependent on proper nutrition, calcium, and vitamin D. Other underlying diseases such as diabetes or osteoporosis and smoking may prevent appropriate healing and cause complications. If the fracture heals adequately with proper alignment and immobilization, optimal healing occurs. In complicated fractures, internal fixation may be performed to ensure proper alignment. Understanding the pathophysiology of load-bearing tissues is important for developing appropriate and effective prevention strategies and treatments for osteoarthritis, tendonitis, and fractures. This will prevent early involvement and sensible management by helping to preserve joint function, reduce pain, and improve the quality of life of affected people [109].

2.9 Conclusion

Biotribology, the science of the surfaces of biological entities, has been increasingly studied to understand how these living organisms work under mechanical stress to manage friction, wear, and lubrication. The insights gained thus far have shed light on how such complexity is designed naturally and how it adjusts under mechanical stress. To round out this work, the following summary of key issues discussed during this text:

Understanding biological surfaces: Biological surfaces represent the very first interfaces both between the tissues and between the body and its surroundings. In these surfaces, it protects the inner structures, allows for movement, and manages the mechanical load. Therefore, their structure and function are very closely related to friction and wear minimization under biological conditions. This refers to cellular and extracellular contributions that involve contributions from cellular components such as epithelial cells and fibroblasts as well as extracellular components including the ECM in their contribution to maintaining the integrity and functions of biological surfaces. These consist of other ECM components that consist of collagen, elastin, and other structural forms that provide structural support while cells provide tissue repair and adaptation.

Interfacing of tissues: Epithelial-mesenchymal junctions, tendon-bone interfaces, and synovial joints are specialized interfaces that ensure the presence of interfacing tissues with different mechanical properties and facilitate smooth movement by effective transfer of forces with prevention of damage.

Biological mechanisms of lubrication: Many biological lubricants, such as synovial fluid, mucus, and saliva, assist in the lubrication of two moving surfaces. Such fluids are described to be compatible with a given environment, such as a joint cavity or mucous surface; thus, they can fluidly move while also providing protection against wear.

Biomechanics of load-bearing tissues: Bones, cartilage, tendons, and ligaments are basic tissues that are important for understanding the mechanical properties of the body. The tissue is adapted to physical forces and viscoelastic forces with well-defined stress-strain relationships so that it can experience repeat loading without failure.

Pathophysiology of joint degeneration: Osteoarthritis, tendonitis, and other degenerative conditions underpin the need for the maintenance of appropriate biotribological function. These changes occur due to disruptions in normal lubrication mechanisms, wear of articular cartilage, or damage to load-bearing structures, leading to pain and loss of mobility. Clinical and engineering applications of biotribology include prosthetic joint design, tissue engineering, and regenerative medicine. Understanding the mechanisms of natural lubrication and load bearing is important for developing implants, biomaterials, and therapies that restore or mimic the properties of native tissue.

Overall, the area of biotribology provides a kind of multidisciplinary prospective (biology, mechanics, and materials science) on how living systems cope with physical forces. By understanding this, we can expand our understanding of human health and disease while guiding innovations in medical technology. Such grand landmarks in principles understanding through biotribology would always hold the full promise of open opportunities for subsequent research and innovations to push the self-advance ability through various channels. Combined with computational models, they might provide greater insight into tissue behavior under mechanical stress and better understanding of friction and wear in real time. Biomimetic materials and artificial lubricants: Biomimetic materials are developed to mimic natural tissue and interface properties. The development of synthetic cartilage and hydrogels will be oriented toward the artificial reproduction of the viscoelasticity and lubrication properties of natural cartilage, which are critical for joint replacement and repair. Artificial lubricants simulating synovial fluid or mucus would work much better in joint treatments and implants. The restoration of proper lubrication and mechanical function may delay or prevent surgical interventions such as joint replacement in patients with osteoarthritis or other joint diseases. Mechanobiology refers to the research area focused on how cells sense and respond to mechanical stimuli, an area of crucial research into how tissues adapt to load and stress. The regulation of signaling pathways in tissue remodeling and ECM production may offer new therapeutic approaches to help strengthen tissues by augmenting the body's natural repair process. Wearable technology and biotribological monitoring: Wearable sensors and devices could be integrated for real-time monitoring of joint movements, loads, and friction levels, thus allowing assessment of the biomechanical performance of individuals, tracking of degenerative conditions, and guidance through personalized rehabilitation protocols to obtain better outcomes.

Microbiome and surface interactions: Recent studies suggest that the microbiome, most notably in the gut or the skin, for example, influences surface properties and their interaction with lubricants. Further insights into the roles of microbes in sustaining or changing surface or fluid properties may thus lead to novel approaches for treating conditions such as inflammatory joint disease or skin disorders. The biotribological adaptability of other organisms has evolved in incredible ways to enable specialized functions, which can be inspiration for novel engineering solutions. Studying the surfaces of fish scales, insect wings, and the lubricated joints of mammals may reveal materials and structures applicable in medical and industrial technologies.

Personalized medicine in biotribology: Advancements in genetic analysis and data-driven medicine are expected to lead to the development of more individualized treatments for biotribological interventions. Treatment of joint lubrication, cartilage repair, or tissue engineering may be optimized about effectiveness and outcome by focusing on the patient's genetic profile or specific biomechanical characteristics.

Sustainable and biodegradable materials: The following research stages relate to how the processes of wear and tear may be slowed, thereby increasing the longevity of prosthetic devices and improving living conditions for the aged population. Biology has shown that nature uses surfaces for dealing with friction, wear, and mechanical forces in its maintenance to sustain stability and function within the body. Advances in biotribology continue to assure not only a better understanding of the natural world but also better health for people through innovative therapies, materials, and technologies. The gap between biological mechanisms and engineering applications for more effective treatments for conditions affecting movement and joint function will certainly continue to be bridged, eventually leading to the betterment of individuals at all stages of life.

References

[1] Gehr P. (2018). Interaction of nanoparticles with biological systems. *Colloid Surfaces B: Biointerfaces*, 172, 395–399.

[2] Arzt E, Quan H, McMeeking RM, Hensel R. (2021). Functional surface microstructures inspired by nature–From adhesion and wetting principles to sustainable new devices. *Progress in Materials Science*, 120, 100823.

[3] Mitra SP. (2020). Protein adsorption on biomaterial surfaces: Subsequent conformational and biological consequences – a review. *Journal of Surface Science & Technology*, 36, 7–38.

[4] Yilmaz, H., Abdulazez, I. F., Gursoy, S., Kazancioglu, Y., & Ustundag, C. B. (2025). Cartilage Tissue Engineering in Multilayer Tissue Regeneration. *Annals of Biomedical Engineering*, 53(2), 284–317.

[5] Porzionato A, Stocco E, Barbon S, Grandi F, Macchi V, De Caro R. (2018). Tissue-engineered grafts from human decellularized extracellular matrices: A systematic review and future perspectives. *International Journal of Molecular Sciences*, 19(12), 4117.

[6] Liu WL, Zou MZ, Qin SY, Cheng YJ, Ma YH, Sun YX, Zhang XZ. (2020). Recent advances of cell membrane-coated nanomaterials for biomedical applications. *Advanced Functional Materials*, 30(39), 2003559.

[7] Şenel S. (2021). An overview of physical, microbiological and immune barriers of oral mucosa. *International Journal of Molecular Sciences*, 22(15), 7821.

[8] Prescott SL, Larcombe DL, Logan AC, West C, Burks W, Caraballo L, Campbell DE. (2017). The skin microbiome: Impact of modern environments on skin ecology, barrier integrity, and systemic immune programming. *World Allergy Organisation of Journal*, 10, 1–16.

[9] Yu F, Nivasini D, Kumar OS, Choudhury D, Foo LC, Ng SH. (2018). Microfluidic platforms for modeling biological barriers in the circulatory system. *Drug Discovery Today: Technologies*, 23(4), 815–829.

[10] Tee JK, Yip LX, Tan ES, Santitewagun S, Prasath A, Ke PC, Leong DT. (2019). Nanoparticles' interactions with vasculature in diseases. *Chemical Society Reviews*, 48(21), 5381–5407.

[11] Wells JM, Brummer RJ, Derrien M, MacDonald TT, Troost F, Cani PD, Garcia-Rodenas CL. (2017). Homeostasis of the gut barrier and potential biomarkers. *American Journal of Physiology - Gastrointestinal and Liver Physiology*, 312(3), G171–G193.

[12] Ruprecht V, Monzo P, Ravasio A, Yue Z, Makhija E, Strale PO, Viasnoff V. (2017). How cells respond to environmental cues–insights from bio-functionalized substrates. *Journal of Cell Science*, 130(1), 51–61.

[13] Javaid M, Haleem A, Singh RP, Suman R. (2023). Sustaining the healthcare systems through the conceptual of biomedical engineering: A study with recent and future potentials. *Biomedical Technology*, 1, 39–47.

[14] Bayer IS. (2018). Advances in tribology of lubricin and lubricin-like synthetic polymer nanostructures. *Lubricants*, 6(2), 30.

[15] Siddaiah A, Menezes PL. (2016). Advances in bio-inspired tribology for engineering applications. *Journal of Bio- and Tribo-Corrosion*, 2, 1–19.
[16] Xie C. (2019). Bio-inspired nanofunctionalisation of biomaterial surfaces: A review. *Biosurface and Biotribology*, 5(3), 83–92.
[17] Wagner RM, Maiti R, Carré MJ, Perrault CM, Evans PC, Lewis R. (2021). Bio-tribology of vascular devices: A review of tissue/device friction research. *Biotribology*, 25, 100169.
[18] Krakowski P, Rejniak A, Sobczyk J, Karpiński R. (2024). Cartilage integrity: A review of mechanical and frictional properties and repair approaches in osteoarthritis. *Healthcare*, 12(16), 1648.
[19] Lanza A, Ruggiero A, Sbordone L. (2019). Tribology and dentistry: A commentary. *Lubricants*, 7(6), 52.
[20] Yuan, H., & Cui, W. (2024). The lubricated matter in body. *Progress in Materials Science*, 146, 101334.
[21] Austin W, Hdeib M, Fraser P, Goldchtaub M, Shams E, Han T, Adibnia V. (2024). Oral lubrication, xerostomia, and advanced macromolecular lubricants for treatment of dry mouth. *Lubricants*, 12(4), 126.
[22] Käsdorf BT, Weber F, Petrou G, Srivastava V, Crouzier T, Lieleg O. (2017). Mucin-inspired lubrication on hydrophobic surfaces. *Biomacromolecules*, 18(8), 2454–2462.
[23] Mang T, Busch C. (2017). Lubricants in the tribological system. *Lubricants Lubrication*, 11–30.
[24] Kruzic JJ, Arsecularatne JA, Tanaka CB, Hoffman MJ, Cesar PF. (2018). Recent advances in understanding the fatigue and wear behavior of dental composites and ceramics. *Journal of the Mechanical Behavior of Biomedical Materials*, 88, 504–533.
[25] Hu C, Ashok D, Nisbet DR, Gautam V. (2019). Bioinspired surface modification of orthopedic implants for bone tissue engineering. *Biomaterials*, 219, 119366.
[26] Vladkova TG, Staneva AD, Gospodinova DN. (2020). Surface engineered biomaterials and ureteral stents inhibiting biofilm formation and encrustation. *Surface and Coatings Technology*, 404, 126424.
[27] Karamanos NK, Theocharis AD, Piperigkou Z, Manou D, Passi A, Skandalis SS, Onisto M. (2021). A guide to the composition and functions of the extracellular matrix. *FEBS Journal*, 288(24), 6850–6912.
[28] Poole JJ, Mostaço-Guidolin LB. (2021). Optical microscopy and the extracellular matrix structure: A review. *Cells*, 10(7), 1760.
[29] Crossley RM, Johnson S, Tsingos E, Bell Z, Berardi M, Botticelli M, Shuttleworth R. (2024). Modeling the extracellular matrix in cell migration and morphogenesis: A guide for the curious biologist. *Frontiers in Cell and Developmental Biolo Gy*, 12, 1354132.
[30] Honda H. (2017). The world of epithelial sheets. *Development, Growth & Differentiation.*, 59(5), 306–316.
[31] Xu H, Ghishan FK. (2018). Molecular physiology of gastrointestinal function during development. *Physiology of the Gastrointestinal Tract*, 235–269.
[32] Cherkas OA, Kobeza PA, Marchenko DG. (2023). Basic principles of the structure and organization of connective tissue. *Morphologia*, 1, 77–85.
[33] Mukund K, Subramaniam S. (2020). Skeletal muscle: A review of molecular structure and function, in health and disease. *Wiley Interdisciplinary Reviews-Systems Biology and Medicine*, 12(1), e1462.
[34] Lee KH, Cha M, Lee BH. (2021). Crosstalk between neuron and glial cells in oxidative injury and neuroprotection. *International Journal of Molecular Sciences*, 22(24), 13315.
[35] Borst K, Dumas AA, Prinz M. (2021). Microglia: Immune and non-immune functions. *Immunity*, 54(10), 2194–2208.
[36] Owczarzy A, Kurasiński R, Kulig K, Rogóż W, Szkudlarek A, Maciążek-Jurczyk M. (2020). Collagen-structure, properties and application. *Biomaterials Science & Engineering*, Vol 23, pg. 156.
[37] Manou D, Caon I, Bouris P, Triantaphyllidou IE, Giaroni C, Passi A, Theocharis AD. (2019). The complex interplay between extracellular matrix and cells in tissues. *The Extracellular Matrix: Methods and Protocols*, 1–20.
[38] Pomin VH, Mulloy B. (2018). Glycosaminoglycans and proteoglycans. *Pharmaceuticals*, 11(1), 27.

[39] Vijayavenkataraman S, Yan WC, Lu WF, Wang CH, Fuh JYH. (2018). 3D bioprinting of tissues and organs for regenerative medicine. *Advanced Drug Delivery Reviews*, 132, 296–332.

[40] Mostafa HK. (2022). Different cells of the human body: Categories and morphological characters. *Journal of Microscopy and Ultrastructure*, 10(2), 40–46.

[41] Burlew BS, Weber KT. (2000). Connective tissue and the heart: Functional significance and regulatory mechanisms. *Clinical Cardiology*, 18(3), 435–442.

[42] Vogt G. (2021). Synthesis of digestive enzymes, food processing, and nutrient absorption in decapod crustaceans: A comparison to the mammalian model of digestion. *Zoology*, 147, 125945.

[43] Rahmati M, Silva EA, Reseland JE, Heyward CA, Haugen HJ. (2020). Biological responses to physicochemical properties of biomaterial surface. *Chemical Society Reviews*, 49(15), 5178–5224.

[44] Abarca-Cabrera L, Fraga-García P, Berensmeier S. (2021). Bio-nano interactions: Binding proteins, polysaccharides, lipids and nucleic acids onto magnetic nanoparticles. *Biomaterials Research*, 25(1), 12.

[45] LaFoya B, Munroe JA, Miyamoto A, Detweiler MA, Crow JJ, Gazdik T, Albig AR. (2018). Beyond the matrix: The many non-ECM ligands for integrins. *International Journal of Molecular Sciences*, 19(2), 449.

[46] Lahir YH. (2016). Understanding the basic role of glycocalyx during cancer. *Journal of Radiation Research*, 7(3), 79–84.

[47] Wainwright DK, Lauder GV, Weaver JC. (2017). Imaging biological surface topography in situ and in vivo. *Methods in Ecology & Evolution*, 8(11), 1626–1638.

[48] Chen L, Yan C, Zheng Z. (2018). Functional polymer surfaces for controlling cell behaviors. *Mat Today: Proceedings*, 21(1), 38–59.

[49] Hasegawa M, Saruta J, Hirota M, Taniyama T, Sugita Y, Kubo K, Ogawa T. (2020). A newly created meso-, micro-, and nano-scale rough titanium surface promotes bone-implant integration. *International Journal of Molecular Sciences*, 21(3), 783.

[50] Otzen DE. (2017). Biosurfactants and surfactants interacting with membranes and proteins: Same but different?. *Biochimica Et Biophysica Acta(BBA)-Biomembranes*, 1859(4), 639–649.

[51] Sharma SK, Grewal HS. (2023). Tribological behavior of bioinspired surfaces. *Biomimetics*, 8(1), 62.

[52] Fabbrizzi L. (2022). Beyond the molecule: Intermolecular forces from gas liquefaction to X– H··· π hydrogen bonds. *ChemPlusChem*, 87(1), e202100243.

[53] Reines BP, Ninham BW. (2019). Structure and function of the endothelial surface layer: Unraveling the nanoarchitecture of biological surfaces. *Quarterly Reviews of Biophysics*, 52, e13.

[54] Owen MC, Gnutt D, Gao M, Wärmländer SK, Jarvet J, Gräslund A, Strodel B. (2019). Effects of in vivo conditions on amyloid aggregation. *Chemical Society Reviews*, 48(14), 3946–3996.

[55] Cai S, Wu C, Yang W, Liang W, Yu H, Liu L. (2020). Recent advance in surface modification for regulating cell adhesion and behaviors. *Nanotechnology Reviews*, 9(1), 971–989.

[56] Singh AK, Malviya R. (2023). Graft copolymers of polysaccharide: Synthesis methodology and biomedical applications in tissue engineering. *Current Pharmaceutical Biotechnology*, 24(4), 510–531.

[57] Levato R, Jungst T, Scheuring RG, Blunk T, Groll J, Malda J. (2020). From shape to function: The next step in bioprinting. *Advances in Materials*, 32(12), 1906423.

[58] Rout-Pitt N, Farrow N, Parsons D, Donnelley M. (2018). Epithelial mesenchymal transition (EMT): A universal process in lung diseases with implications for cystic fibrosis pathophysiology. *Respiratory Research*, 19, 1–10.

[59] Popper HH. (2016). Progression and metastasis of lung cancer. *Cancer & Metastasis Reviews*, 35, 75–91.

[60] Luo W, Wang Y, Han Q, Wang Z, Jiao J, Gong X, Wu M. (2022). Advanced strategies for constructing interfacial tissues of bone and tendon/ligament. *Journal of Tissue Engineering*, 13, 20417314221144714.

[61] Gaut L, Duprez D. (2016). Tendon development and diseases. *Wiley Interdisciplinary Reviews. Developmental Biology*, 5(1), 5–23.
[62] Coglianese D. (2024). Individuals with localized musculoskeletal and connective tissue disorders. In *Clin Exerc Pathophysiol Phys Ther* (pp. 385–431).
[63] Nimni ME, Harkness RD. (2018). Molecular structure and functions of collagen. In *Collagen* (pp. 1–78).
[64] Boppart MD, Mahmassani ZS. (2019). Integrin signaling: Linking mechanical stimulation to skeletal muscle hypertrophy. *American Journal of Physiology. Cell Physiology*, 317(4), C629–C641.
[65] Guo L, Li PC, Rong XQ, Wei XC. Key roles of the superficial zone in articular cartilage physiology, pathology, and regeneration. Chin Med J 2025; 138: 1399–1410. doi: 10.1097/CM9.0000000000003319
[66] Dowson D. (2012). Bio-tribology. *Faraday Discussion*, 156(1), 9–30.
[67] Yakut Y, Tuncer A. (2020). Architecture of human joints and their movement. In *Comp Kinesiol Hum Body* (pp. 47–57).
[68] Jin Z, Dowson D. (2013). Bio-friction. *Friction*, 1, 100–113.
[69] Cyriac F, Akchurin A. (2020). Thin film lubrication, lubricants and additives. In *Tribology in Materials and Applications* (pp. 33–75).
[70] Diosa JG, Moreno R, Chica EL, Villarraga JA, Tepole AB. (2021). Changes in the three-dimensional microscale topography of human skin with aging impact its mechanical and tribological behavior. *PLoS One*, 16(7), e0241533.
[71] Vedadghavami A, Zhang C, Bajpayee AG. (2020). Overcoming negatively charged tissue barriers: Drug delivery using cationic peptides and proteins. *Nano Today*, 34, 100898.
[72] Boettcher K, Winkeljann B, Schmidt TA, Lieleg O. (2017). Quantification of cartilage wear morphologies in unidirectional sliding experiments: Influence of different macromolecular lubricants. *Biotribology*, 12, 43–51.
[73] Marian M, Shah R, Gashi B, Zhang S, Bhavnani K, Wartzack S, Rosenkranz A. (2021). Exploring the lubrication mechanisms of synovial fluids for joint longevity–a perspective. *Colloid Surfaces B: Biointerfaces*, 206, 111926.
[74] Muralidharan K, Vignesh V, RV V, Govindaraju M, Baghad A, Narassima MS, Siengchin S. (2024). Comprehensive overview of nano, micro, and macro tribometers in practice. *Journal of Bio- and Tribo-corrosion.*, 10(3), 44.
[75] Rouf S, Raina A, Ul Haq MI, Naveed N. (2022). Sensors and tribological systems: Applications for industry 4.0. *Ndustrial Robot: The International Journal of Robotics Research and Application*, 49(3), 442–460.
[76] Camarero-Espinosa S, Rothen-Rutishauser B, Foster EJ, Weder C. (2016). Articular cartilage: From formation to tissue engineering. *Biomaterial Science*, 4(5), 734–767.
[77] Gahunia HK, Pritzker KP. (2020). Structure and function of articular cartilage. In *Articular Cartilage of the Knee: Health, Disease and Therapy* (pp. 3–70).
[78] Panwar H, Rokana N, Aparna SV, Kaur J, Singh A, Singh J, Puniya AK. (2021). Gastrointestinal stress as innate defence against microbial attack. *Journal of Applied Microbiology*, 130(4), 1035–1061.
[79] Hariri BM, Cohen NA. (2016). New insights into upper airway innate immunity. *American Journal of Rhinology and Allergy*, 30(5), 319–323.
[80] Pedersen AML, Sørensen CE, Proctor GB, Carpenter GH, Ekström J. (2018). Salivary secretion in health and disease. *Journal of Oral Rehabilitation*, 45(9), 730–746.
[81] Ha P, Liu TP, Li C, Zheng Z. (2023). Novel strategies for orofacial soft tissue regeneration. *Advances in Wound Care*, 12(6), 339–360.
[82] Pradal C, Yakubov GE, Williams MA, MA M, Stokes JR. (2019). Lubrication by biomacromolecules: Mechanisms and biomimetic strategies. *Bioinspiration and Biomimetics*, 14(5), 051001.

[83] Ambrogi M, Vezina CM. (2024). Roles of airway and intestinal epithelia in responding to pathogens and maintaining tissue homeostasis. *Frontiers in Cellular Infection and Microbiology*, 14, 1346087.

[84] Al-Atawi NO, Mashat DS. (2022). Synovial joint study. *American Journal of Computational Mathematics*, 12(1), 7–24.

[85] Gelberg HB. (2017). Alimentary system and the peritoneum, omentum, mesentery, and peritoneal cavity. *Pathologic Basis of Veterinary Disease*, 324.

[86] Cooper BG, Bordeianu C, Nazarian A, Snyder BD, Grinstaff MW. (2018). Active agents, biomaterials, and technologies to improve biolubrication and strengthen soft tissues. *Biomaterials*, 181, 210–226.

[87] Lin W, Klein J. (2021). Recent progress in cartilage lubrication. *Advances in Materials*, 33(18), 2005513.

[88] Lin W, Klein J. (2022). Hydration lubrication in biomedical applications: From cartilage to hydrogels. *Accounts of Materials Research*, 3(2), 213–223.

[89] Kaya DÖ. (2020). Architecture of tendon and ligament and their adaptation to pathological conditions. In *Comparative Kinesiology of the Human Body* (pp. 115–147).

[90] Isojima T, Sims NA. (2021). Cortical bone development, maintenance and porosity: Genetic alterations in humans and mice influencing chondrocytes, osteoclasts, osteoblasts and osteocytes. *Cellular and Molecular Life Sciences*, 78(15), 5755–5773.

[91] Hart NH, Nimphius S, Rantalainen T, Ireland A, Siafarikas A, Newton RU. (2017). Mechanical basis of bone strength: Influence of bone material, bone structure and muscle action. *Journal of Musculoskeletal and Neuronal Interactions*, 17(3), 114.

[92] Żylińska B, Sobczyńska-Rak A, Lisiecka U, Stodolak-Zych E, Jarosz Ł, Szponder T. (2021). Structure and pathologies of articular cartilage. *In Vivo*, 35(3), 1355–1363.

[93] Pathria MN, Chung CB, Resnick DL. (2016). Acute and stress-related injuries of bone and cartilage: Pertinent anatomy, basic biomechanics, and imaging perspective. *Radiology*, 280(1), 21–38.

[94] Sharabi M. (2022). Structural mechanisms in soft fibrous tissues: A review. *Frontiers in Materials*, 8, 793647.

[95] Lim WL, Liau LL, Ng MH, Chowdhury SR, Law JX. (2019). Current progress in tendon and ligament tissue engineering. *Tissue Engineering and Regenerative Medicine*, 16, 549–571.

[96] Blazevich AJ, Fletcher JR. (2023). More than energy cost: Multiple benefits of the long Achilles tendon in human walking and running. *Biology Review*, 98(6), 2210–2225.

[97] Gómez-González M, Latorre E, Arroyo M, Trepat X. (2020). Measuring mechanical stress in living tissues. *Nature Reviews Physics*, 2(6), 300–317.

[98] Holmes DP. (2019). Elasticity and stability of shape-shifting structures. *Current Opinion in Colloid and Interface Science*, 40, 118–137.

[99] DhiviyaLakshmi L, Muthukumaran V. (2022). Overview of biological tissue properties for biotribology. In *Tribol Charact Surf Coatings* (pp. 253–280).

[100] Gracey E, Burssens A, Cambré I, Schett G, Lories R, McInnes IB, Elewaut D. (2020). Tendon and ligament mechanical loading in the pathogenesis of inflammatory arthritis. *Natural Review Rheumatology*, 16(4), 193–207.

[101] Pagar KR, Khandbahale SV, Kasar PM. (2019). Osteoarthritis: Pathophysiology and current treatment modalities. *Asian Journal of Pharmaceutical Research*, 9(4), 289–298.

[102] Peng Z, Sun H, Bunpetch V, Koh Y, Wen Y, Wu D, Ouyang H. (2021). The regulation of cartilage extracellular matrix homeostasis in joint cartilage degeneration and regeneration. *Biomaterials*, 268, 120555.

[103] Megha KB, Joseph X, Akhil V, Mohanan PV. (2021). Cascade of immune mechanism and consequences of inflammatory disorders. *Phytomedicine*, 91, 153712.

[104] Aicale R, Tarantino D, Maffulli N. (2018). Overuse injuries in sport: A comprehensive overview. *Journal of Orthopaedic Surgery and Research*, 13, 1–11.

[105] Urits I, Smoots D, Franscioni H, Patel A, Fackler N, Wiley S, Viswanath O. (2020). Injection techniques for common chronic pain conditions of the foot: A comprehensive review. *Pain and Therapy*, 9, 145–160.
[106] Patterson JT, Tangtiphaiboontana J, Pandya NK. (2018). Management of pediatric femoral neck fracture. *Journal of the American Academy of Orthopaedic Surgeons*, 26(12), 411–419.
[107] Runyan CM, Gabrick KS. (2017). Biology of bone formation, fracture healing, and distraction osteogenesis. *The Journal of Craniofacial Surgery*, 28(5), 1380–1389.
[108] Wang L, You X, Zhang L, Zhang C, Zou W. (2022). Mechanical regulation of bone remodeling. *Bone Research*, 10(1), 16.
[109] Foster AL, Moriarty TF, Zalavras C, Morgenstern M, Jaiprakash A, Crawford R, Metsemakers WJ. (2021). The influence of biomechanical stability on bone healing and fracture-related infection: The legacy of Stephan Perren. *Injury*, 52(1), 43–52.

Annu Rani, T.P. Singh, and Raushan Singh

3 Friction and wear mechanisms in biological systems

Abstract: This chapter explores the mechanisms of friction and wear in biological interfaces, focusing on the complexities of interactions between tissues and biomaterials. It provides a detailed overview of frictional behavior in biological systems, wear processes in natural tissues, and the key factors that influence tribological performance. Emphasis is placed on the role of biological fluids, surface characteristics, and the dynamic nature of living systems, with applications to fields such as orthopedics, dental tribology, and biomedical device design.

Keywords: Biotribology, friction, wear, biological interfaces, synovial fluid, viscoelasticity, tissue engineering, biomedical implants

3.1 Introduction

Because of the special combination of softness, viscoelasticity, and intricate surface interactions seen in biological systems, studying wear and friction in these systems is both difficult and crucial for biomedical applications. Understanding joint mechanics, implant lifetime, and the wear properties of tissues like skin and cartilage all depend on biotribology or tribology in biological contexts [1]. The fundamental causes of wear and friction in biological systems are reviewed in this chapter, along with their implications for tissue engineering and the design of medical devices.

3.2 Mechanisms of friction in biological interfaces

3.2.1 Overview of friction in biological systems

A key component of biotribology is friction in biological systems, where interactions between moist and soft surfaces – such as those in joints, skin, and other tissues – are essential to preserving regular physiological function [2]. The intricate microstructures, viscoelastic characteristics, and the presence of natural lubricants like mucous and synovial fluid distinguish biological interfaces from conventionally designed surfaces. These elements work together to produce distinct frictional behavior that differs greatly from that of inanimate materials [3].

https://doi.org/10.1515/9783111563176-003

3.3 Friction mechanisms in biological interfaces

The resistance to movement between surfaces in contact, such as the rubbing of skin against outside objects or the sliding of bones at a joint, is the source of friction in biological systems. This friction falls into a number of categories.

3.3.1 Static friction

Static friction is the amount of force required to start movement between two surfaces that are in contact. This could be seen in biological systems as skin adhesion to other surfaces or stiff joints.

3.3.2 Kinetic (dynamic) friction

Kinetic friction is the resistance encountered when touching surfaces move relative to one another. For instance, synovial fluid in joints facilitates smooth articulation during movement by lowering kinetic friction [4].

3.3.3 Viscoelastic friction

This is only found in biological materials, where the time-dependent deformation of viscoelastic tissues, such as skin, cartilage, and tendons, affects the frictional response. As biological interfaces react to varying loading circumstances, speeds, and hydration levels, they frequently move between distinct friction regimes [5].

3.4 Factors influencing friction in biological systems

The presence of biological lubricants, surface conditions, and material characteristics all affect the frictional behavior in biological systems. Important elements consist of the following.

3.4.1 Surface roughness and microstructure

The contact area between sliding surfaces is influenced by the surface roughness of tissues such as the articular cartilage in joints [6]. Because they interlock more, rougher surfaces tend to increase friction, whereas smoother surfaces tend to decrease it [7].

3.4.2 Elasticity and deformation

Because biological tissues are frequently pliable and soft, friction and contact mechanics are impacted. Articular cartilage, for instance, deforms when pressure is applied, spreading weight and producing a smoother sliding surface that lowers friction [8].

3.4.3 Hydration and lubrication

Water content is essential for lowering friction in tissues such as mucous membranes and cartilage. Tissues can swell due to hydration levels, forming a hydrated gel layer that makes sliding easier [3]. By creating a lubricating layer between surfaces, natural lubricants such as saliva in the mouth or synovial fluid in joints further lessen friction [9].

3.4.4 Role of natural lubricants

Low friction at biological contacts requires the use of natural lubricants. Mucus, sebum, and synovial fluid are examples of substances that work by creating thin, protective coatings between the body's moving surfaces [10]. By minimizing direct surface-to-surface contact, these coatings considerably lower mechanical wear and friction. One excellent illustration is the synovial fluid found in joints, which lubricates and cushions the joint surfaces to promote smooth articulation [11].

3.5 Synovial fluid

The viscoelastic fluid known as synovial fluid, which is found in synovial joints like the knees and hips, has two purposes: it absorbs stress and provides lubrication. It is abundant in compounds like lubricin and hyaluronic acid, which reduce friction between articulating surfaces and are essential for boundary lubrication, which is most successful at low sliding speeds [12, 13].

3.5.1 Mucous

Mucus creates a lubricating layer on mucosal surfaces, such as those in the gastrointestinal tract, which facilitates food passage while reducing mechanical stress and friction between tissues [14].

3.5.2 Sebum

Sebum functions as a natural emollient on the skin, lowering friction between the layers of skin and outside objects. These lubricants facilitate smooth movements and interactions by lowering friction and shielding sensitive tissues from deterioration [15].

3.6 Friction in specific biological contexts

Different parts of the body experience unique frictional conditions based on their function and environment:

3.6.1 Joint articulation

Low friction during movement is ensured in synovial joints via the interaction of bone geometry, synovial fluid, and cartilage. An almost frictionless interface made possible by articular cartilage is essential for joint function [16].

3.6.2 Skin friction

The frictional behavior of skin contacts changes depending on surface texture, oil content, and hydration levels. Understanding skin friction can assist in reducing irritation and enhancing comfort in applications such as prosthetic design [17].

3.6.3 Dental contacts

Friction plays a role in the mastication process, where teeth interact with food and with each other. Saliva provides a lubricating film that aids in reducing friction, facilitating smooth chewing, and protecting the enamel from excessive wear [3].

3.7 Challenges in studying friction in biological systems

Studying friction in biological systems presents unique challenges due to the dynamic nature of living tissues and the complexity of interactions at the molecular level. Biological tissues constantly change in response to environmental stimuli, loading conditions, and aging, which can alter their frictional properties over time [10]. Additionally, replicating the conditions of in vivo friction in laboratory settings requires advanced techniques like atomic force microscopy (AFM) and tribometers that can mimic the softness and hydration of biological tissues. Understanding friction in biological systems is not only essential for studying natural processes but also for developing medical devices, prosthetics, and implants that need to interact seamlessly with human tissues. By analyzing the mechanisms and factors influencing friction, biotribology can provide insights into the design of artificial joints, skin-contact materials, and other biomedical applications [11].

3.8 Boundary lubrication in biological systems

Boundary lubrication is a crucial mechanism in biological systems, where thin molecular layers reduce friction between surfaces in close contact. Unlike hydrodynamic lubrication, which relies on a thicker film of fluid to keep surfaces separated, boundary lubrication operates when surfaces are so close that fluid layers are thin, often on the order of nanometers. This type of lubrication is vital for maintaining low friction and minimizing wear in many biological interfaces, particularly in load-bearing joints like knees and hips as well as in softer tissues like skin and mucosal surfaces [12].

3.8.1 Definition and principles of boundary lubrication

When a molecular film develops between two surfaces, it prevents direct solid-solid contact, and boundary lubrication takes place, thereby lowering the friction. In biological systems, this molecular film usually comprises proteins, glycoproteins, lipids, or other biopolymers that cling to surfaces and produce a slick interface [3]. Boundary lubrication's efficacy depends on the molecular composition, surface adhesion capacity, and water retention capacity of the molecules involved.

3.8.1.1 Adhesion of molecules

Boundary lubricants cling to the surfaces they lubricate, so forming a monolayer or thin film that separates the surfaces. These molecules' adhesive qualities help to lower friction by means of minimizing direct interactions between the contacting surfaces.

3.8.1.2 Hydration layers

Many boundary lubricants in biological systems are hydrophilic, which lets them draw water molecules and create a hydration layer. This layer not only lubricates but also cushions the surfaces, therefore lowering friction by means of hydration and molecular sliding [10].

3.8.2 Key molecules in biological boundary lubrication

Many molecules in different biological settings provide boundary lubrication by offering a slick interface that reduces wear and friction. Among the most researched bodily lubricants are the following.

3.8.2.1 Lubricin [proteoglycan 4]

Lubricin is a glycoprotein in synovial fluid that plays a key role in lowering friction in specific joints. Lubricin creates a boundary layer on the articular cartilage surface, producing a smooth interface that avoids direct contact between cartilage surfaces during motion. The structure of lubricin enables it to adhere to the cartilage surface, producing a brush-like layer that retains water and lowers shear forces [18].

3.8.2.2 Hyaluronic acid

Hyaluronic acid is a second major synovial fluid component that helps to lubricate boundaries. Hyaluronic acid is a high-molecular-weight, negatively charged polysaccharide with a high capacity to hold water and form a viscoelastic gel that serves to cushion as well as lubricate. Hyaluronic acid increases the boundary lubricating effectiveness of synovial fluid by keeping the joints hydrated and presenting a viscous film between gliding surfaces [19].

3.8.2.3 Mucins

In mucosal tissues, for example, in the gastrointestinal tract or ocular surface, mucins are boundary lubricants. Mucins are glycoproteins of high-molecular weight that create a gel-like protective covering on epithelial surfaces. They retain water and produce a smooth, slippery surface that minimizes friction during food movement through the digestive tract or eyelid blinking over the eye surface [26].

3.8.2.4 Phospholipids

Along with proteins, boundary lubrication is facilitated by lipids in the case of pulmonary systems (e.g., surfactants in lungs) and skin. Phospholipids can create bilayers or monolayers on tissue surfaces, coating them with a lubricating film that minimizes friction between tissues or between tissues and the outside surfaces. Boundary lubrication in synovial joints is mainly facilitated by two molecules: lubricin and hyaluronic acid. As indicated in Fig. 3.1, lubricin binds to cartilage surfaces, forming a low-friction interface, whereas hyaluronic acid gives structural support and keeps synovial fluid viscosity [11].

3.8.3 Boundary lubrication in synovial joints

Synovial joints, like those of the knee, shoulder, and hip, depend strongly on boundary lubrication for reducing friction in movement and for conserving cartilage from wear [20]. The combination of lubricin, hyaluronic acid, and other synovial fluid proteins provides a situation in which cartilage surfaces can glide easily against one another with little wear [12].

3.8.3.1 Mechanism in cartilage

Articular cartilage that protects the bone ends in synovial joints possesses a smooth surface that is covered by boundary lubricants. Such lubricants, especially lubricin, bond to the cartilage surface, forming a hydrated layer that decreases friction during joint movement [3].

3.8.3.2 Role of synovial fluid

Synovial fluid is the vehicle upon which these lubricants are suspended. It keeps the cartilage surfaces well-lubricated even under changing load conditions, shifting between boundary and mixed lubrication as the joint articulates [19].

3.8.3.3 Shear-thinning behavior

The viscoelastic nature of synovial fluid means that it can exhibit shear-thinning properties – becoming less viscous under high shear rates, such as during rapid joint movement. This allows the fluid to flow more easily and maintain lubrication, reducing resistance to motion:

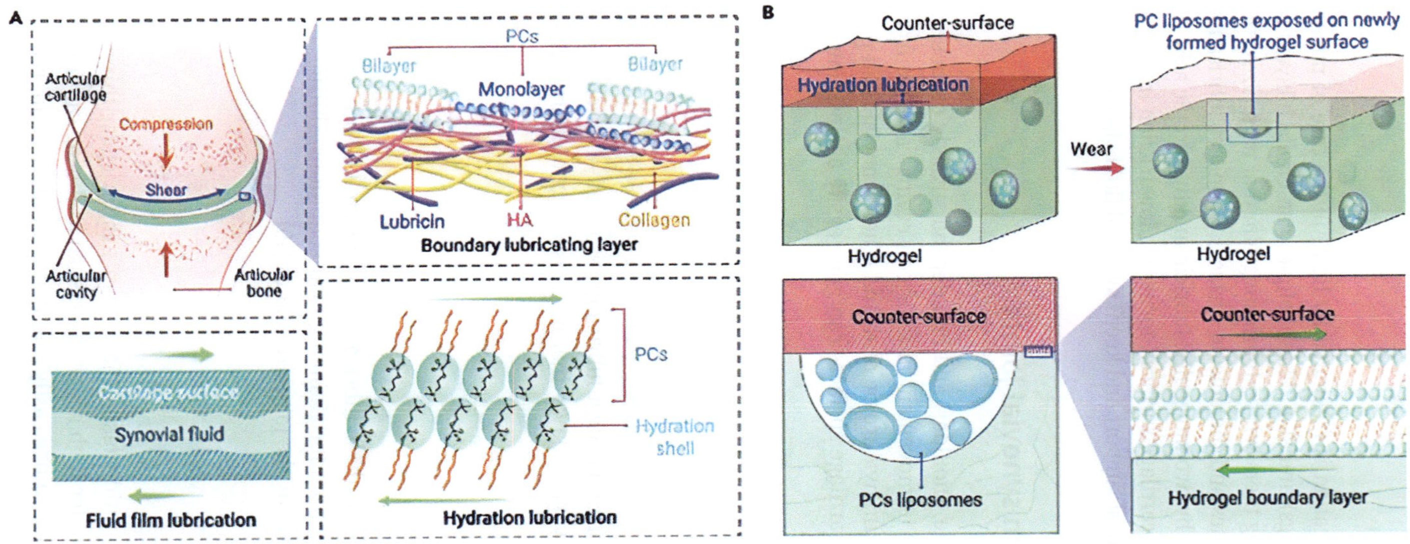

Fig. 3.1: Diagram showing the role of lubricin and hyaluronic acid in boundary lubrication at the cartilage surface within a synovial joint (reprint with permission [19]).

$$\mu = \frac{T}{\sigma_n} \tag{3.1}$$

Approximation of the friction coefficient in boundary lubrication μ is the coefficient of friction, τ is the shear stress, and σ_n is the normal stress (eq. (3.1)). In biological systems, lower shear stress due to effective boundary lubrication leads to a lower friction coefficient [20, 21].

3.8.4 Boundary lubrication in soft tissues

Boundary lubrication is also important in soft tissues like skin and mucosal surfaces, where it helps to reduce irritation and wear caused by friction.

3.8.4.1 Skin

The stratum corneum, the outermost skin layer, is covered with natural lubricants such as sebum, which are boundary lubricants. This layer reduces friction between skin and other surfaces including clothing or prosthetics. The efficiency of this lubrication depends on hydration levels, temperature, and skin care products.

3.8.4.2 Mucosal surfaces

Within the mouth, mucins form a boundary lubrication layer that facilitates easy movement of the tongue and minimizes tissue-to-tissue friction. It is also essential in the gastrointestinal tract, where a lubricating layer of mucous enables food to move freely through the digestive system. Figure 3.2 demonstrates the process by which mucins form a boundary lubrication layer. The diagram shows how mucin glycoproteins interact with water molecules to form a hydrated, low-friction film on the surface of mucosal tissue. This hydrated film causes tissues to slide against each other smoothly, thereby decreasing friction and mechanical abrasion during physiological activity [22].

3.9 Clinical relevance and applications of boundary lubrication

Understanding boundary lubrication in biological systems has direct implications for the design of medical devices, implants, and therapeutic agents:

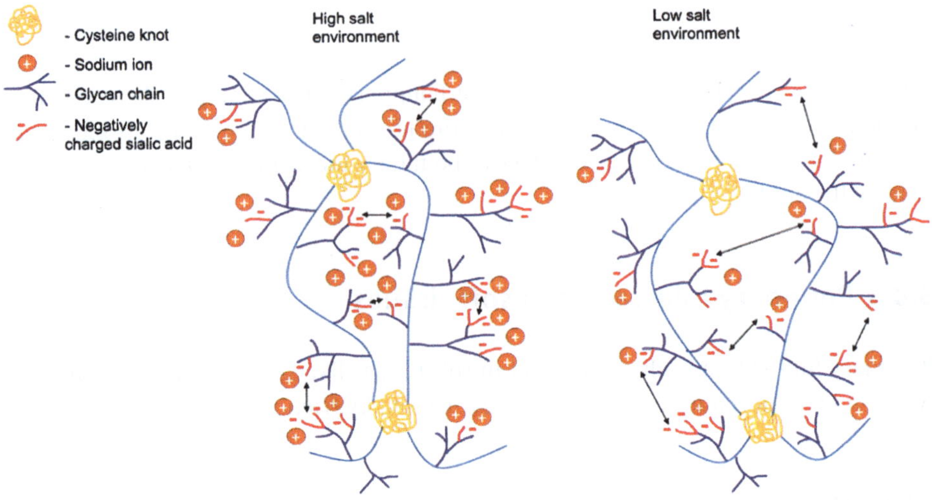

Fig. 3.2: Illustration of the role of mucins in creating a boundary lubrication layer in mucosal tissues, highlighting the interaction between mucins and water molecules (reprint with permission [22]).

3.9.1 Artificial joints

The production of materials that replicate the boundary lubrication effect of lubricin and hyaluronic acid can enhance the performance and lifespan of artificial joints. The materials are expected to minimize wear particles caused by friction, which leads to implant loosening and inflammation.

3.9.2 Dry eye treatments

For dry eye syndrome, for instance, artificial gels or tears with molecules that replicate the surface lubricating property of natural mucins may reestablish lubrication to the surface of the eye, easing pain and friction.

3.9.2.1 Skin and prosthetic interfaces

Designing prosthetic limbs or devices that communicate with skin demands knowledge of boundary lubrication to avoid skin irritation. Adding biocompatible lubricating films will enhance comfort and minimize the potential for skin breakdown. Figure 3.3 shows the role of synovial fluid constituents, including hyaluronic acid and lubricin, in the creation of a protective film on the cartilage surfaces. This movie reduces direct contact between layers of cartilage and hence friction and avoids damage to joints during movement. Figure 3.3 is the schematic of boundary lubrication mechanisms in a hip joint, wherein the role of these synovial fluid components has been shown that maintains low friction and avoids wear under physiological conditions [23].

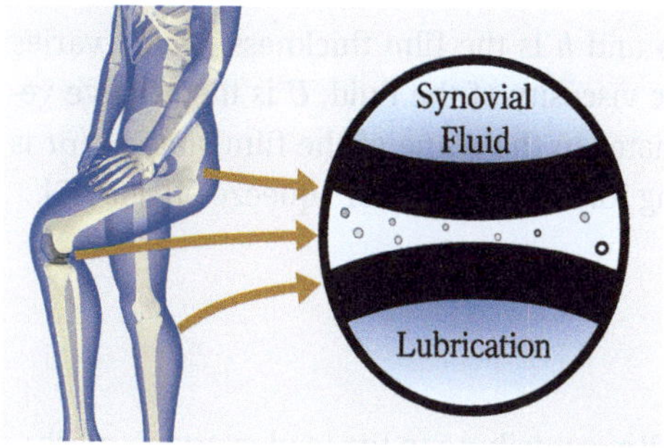

Fig. 3.3: Schematic of boundary lubrication mechanisms in a hip joint, showing the role of synovial fluid components in reducing friction (reprint with permission [23]).

3.9.2.2 Challenges and future directions in boundary lubrication research

Boundary lubrication research in biological systems is hindered by the intricate interactions between molecular layers and dynamic tissue environments. Sophisticated instruments such as AFM and high-resolution imaging technologies are employed to investigate these interactions at the molecular level. Synthetic lubricants that can replicate or surpass natural boundary lubrication are envisioned in future research, with the potential to provide more effective treatments for joint diseases and higher-performing biomedical devices [19].

This section gives a comprehensive overview of the principles, major molecules, and applications of boundary lubrication in biological systems, emphasizing its role in maintaining low friction and tissue protection [17]. It lays the groundwork for discussing other mechanisms of lubrication and their interactions in biological systems. Please let me know if you would like to add any additional information or changes in any area [24].

3.10 Hydrodynamic and mixed lubrication in joints

Hydrodynamic lubrication requires the development of a fluid film between contact surfaces in motion, whose viscosities and joint geometry are important factors. Mixed lubrication exists if fluid film and surface contacts both occur.

3.10.1 Reynolds equation for joint lubrication

$$\frac{\partial}{\partial x}\left(h^3\frac{\partial p}{\partial x}\right)+\frac{\partial}{\partial y}\left(h^3\frac{\partial p}{\partial y}\right)=6\eta U\frac{\partial h}{\partial x}+12\eta\frac{\partial h}{\partial t} \tag{3.2}$$

where p is the pressure within the fluid film and h is the film thickness, which varies with surface shape and load, η\is the dynamic viscosity of the fluid, U is the relative velocity of the surfaces, x and y are the coordinates in the plane of the film, and $\partial h/\partial t$ is the rate of change in film thickness, accounting for compression or squeeze effects [15].

3.10.2 Mixed lubrication regimes

When both direct surface contacts and fluid film contribute to the load-carrying mechanism it is termed as Mixed lubrication regimes. This regime generally occurs in dynamic loading situations, for example, walking, running, or abrupt movements that involve joints like the knee or hip. For these situations, the synovial fluid (a non-Newtonian fluid rich in proteins, lubricin, and hyaluronic acid, whose viscosity is typically high) gets partially squeezed out because of heavy loads or varying velocities. Although some regions of the cartilage surfaces are kept apart by a thin film of fluid (hydrodynamic effect), surface asperities (microscopic roughness) in other locations do make contact and exhibit boundary lubrication behavior. This regime is particularly significant in the knee joint due to the complex motions (flexion-extension, rotation, and translation) that establish highly variable loading environments. Similarly, the hip joint, despite its ball-and-socket stability, undergoes high compressive forces during locomotion, making mixed lubrication essential for joint longevity [20].

3.10.3 Clinical implications

Comprehension of the lubrication processes in natural joints – that is, boundary, hydrodynamic, and mixed regimes – is essential to prosthetic joint design and performance as well as artificial hips and knees. Mimicking the natural environment of lubrication can minimize friction and wear in implants, thus maximizing their functional lifetime and enhancing patient outcomes. Newer prosthetic designs feature biomimetic surfaces, low-friction coatings, and synthetic lubricants that seek to mimic the lubrication response of natural joints. Elucidating these mechanisms also helps with postoperative rehabilitation. Predicting wear during implantation and designing the next generation of joint replacements with increased mobility and patient comfort, Figure 3.4 shows a schematic of hydrodynamic and mixed lubrication modes in the knee joint. It describes how geometry of the joint, synovial fluid flow, and topography of cartilage surfaces determine these lubrication modes, making movement smooth and wear minimal [24].

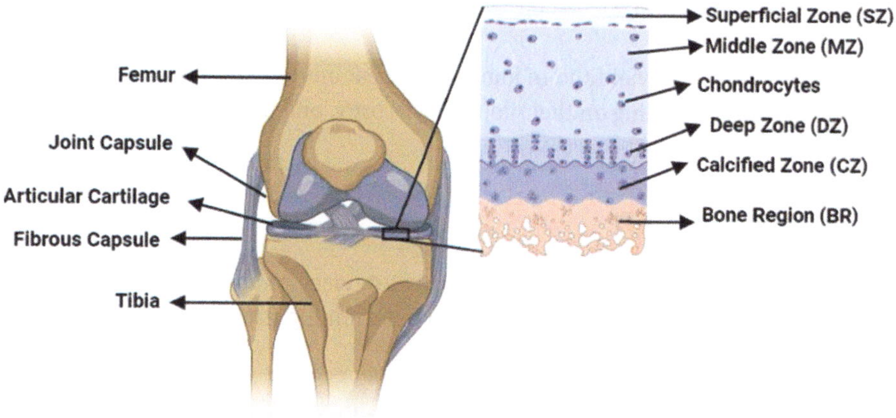

Fig. 3.4: Schematic of hydrodynamic and mixed lubrication modes in a knee joint (reprint with permission [24]).

3.11 Friction at soft-tissue interfaces

Friction between soft-tissue interfaces is the resistance that is experienced when two soft tissue surfaces or a tissue and an external material move past one another. The investigation of this friction force is important in biomechanics, biomaterials, and medical device development [25]. Knowledge of how friction functions in biological systems provides insight into the creation of more efficient prosthetics, eliminating discomfort, avoiding injury, and improving the performance of artificial implants [26]. Soft tissues like skin, muscle, and fat possess intricate characteristics, including nonlinear elasticity, viscoelasticity, and moisture sensitivity, all of which affect frictional behavior [27].

3.11.1 Frictional forces at these interfaces arise due to several factors

3.11.1.1 Mechanical properties of tissues

Soft tissues are highly variable in their stiffness, elasticity, and surface texture. For example, skin, which is a barrier, possesses special properties relative to deeper tissues such as muscles or tendons.

3.11.1.2 Surface lubrication

Body fluids, including synovial fluid in joints and sweat on the skin, can either increase or decrease friction depending on their volume, viscosity, and distribution.

3.11.1.3 Material interactions

When soft tissues come in contact with medical devices or implants, the material properties of those devices play a major role in determining friction levels.

3.11.2 Mechanical behavior and frictional properties of soft tissues

Soft tissues exhibit viscoelastic behavior, such that response to friction forces is a function of applied load rate and time. This is in contrast to conventional engineering materials, which typically respond in linear, predictable fashions. For instance, skin is layered, with the outermost stratum corneum being the chief location for friction interaction. Thickness, hydration, and lipid composition of this stratum significantly influence friction levels. When the skin is well-hydrated, the skin softens and is more flexible, lowering frictional resistance to a certain degree. Excess moisture, however, can cause maceration, raising friction and the risk of skin breakdown [28].

3.12 Factors influencing friction at the skin interface

3.12.1 Hydration and skin condition

The degree of skin hydration alters its frictional properties. Studies show that dry skin often has higher friction due to increased roughness, while hydrated skin can exhibit lower friction [25].

3.12.2 Aging and skin health

Age and health conditions, such as diabetes, influence skin's elasticity, hydration, and overall frictional behavior [26].

3.12.3 Environmental conditions

External factors, such as temperature and humidity, alter skin's frictional response by affecting its moisture level and elasticity [27].

3.13 Muscle and joint friction

So, when it comes to muscle and joint connections, low friction is quite significant. That is where lubricants such as synovial fluid enter the picture. This material is very effective at minimizing wear and tear on the cartilage that caps our joints, which, to be honest, is vital for smooth movement. But here's the catch: if that synovial fluid gets compromised or dwindles – like it does in osteoarthritis – friction can skyrocket. And no one likes that because it equals more joint pain and possible harm. Understanding how friction behaves in our joints is a big deal. It might seriously aid in discovering new treatments or even designing synthetic lubricants that make our joints feel fine [29].

3.14 Medical devices and friction at tissue interfaces

Medical devices such as prosthetics, implants, and catheters have this curious interaction with soft tissues – basically rub against them, yeah? And if there is too much friction, well, that is a problem. Discomfort, possible tissue damage, even the possibility of infection. Not a pleasant thing at all! And conversely, if there is not enough friction that also causes problems such as slippage. And we certainly do not want an unstable device, right? It is all about getting that sweet spot [30].

3.14.1 Friction in prosthetics

Prosthetic limbs actually require that sweet spot of friction. Consider lower-limb prosthetics – if there is too much friction at the socket, it can cause some serious skin irritation or even ulcers, which nobody wants. Conversely, if there is not enough friction, it can make things rather unstable and uncomfortable. So, how do we address this? Well, materials such as silicone are well-liked when it comes to prosthetic liners because they have low friction, which keeps irritation at bay. But the catch is that these materials also must be flexible enough to accommodate movement and pressure that comes with doing things like walking or running. It is a balancing act, for sure! [31]

3.14.2 Implants and joint friction

Knee and hip implants operate in a high-friction environment so that the implant material has to be extremely wear-resistant. Cobalt-chromium alloys and titanium are favored options because they are strong and have relatively low friction. Low-friction substitutes that mimic natural cartilage are under investigation that may extend the life of the implant and decrease wear using advanced polymers, hydrogels, and other technologies [32–34].

3.14.3 Catheters and tissue friction

In insertion and movement maneuvers, catheters encounter sensitive internal structures such as blood vessels or the bladder wall where friction must be minimized to prevent damage [35, 36]. Antimicrobial and hydrophilic coatings that provide frictional resistance enhance ease of insertion and reduce infection risk. However, these coatings are removed with use, and this underscores the necessity for more lasting low-friction alternatives [37].

3.15 Research and advances in reducing friction at soft-tissue interfaces

Recent studies have been devoted to optimizing the tribological features of materials and coatings for particular applications.

3.15.1 Biolubricants and synthetic fluids

Synthetic lubricants mimicking synovial fluid are being developed to reduce friction in joint prosthetics, prolonging their life and enhancing patient comfort.

3.15.2 Hydrogels

Synovia-like synthetic lubricants are also being designed to lubricate joint implants, to decrease wear and tear on these devices and to make them more comfortable for the patient.

3.15.3 Surface modification techniques

These water-based polymers are being investigated for their similarities to natural tissue properties. Hydrogels have low friction and are capable of the adjust to varying mechanical loads, which has applications such as joint replacements and soft tissue repairs.

3.16 Challenges and future directions

Despite advances, there are many challenges still ahead in understanding and controlling friction in soft-tissue interfaces. Uncertainty in patient anatomy, motion patterns, and tissue health make it challenging to standardize material and solutions in populations. Second, the long-term consequences of friction-reducing coatings and materials need additional research to make them safe and effective over long periods of time. Future work will involve the creation of "smart" materials that are sensitive to environmental conditions, self-lubricating implants, and biomimetic coatings capable of better mimicking natural tissue properties [38]. Additional interdisciplinary effort in biomechanics, materials science, and medical engineering will also be essential in progressing the friction management at soft-tissue interfaces. Soft-tissue interface friction is a complex phenomenon with biomechanics, prosthetic, and implant design implications. By learning the distinct characteristics of soft tissues and their interactions with materials, researchers and engineers can design solutions that optimize patient outcomes, comfort, and medical device life. Satisfying friction without causing discomfort, injury, or device failure will continue to be a fundamental challenge, but developments in biomaterials and surface technology hold significant promise for resolution of these challenging issues [19].

3.17 Micro- and nanoscale friction

3.17.1 Friction at small scales

Friction, motion resistance between colliding surfaces, exhibits distinctly at micro- and nanoscale sizes in contrast to the higher macroscopic interaction levels. At micro- and nanoscale sizes, special physical effects, including surface forces, adhesion, and atomic interactions, contribute significantly to the frictional behavior as a whole [39]. An understanding of friction at the nanoscale and microscale is imperative to push microelectromechanical systems (MEMS) and nanoelectromechanical systems (NEMS) applications, data storage devices, and biomedical implants further, as severe friction and wear can critically compromise performance and lifespan [40].

3.17.2 Fundamentals of micro- and nanoscale friction

Common macroscopic friction laws, like Amontons' laws, hold little well at micro- and nanoscale sizes. Friction in these regimes is frequently dominated by surface interactions rather than material bulk properties. The surface-volume ratio grows exponentially as dimensions become smaller, resulting in the more significant impact of surface forces such as van der Waals interactions, capillary forces, and electrostatic forces [41].

Adhesion and atomic stick-slip behavior are two key concepts in nanoscale friction. Adhesion is the stickiness of surfaces in contact, which becomes more important as the contacting surface area gets smaller. Stick-slip behavior is the phenomenon of surfaces alternatingly sticking and slipping at the atomic level, which results in intermittent, jerky motion and frictional force fluctuations. This atomic-scale interaction can be seen and quantified by methods such as AFM and lateral force microscopy (LFM) [39].

3.17.3 Measurement techniques

Measuring friction at the nanoscale and microscale demands accurate instrumentation capable of measuring very small forces and displacements. AFM is an anchor technique for such measurements, enabling the quantification of frictional forces at the atomic level by imaging a sharp tip over a sample surface and measuring lateral (frictional) forces. In LFM, a specific AFM mode, torsion of the cantilever is employed to measure frictional forces more directly. Another useful instrument is the quartz crystal microbalance, in which it is possible to measure changes in oscillation frequency of a quartz crystal due to frictional forces generated by thin films or molecular layers [39].

3.17.4 Physical mechanisms at play

The unique behavior of friction at the micro- and nanoscale arises from the interplay of several mechanisms, often absent or negligible in bulk materials.

3.17.4.1 Adhesion

As surface roughness diminishes at small scales, the contact area tends toward atomic dimensions, and adhesion forces become significant. This can result in a phenomenon called "stiction," where surfaces stick together because of strong adhesive forces [40].

3.17.4.2 Atomic stick-slip motion

On the atomic scale, surfaces do not move smoothly but rather undergo stick-slip motion. Atomic stick-slip is caused by atoms becoming stuck into energy minima on the other surface until enough force is applied to break this binding, leading to an abrupt slip [42].

3.17.4.3 Wear and plasticity

Materials can show plastic and elastic deformation at small length scales, influencing how energy is dissipated during frictional contact. The greater effect of surface roughness at such length scales also implies that surface topography plays a large role in frictional forces [43].

3.17.5 Applications of micro- and nanoscale friction

Micro- and nanoscale friction studies are vital in fields where tiny, precise motions are essential.

3.17.5.1 MEMS and NEMS

In such systems, friction can cause energy losses, stiction, and wear, which reduce device efficiency and lifespan. Minimizing friction using engineered coatings or lubrication is essential for reliable operation [41].

3.17.5.2 Data storage and electronics

The read-write heads in hard drives operate at the nanoscale, where friction must be carefully controlled to prevent damage and data loss. Research in nanoscale friction is pivotal for improving the durability and density of storage technologies [44].

3.17.5.3 Biomedical devices

Implants, stents, and other biomedical devices often operate in dynamic environments where low friction is essential to prevent tissue damage and ensure smooth functioning.

3.17.6 Future directions

Research into micro- and nanoscale friction continues to evolve, with advances in both experimental methods and theoretical understanding. One promising area is superlubricity, a phenomenon where friction approaches near-zero values. This effect has been observed in specific materials, such as graphene, and holds promise for developing wear-resistant coatings and next-generation nanodevices [45]. Additionally, molecular dynamics simulations are becoming a powerful tool in studying and predicting frictional behavior at atomic and molecular levels, complementing experimental studies and paving the way for innovative designs in nanotribology. Micro- and nanoscale friction remains a challenging yet exciting area of study, with far-reaching implications across various advanced technologies. Continued advancements in this field could unlock new possibilities for creating highly efficient, durable, and miniaturized devices in sectors ranging from consumer electronics to medical technology [46].

3.18 Wear processes in natural tissues and biomaterials

3.18.1 Types of wear in biological systems

In biological systems, wear refers to the progressive damage or removal of material from biological surfaces, often due to mechanical interactions with other surfaces. This phenomenon is especially important in joints, teeth, and implants, where wear impacts functionality and longevity. Unlike in engineered systems, biological wear processes are complex and influenced by various factors including the mechanical properties of tissues, biological lubrication, and cellular responses to damage. Here, we explore the primary types of wear observed in biological systems: abrasive, adhesive, fatigue, and corrosive wear [47].

3.18.1.1 Abrasive wear

Abrasive wear occurs when a harder surface or particle scratches or cuts into a softer surface, removing material through a process of microcutting or plowing. In biological systems, abrasive wear is common in joints and teeth, where foreign particles or irregularities on opposing surfaces act as abrasives. In joint cartilage, for example, abrasive wear can be induced by microscopic debris or wear particles that enter the joint space, particularly in conditions like osteoarthritis. These particles, often formed from degraded cartilage or bone, become embedded in the cartilage surface, leading to further wear during joint movement [48]. In teeth, abrasive wear can result from food particles or environmental grit and is influenced by factors such as diet and chewing habits. In dentistry, abrasive wear [*attrition*] is often categorized by its impact on the enamel surface and the role of hard foods, bruxism [*grinding*], or even acidic environments that soften enamel, making it more susceptible to abrasion [49].

Biological systems have natural defenses against abrasive wear, such as synovial fluid in joints, which acts as a lubricant and helps to reduce friction. In dental applications, fluoride treatments strengthen enamel against abrasive forces. Medical interventions can also include dietary adjustments or therapeutic approaches aimed at reducing debris within joints or improving lubrication [50].

3.18.1.2 Adhesive wear

Adhesive wear occurs when two surfaces come into contact, adhere at microscopic points, and then detach, transferring material from one surface to the other. This type of wear is typically due to the formation of microjunctions between surfaces under high contact pressures, followed by subsequent rupture when the surfaces move relative to each other. In biological joints, adhesive wear is influenced by the composition of cartilage and synovial fluid. Cartilage is designed to minimize adhesion through a highly organized structure and low-friction coating provided by synovial fluid, which contains lubricating molecules such as lubricin [11]. However, in pathological conditions, such as rheumatoid arthritis, the breakdown of cartilage structure and loss of lubricin can lead to increased adhesion between articulating surfaces, enhancing wear [51].

Synovial fluid naturally provides a low-friction environment to counteract adhesive wear. Emerging therapies focus on enhancing or supplementing the joint's natural lubricating properties such as injecting hyaluronic acid or synthetic lubricants into arthritic joints. In prosthetic implants, materials like ultra-high-molecular-weight polyethylene (UHMWPE) are used to reduce adhesive wear between metal and polymer surfaces, prolonging implant longevity [52].

3.18.1.3 Fatigue wear

Fatigue wear occurs due to repeated stress cycles on a surface, leading to microcracking and, eventually, material breakdown. This type of wear is particularly relevant in load-bearing biological structures, such as articular cartilage and bone, where repeated movements cause cumulative damage over time. In joints, fatigue wear is associated with repetitive loading during normal daily activities such as walking or running. Over time, microscopic cracks form within the cartilage or underlying bone due to cyclic stresses, and these can propagate to create larger areas of damage [53]. Cartilage, which has limited regenerative capacity, is especially vulnerable to fatigue wear.

This is a primary factor in degenerative joint diseases like osteoarthritis, where repetitive load cycles lead to progressive cartilage thinning and the formation of rough, uneven surfaces that exacerbate wear [54].

Biological tissues adapt to manage fatigue wear through cellular remodeling and self-repair. In bone, microdamage can trigger a remodeling response where osteoclasts remove damaged areas, and osteoblasts deposit new bone tissue [55]. However, the repair capability of cartilage is limited, and therapies are primarily focused on load management such as weight reduction and physical therapy to reduce joint stress. In cases of severe wear, joint replacement surgery may be required [56].

3.18.1.4 Corrosive wear

Corrosive wear (tribocorrosion) in biological systems involves material degradation due to chemical reactions with bodily fluids, often in combination with mechanical action. Although more commonly discussed in relation to metal implants, corrosive wear can also affect natural tissues. For instance, metal implants in orthopedic and dental applications are exposed to corrosive wear when bodily fluids interact with the metal surface, potentially releasing ions into surrounding tissues. This type of wear can lead to inflammatory responses and even implant loosening or failure [57, 58]. In natural teeth, corrosive wear occurs when acids from foods or bacterial metabolism degrade the enamel or dentin, making the surface more susceptible to mechanical wear (a phenomenon often called erosion) [59, 60].

Corrosive wear in implants is minimized through the use of corrosion-resistant materials such as titanium alloys and ceramics [61]. Dental care practices, such as avoiding acidic foods and using fluoride treatments, also help reduce the effects of corrosive wear [62]. New developments in coating technologies for implants aim to create barrier layers that protect against both mechanical and chemical wear, enhancing the longevity of implanted devices [63].

Wear in biological systems is multifaceted, with types such as abrasive, adhesive, fatigue, and corrosive wear each contributing uniquely to tissue damage and degeneration. While biological systems have evolved adaptive responses to some forms of wear, modern medical interventions and materials engineering play a critical role in managing wear in cases where natural defenses are inadequate. Understanding the mechanisms behind each type of wear in biological systems informs both preventative strategies and therapeutic innovations, ultimately contributing to better health outcomes and the longevity of natural and artificial biological structures. Figure 3.5 presents illustrations of different wear mechanisms observed in cartilage and dental enamel. In cartilage, wear can occur through abrasive, adhesive, and fatigue mechanisms, especially under insufficient lubrication or excessive loading. Dental enamel, being the hardest tissue in the human body, primarily undergoes abrasive and erosive wear, often exacerbated by diet and oral hygiene habits. These insights are crucial for both orthopedic and dental implant design, as they guide material selection, surface engineering, and long-term durability assessments [64].

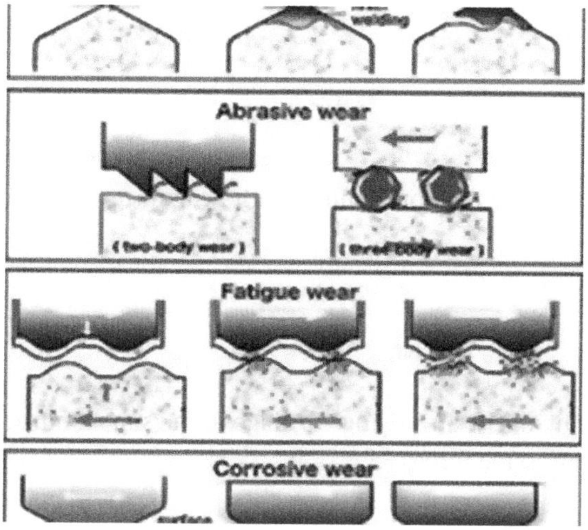

Fig. 3.5: Illustrations of different wear mechanisms in cartilage and dental enamel (reprint with permission [65]).

3.19 Wear in articular cartilage

Articular cartilage covers the ends of bones within synovial joints, providing a smooth, lubricated surface that enables frictionless movement and load distribution. Despite its resilience, cartilage is susceptible to wear over time due to various mechanical, biochemical, and biological factors. Unlike other tissues, cartilage lacks blood vessels and nerves, limiting its regenerative capacity, so wear can lead to irreversible damage, which may contribute to degenerative diseases like osteoarthritis [65]. This chapter examines the mechanisms of cartilage wear, the types of wear that impact joint health, and the implications for managing joint health and disease.

3.19.1 Structure and function of articular cartilage

Articular cartilage is a highly specialized tissue primarily composed of chondrocytes embedded within an extracellular matrix (ECM) rich in collagen fibers, proteoglycans, and water. The ECM enables cartilage to endure compressive loads and maintain structural integrity [66, 67]. This tissue has a unique layered structure that is tailored to resist different forms of mechanical stress:

3.19.1.1 Superficial zone

The outermost layer is rich in collagen and organized to resist shear forces, contributing to low-friction articulation [68].

3.19.1.2 Middle zone

This layer contains a random collagen fiber network, allowing the cartilage to withstand compressive forces [69].

3.19.1.3 Deep zone

The collagen fibers here align perpendicularly, anchoring cartilage to the subchondral bone, enhancing its resistance to compressive forces.

3.19.1.4 Calcified zone

This region directly interfaces with subchondral bone, securing the cartilage to the bone surface.

The zonal arrangement of collagen fibers and proteoglycans makes articular cartilage well-suited to withstand repetitive mechanical loading. However, due to limited cellularity and lack of vasculature, cartilage has limited self-repair capabilities, making it vulnerable to progressive wear over time [70].

3.20 Mechanisms underlying cartilage wear

The wear of articular cartilage involves multiple interrelated mechanisms including loss of lubrication, structural degradation of the ECM, and biological responses to wear particles.

3.20.1 Loss of lubrication

Synovial fluid plays a critical role in reducing friction between cartilage surfaces. It contains lubricin, hyaluronic acid, and phospholipids, which work together to provide boundary and fluid-film lubrication. When lubrication is compromised, friction between surfaces increases, accelerating abrasive and adhesive wear. Lubricin, in particular, coats cartilage surfaces to prevent adhesion, but aging, injury, or inflammation can reduce lubricin levels, making the joint more vulnerable to wear [51].

3.20.2 Extracellular matrix degradation

Cartilage relies on its ECM for structural integrity. Collagen provides tensile strength, while proteoglycans, such as aggrecan, retain water to offer compressive resilience. During wear, the ECM structure weakens due to the degradation of collagen and proteoglycans, often initiated by chondrocyte activity in response to mechanical stress. This degradation exposes the cartilage to further mechanical stress, creating a feedback loop that accelerates wear. The reduced ECM integrity also alters the load distribution within the joint, exacerbating stress on other cartilage areas [71].

3.20.3 Inflammatory response to wear particles

Wear particles generated during cartilage breakdown can trigger an inflammatory response within the joint, known as synovitis. Immune cells recognize wear particles as foreign bodies, leading to the release of proinflammatory cytokines like IL-6, IL-8, and TNF-α. These cytokines amplify cartilage degradation by stimulating chondrocytes to produce matrix-degrading enzymes, accelerating cartilage breakdown. In osteoarthritis, this inflammatory process becomes chronic, exacerbating cartilage wear and contributing to pain and dysfunction [72].

3.21 Implications for managing cartilage wear

Given the limited regenerative ability of articular cartilage, understanding and managing wear mechanisms are crucial for preserving joint health and mitigating degenerative diseases. Several strategies have been explored to address cartilage wear.

3.21.1 Lubricant supplementation

Injectable therapies such as hyaluronic acid aim to replenish joint lubrication and reduce friction. Synthetic lubricants and lubricin-mimetic compounds are also being investigated to improve joint lubrication in patients with arthritis [73].

3.21.2 Anti-inflammatory treatments

Anti-inflammatory medications, including NSAIDs and corticosteroids, are commonly used to control joint inflammation. Newer therapies, such as biologics that target specific inflammatory pathways, can be more effective in reducing joint inflammation and protecting cartilage from corrosive wear [74].

3.21.3 Cartilage regeneration and repair

Advances in tissue engineering and regenerative medicine offer potential solutions for repairing damaged cartilage. Techniques such as microfracture surgery, autologous chondrocyte implantation, and mesenchymal stem cell therapy aim to stimulate new cartilage growth. While these approaches show promise, regenerating durable, load-bearing cartilage that can withstand wear remains a challenge [75].

3.21.4 Physical therapy and load management

Lifestyle modifications, including physical therapy, weight management, and low-impact exercise, can reduce the mechanical load on joints, slowing the progression of wear [76]. Strengthening muscles around the joint can also help stabilize the joint and minimize stress on the cartilage. Wear in articular cartilage is a complex, multifaceted process driven by mechanical, biochemical, and inflammatory factors. The types of wear – abrasive, adhesive, fatigue, and corrosive – each contribute uniquely to cartilage degradation and have significant implications for joint health. While cartilage's limited regenerative capacity makes wear management challenging, advances in therapies targeting lubrication, inflammation, and tissue repair hold promise for preserving joint function and improving outcomes for individuals with joint degeneration. Further research into the mechanisms of cartilage wear and the development of targeted interventions will be crucial in advancing our understanding and treatment of joint diseases such as osteoarthritis.

3.22 Dental wear: enamel and dentin

Dental wear is a natural process where the outermost structures of teeth, primarily enamel and dentin, gradually erode or degrade over time. While some degree of wear is expected as part of the aging process, excessive or accelerated dental wear can lead to tooth sensitivity, pain, and even tooth loss. Dental wear results from mechanical, chemical, and sometimes biological factors, and understanding the mechanisms and patterns of wear in enamel and dentin is essential for effective dental care and restoration [77, 78].

3.23 Structure and function of enamel and dentin

Enamel is the hard, outer layer of the tooth and is the most mineralized tissue in the human body. Composed primarily of hydroxyapatite crystals, enamel is exceptionally strong, protecting the underlying dentin and pulp from mechanical forces, temperature extremes, and chemical erosion. However, enamel is also brittle and lacks regenerative ability; once it is worn or eroded, it cannot self-repair. The thickness and density of enamel vary across the tooth surface, being thicker at the cusp tips where chewing forces are highest and thinner near the cervical region of the tooth. Dentin, located beneath the enamel, is a more resilient but less-mineralized tissue. Unlike enamel, dentin is composed of a network of microscopic tubules, giving it a degree of flexibility that helps absorb the mechanical stresses transferred through enamel. The dentin also houses odontoblastic cells capable of limited reparative action, allowing some reactionary dentin to form in response to certain types of wear or injury. However, dentin is more susceptible to chemical dissolution and mechanical wear than enamel.

Together, enamel and dentin form a composite structure that provides both strength and resilience, allowing teeth to withstand considerable force over a lifetime. However, various factors can cause the gradual erosion or degradation of these tissues, compromising tooth function and health.

3.23.1 Types of dental wear

Dental wear can be broadly categorized into attrition, abrasion, erosion, and abfraction. Each of these types of wear affects enamel and dentin differently, and they often coexist, compounding the overall impact on dental health.

3.23.1.1 Attrition

Attrition is the mechanical wear that occurs due to direct tooth-to-tooth contact, typically from chewing or grinding (bruxism). This type of wear primarily affects enamel, leading to the flattening of occlusal surfaces and the reduction of cusp height. As attrition progresses, the enamel layer may wear down entirely, exposing the underlying dentin. Since dentin is less resistant to wear than enamel, the process accelerates once the enamel is lost. Over time, attrition can lead to significant tooth structure loss and a flattened bite, potentially impairing the functionality of the bite and leading to temporomandibular joint (TMJ) disorders.

3.23.1.2 Abrasion

Abrasion results from mechanical wear caused by an external substance such as toothbrush bristles, abrasive toothpaste, or oral habits like nail biting or using teeth as tools. Abrasive wear commonly occurs on the cervical region of the teeth near the gum line, where the enamel layer is thinner. Excessive brushing with abrasive toothpaste or using a hard-bristled toothbrush can lead to the gradual erosion of enamel at this region, exposing the dentin and leading to increased sensitivity and a heightened risk of decay. The progression of abrasion highlights the need for proper brushing techniques and the use of nonabrasive oral care products to protect enamel integrity.

3.23.1.3 Erosion

Erosion is a chemical process that involves the dissolution of dental hard tissues by acids not produced by bacteria. Sources of these acids can be dietary (such as acidic foods and beverages) or intrinsic, as seen in conditions like gastroesophageal reflux disease (GERD) or chronic vomiting. Erosion initially affects the enamel surface, causing it to soften and demineralize, which makes it more susceptible to mechanical wear. As enamel wears away, the underlying dentin becomes exposed to acidic environments and continues to erode, often at a faster rate due to its lower mineral content. Erosive wear is commonly seen on the palatal surfaces of anterior teeth in individuals with GERD and on the occlusal surfaces of molars in those with high-acidic food intake. Managing erosive wear often requires dietary modifications, neutralizing agents, and protective dental treatments [77].

3.23.1.4 Abfraction

Abfraction is a mechanical wear phenomenon associated with biomechanical loading forces that cause flexure and microfractures in the enamel, particularly around the cervical areas of teeth. This type of wear is thought to result from occlusal forces that concentrate stress along the tooth neck, where enamel is thinner and more prone to cracking. Repeated stress in these areas can lead to chipping and loss of enamel, creating wedge-shaped lesions near the gum line. Once the enamel is lost, dentin becomes more vulnerable to further wear and chemical degradation. Abfraction lesions are often managed through occlusal adjustments or the use of dental restorations to protect the affected areas [79].

3.24 Implications of dental wear on oral health

The progressive wear of enamel and dentin has significant implications for oral health. Enamel wear increases susceptibility to dentin exposure, leading to increased tooth sensitivity and vulnerability to cavities. Once the enamel is compromised, the tooth loses its natural protective barrier, allowing acids, bacteria, and external stimuli to more easily reach the inner structures of the tooth including the pulp. In severe cases, excessive wear can result in pulpitis (inflammation of the pulp), root canal exposure, and eventual tooth loss [74].

Tooth wear also affects aesthetics and functionality. Excessive wear on the occlusal surfaces can alter the shape and alignment of the teeth, leading to bite changes that affect mastication (chewing) efficiency and can contribute to TMJ disorders. The shortening and flattening of teeth from attrition can impact facial structure and lead to an older appearance, as teeth help maintain the height and structure of the lower face [80].

3.25 Management and prevention of dental wear

Managing dental wear involves identifying the underlying cause and taking preventive and restorative measures to preserve tooth structure. For attrition, treatment may include stress management techniques, occlusal splints, and bite adjustments to minimize grinding and clenching forces. In cases of abrasion, education on proper brushing techniques and the use of soft-bristled brushes and non-abrasive toothpaste are essential [81].

For patients with erosive wear, dietary counseling to reduce acidic intake and neutralizing agents (such as fluorides) can help protect enamel. Restorative materials like dental bonding or veneers are sometimes applied to cover eroded areas, particularly in the case of advanced wear. Abfraction lesions are managed by reducing excessive occlusal forces, sometimes through bite realignment, and by using dental materials to reinforce the weakened areas [82].

In all cases, routine dental check-ups and early intervention play a crucial role in managing wear and preventing severe damage. Dentists often use fluoride treatments, sealants, and other protective coatings to strengthen enamel and shield dentin from further wear. By understanding the types and causes of dental wear, individuals can adopt preventative measures to preserve the health and functionality of their teeth well into their later years [83].

3.26 Wear of biomedical implants

Biomedical implants have revolutionized healthcare by enhancing the quality of life for millions of individuals worldwide. These implants serve diverse functions, ranging from joint replacements and dental fixtures to cardiovascular stents. With the growing reliance on these medical devices, understanding the wear mechanisms that impact their durability and function becomes essential. Wear in biomedical implants – primarily through friction, abrasion, and corrosion – can lead to device failure, adverse biological reactions, and the necessity for replacement surgeries [84, 86]. This chapter explores the primary causes, implications, and advancements in materials and surface engineering aimed at mitigating wear in biomedical implants.

3.26.1 Materials used in biomedical implants and their wear properties

Material selection for implants is crucial in minimizing wear while ensuring biocompatibility, strength, and mechanical compatibility with body tissues. Common materials include the following.

3.26.1.1 Metals and alloys

Titanium, cobalt-chromium, and stainless-steel alloys are prevalent in orthopedic and dental implants due to their strength and corrosion resistance. However, wear in metal-on-metal implants has raised concerns regarding the release of metal ions, leading to inflammatory reactions [87, 88]. Innovations like surface coatings aim to enhance the wear resistance of these metals [91].

3.26.1.2 Ceramics

Known for their hardness and low wear rates, ceramics such as zirconia and alumina are used in joint replacements, particularly in load-bearing applications [96]. The smooth surfaces of ceramics reduce friction and wear, but they can be brittle and prone to sudden failure under impact [85].

3.26.1.3 Polymers

UHMWPE is a widely used polymer in joint implants due to its low-friction properties. Although it exhibits relatively low wear rates, the debris generated from polymer wear can cause osteolysis – a process in which bone resorption occurs around the implant, leading to loosening [89, 95].

3.26.1.4 Composite materials

Combining two or more materials, such as ceramic and polymer composites, allows for improved wear resistance and toughness. Composite materials are increasingly used to address the limitations of individual materials and enhance implant longevity [94]. Advances in surface engineering, such as the use of diamond-like carbon (DLC) coatings and hydroxyapatite coatings, have significantly improved the wear resistance and biocompatibility of implant materials [93].

3.27 Implications of wear on biomedical implants

The wear of biomedical implants has both mechanical and biological implications.

3.27.1 Mechanical implications

Wear affects the structural integrity and function of the implant. As materials degrade over time, implants may fail to support the required loads or function optimally [92]. Joint implants, for example, may become misaligned due to wear, leading to altered biomechanics and pain.

3.27.2 Biological implications

Wear particles released from implants can lead to adverse biological responses. In orthopedic implants, wear particles – especially from metals and polymers – can accumulate in surrounding tissues and cause inflammation, leading to osteolysis and implant loosening [88, 90]. This response is a significant cause of implant failure, particularly in younger and more active patients. Mitigating the biological effects of wear requires a careful balance of material properties and design considerations to ensure that implants are durable, biocompatible, and suited to the patient's physiological conditions [86, 91].

3.27.3 Strategies to minimize wear in biomedical implants

Various strategies have been developed to address wear in biomedical implants.

3.27.3.1 Material innovations

Research is focused on developing new alloys, ceramics, and polymer materials with enhanced wear resistance and biocompatibility. For instance, titanium-tantalum and cobalt-chromium-molybdenum alloys have shown promise in reducing wear-related issues.

3.27.3.2 Surface coatings and treatments

Surface engineering techniques, such as plasma spray and laser treatment, are used to apply protective coatings that reduce friction and wear. DLC coatings, for example, create a highly durable surface layer that reduces particle release in metal-on-metal joint replacements [97, 98].

3.27.4 Design modifications

Advanced designs, such as articulating surfaces with reduced friction, can help decrease wear rates. Modular designs, which allow for customized implant shapes and configurations, also contribute to reducing wear by optimizing load distribution [99].

3.27.5 Optimized joint lubrication

Improving lubrication within the joint capsule of implants can minimize direct contact between surfaces, reducing wear. Advances in synthetic lubricants, as well as innovations that mimic natural synovial fluid, are being researched to improve the longevity of joint implants [100]. Understanding and addressing the wear mechanisms in biomedical implants is critical for extending their longevity, minimizing adverse biological responses, and reducing the need for revision surgeries. Ongoing research in material science, surface engineering, and implant design holds promise for further improving the wear resistance of biomedical implants, ultimately enhancing patient outcomes and quality of life. By developing new materials and refining implant designs, the medical community continues to advance toward safer, more reliable, and longer-lasting biomedical devices.

3.28 Factors influencing tribological behavior in biological systems

Tribology, the study of friction, wear, and lubrication, is integral to understanding biological systems, particularly within joints and other load-bearing tissues. In the human body, tribological interactions play a crucial role in functions that require smooth movement such as the articulation of joints, the movement of soft tissues, and the flow of bodily fluids. Unlike mechanical systems, biological systems present unique challenges due to the complexity of living tissues, biological fluids, and their adaptive responses to loading [101]. This chapter examines the primary factors influencing tribological behavior in biological systems including material properties, lubrication mechanisms, biomechanical load, environmental conditions, and biological factors. An understanding of these factors is essential for developing materials and implants that can withstand the demanding conditions of biological environments and for improving therapeutic interventions aimed at restoring or enhancing natural tribological performance.

3.28.1 Material properties of biological tissues and implants

The materials involved in biological tribological systems, both natural and synthetic, directly influence friction, wear, and lubrication behaviors. Key aspects are discussed further.

3.28.1.1 Elasticity and viscoelasticity

Biological tissues, such as cartilage and tendons, exhibit both elastic and viscoelastic properties. This enables them to deform under load and return to their original shape, effectively absorbing and dissipating energy. Elasticity contributes to low friction in biological joints, while viscoelasticity provides resistance to wear by distributing forces over time [102, 103].

3.28.1.2 Surface roughness and texture

Surface topography of both natural and artificial tissues impacts the degree of friction and wear. Natural joint surfaces, such as cartilage, are remarkably smooth, reducing friction. Surface roughness of implants, however, can result in high friction and wear, leading to issues like osteolysis in joint replacements.

3.28.1.3 Hardness and toughness

Cartilage, for example, is soft yet tough, a combination that reduces wear while enduring repetitive stresses. In implants, materials such as ceramics and metal alloys provide high hardness to withstand wear, but they must also be tough to prevent fractures or chipping [104].

3.28.2 Lubrication mechanisms in biological systems

Lubrication is vital for minimizing friction and wear in biological joints. In the human body, lubrication mechanisms are dynamic and complex, involving various fluids and interacting with tissue properties.

3.28.2.1 Boundary lubrication

In boundary lubrication, a thin layer of lubricant, primarily composed of glycoproteins like lubricin, adheres to the articulating surfaces, preventing direct contact. This type of lubrication is crucial in joint movement where loads are high and fluid-film lubrication may not be adequate [105].

3.28.2.2 Fluid-film lubrication

Synovial fluid, rich in hyaluronic acid and lubricating glycoproteins, forms a thin film that separates joint surfaces under motion. Fluid-film lubrication is especially effective under low-load conditions, where it prevents direct cartilage-on-cartilage contact, reducing wear and friction.

3.28.2.3 Mixed lubrication

Biological systems often exhibit mixed lubrication, where boundary and fluid-film lubrication mechanisms work together depending on load conditions and movement. Mixed lubrication helps reduce wear in joints by maintaining a balance between the two types of lubrication, allowing smooth transitions between load-bearing and low-load conditions [106]. Advances in synthetic lubricants and biomimetic coatings in biomedical applications seek to replicate these natural lubrication mechanisms to enhance the tribological performance of implants [107].

3.29 Biomechanical load and motion

The loads exerted on biological systems influence tribological behavior significantly. Biomechanical forces vary based on physical activities, such as walking, running, and jumping, which impose different stresses on joints and tissues:

3.29.1 Normal versus shear load

Biological systems experience both normal *[compressive]* and shear loads. For example, during walking, the knee joint endures compressive forces, while shear forces arise from the sliding motion. These combined loads increase the complexity of wear and frictional behavior, necessitating materials that can endure both types of stresses [108].

3.29.2 Dynamic loading

Unlike static systems, biological systems are constantly in motion, leading to cyclic loading and unloading of joints and tissues. Cyclic loading contributes to fatigue wear in implants, which can lead to microcracks, material degradation, and eventual failure. Fatigue-resistant materials and advanced designs help improve durability under these conditions.

3.29.3 Joint congruency and alignment

Joint alignment and congruency affect the distribution of biomechanical loads across joint surfaces. Misalignment or incongruence can result in uneven wear, increasing friction and potentially leading to joint degeneration. For implants, precise alignment during surgical placement is crucial to reduce wear. Biomechanical load considerations are essential in designing implants that can tolerate various movements and stresses without significant wear or failure.

3.29.4 Environmental conditions in the biological system

The environmental conditions within the body, including temperature, pH, and exposure to biochemical substances, create a unique tribological context:

3.29.4.1 Temperature variations

Human body temperature is relatively stable but can vary with factors such as physical activity or inflammation. Temperature changes can affect the viscosity of synovial fluid and other bodily lubricants, impacting lubrication performance in joints. In implants, temperature changes may influence wear rates and material degradation.

3.29.4.2 pH and biochemical environment

Variations in pH levels, often associated with inflammation or infection, can accelerate corrosion in metallic implants, leading to ion release and adverse biological responses. The biochemical environment, including exposure to oxidative agents, also influences wear behavior, especially in polymer-based implants and coatings.

3.29.4.3 Electrochemical reactions

Metallic implants may undergo electrochemical reactions within the body, leading to corrosion and material degradation. This phenomenon, known as tribocorrosion, can result in increased friction, wear, and release of metal ions, which may cause local or systemic toxicity. Environmental factors must be considered in material selection and surface treatments for implants to ensure durability and biocompatibility within the biological setting [108].

3.30 Biological factors: tissue adaptation and cellular responses

The tribological performance of biological systems is not solely dependent on mechanical factors; biological responses and adaptive behaviors also play a crucial role.

3.30.1 Cartilage adaptation and regeneration

Articular cartilage, for example, adapts to biomechanical loads by remodeling its ECM, enhancing its resistance to wear. However, cartilage has limited regenerative capacity, and excessive wear can lead to irreversible degeneration, as seen in osteoarthritis [109, 110].

3.30.2 Bone remodeling and osteolysis

Implants interact with surrounding bone tissue, triggering remodeling responses. Wear particles generated by implants can lead to osteolysis, where the body resorbs bone tissue around the implant. This reaction can destabilize the implant, necessitating revision surgeries. To mitigate this, implants are designed to minimize particle generation and promote osseointegration, where the bone grows into the implant surface [111].

3.30.3 Immune response to wear particles

Wear particles from implants, especially metal ions and polymer debris, can provoke immune responses, resulting in inflammation and subsequent tissue damage. To reduce these adverse reactions, implant materials are selected for their biocompatibility, and coatings are applied to minimize particle release. Biological adaptation and response to wear are complex and underscore the need for implants that harmonize with the body's natural processes [112].

3.31 Advancements in tribological optimization for biomedical applications

Recent advances in material science, surface engineering, and biolubrication have significantly improved tribological performance in biological systems.

3.31.1 Biomimetic materials

Inspired by natural systems, biomimetic materials replicate the structural and functional properties of tissues like cartilage. For example, hydrogels with high water content and elasticity have shown promise in mimicking the low-friction behavior of cartilage in joint applications.

3.31.2 Surface modifications and coatings

Surface engineering techniques, such as DLC coatings and hydroxyapatite coatings, enhance wear resistance and biocompatibility. These coatings reduce friction, prevent corrosion, and minimize wear particle release.

3.31.3 Advanced lubricants and synthetic synovial fluids

Synthetic lubricants, designed to mimic synovial fluid, improve lubrication in joint implants. These biolubricants can reduce friction and wear, extending implant life and improving joint function.

3.31.4 Additive manufacturing for custom implant design

3D printing and other additive manufacturing techniques enable the creation of custom implants with precise geometries and material properties tailored to patient-specific requirements. Custom designs reduce wear by ensuring optimal fit and load distribution. Tribological behavior in biological systems is shaped by a confluence of factors: material properties, lubrication mechanisms, biomechanical loads, environmental conditions, and biological responses. Advances in understanding these factors have paved the way for innovations that enhance the durability and functionality of biomedical implants, aiming for better integration with biological systems and longer implant lifespans. As research continues, the goal remains to replicate the natural tribological efficiency of biological tissues, contributing to improved therapeutic solutions and patient outcomes in healthcare [113].

3.32 Surface characteristics of biological tissues

The surface characteristics of biological tissues play a crucial role in maintaining the biomechanical functionality of the body. Surface properties, such as texture, roughness, elasticity, and hydration, enable tissues to withstand friction, reduce wear, and facilitate lubrication, especially in load-bearing joints. These properties are fundamental in applications ranging from orthopedic implants to tissue-engineered replacements, where replicating the natural surface attributes of biological tissues is key to ensuring biocompatibility and longevity. This chapter explores the surface characteristics of critical biological tissues, including articular cartilage, skin, and endothelial surfaces, emphasizing their unique properties and tribological relevance [114].

3.32.1 Articular cartilage surface

Articular cartilage is a specialized connective tissue covering the ends of bones within synovial joints such as the knee and hip. This tissue is highly adapted for load-bearing and movement, featuring a surface structure optimized for reducing friction and wear.

3.32.2 Smoothness and roughness

Articular cartilage surfaces are remarkably smooth at the macroscopic level, but microscopically, they present a complex texture composed of collagen fibers and proteoglycans. The collagen matrix provides strength, while the proteoglycan-rich ground substance offers cushioning and low friction [115]. This balance of structure enables cartilage to reduce wear under high-stress conditions.

3.32.3 Hydration and lubrication

Articular cartilage is composed of approximately 70–80% water, which is critical for its lubrication. The water content is bound within the tissue by proteoglycans, creating a pressurized fluid environment that acts as a natural lubricant. Under load, this fluid is exuded to the surface, facilitating fluid-film lubrication, which minimizes direct contact between surfaces [116].

3.32.4 Elasticity and compliance

Cartilage exhibits viscoelastic properties, allowing it to deform under pressure and return to its original shape, distributing loads across the joint surface. This property is essential for preventing wear, as it enables cartilage to absorb and dissipate forces encountered during movement. The surface properties of articular cartilage make it one of the most efficient biological tribological surfaces, and mimicking these characteristics in synthetic implants remains a major goal in biomaterials science [117].

3.33 Skin surface characteristics

The skin serves as a protective barrier and is the largest organ in the human body. It must endure constant interaction with external surfaces, making its surface characteristics critical for its protective function and durability:

3.33.1 Texture and roughness

The skin's surface is composed of a layer of dead skin cells [stratum corneum], which provides a slightly rough texture that aids in grip and abrasion resistance. The roughness of skin varies based on body location and age, with thicker, rougher skin on areas like the palms and soles [118].

3.33.2 Hydration and sebum

Skin hydration is regulated by the stratum corneum and natural moisturizing factors, which help maintain flexibility and prevent cracking. Sebum, an oily secretion from sebaceous glands, lubricates the skin, reducing friction between skin layers and external surfaces, which is particularly important in regions prone to chafing [119].

3.33.3 Elasticity and compliance

Skin elasticity allows it to stretch and conform to various surfaces and movements, which helps reduce wear and tear. Elastic fibers, such as elastin, and a network of collagen fibers provide the skin's ability to resist deformation and recover its shape, making it more resilient against daily mechanical stresses [120].

3.34 Endothelial surfaces in blood vessels

The endothelium, the thin layer of cells lining blood vessels, plays a key role in vascular health and fluid dynamics. Its surface properties are critical for minimizing friction between blood flow and the vessel wall, maintaining efficient circulation.

3.34.1 Smoothness and low friction

Endothelial cells are arranged in a tightly packed, smooth monolayer that minimizes resistance to blood flow. This smooth surface is essential for reducing friction and shear stress, allowing blood to circulate with minimal energy expenditure [121].

3.34.2 Hydration and glycocalyx layer

A thin layer called the glycocalyx coats the endothelial surface, consisting of glycoproteins and proteoglycans. This layer retains water, creating a hydrated, gel-like coating that reduces friction and provides a cushion against blood pressure fluctuations, protecting the endothelium from mechanical stress [122].

3.34.3 Adaptability and compliance

The endothelium is adaptive, adjusting its structure in response to varying blood flow rates and pressures. This adaptability helps reduce wear and damage to the vessel walls, preventing conditions such as atherosclerosis, where excessive mechanical stress can lead to plaque buildup [123, 124]. Understanding endothelial surface properties is essential for designing vascular grafts and stents that are biocompatible, friction-resistant, and durable within the dynamic environment of blood vessels [125].

The surface characteristics of biological tissues are specialized to support their unique biomechanical and protective roles within the body. Articular cartilage provides low-friction surfaces for joint articulation, skin offers protection and flexibility, and the endothelium enables smooth blood flow. Each of these tissues has evolved to meet specific tribological demands through a combination of hydration, elasticity, smoothness, and adaptability [126]. Designing synthetic materials and implants that replicate these surface properties continues to be a significant focus in biomaterials research, aiming to enhance the performance and compatibility of biomedical devices with the body's natural tissues [127].

3.35 Environmental factors influencing tribological behavior in biological systems

In biological systems, environmental factors play a significant role in influencing the tribological behavior of tissues and implants. Unlike mechanical systems, biological environments are complex, variable, and responsive to a range of physiological conditions [128–130]. Factors such as temperature, pH, biochemical composition, and exposure to varying loads create a unique setting that impacts the friction, wear, and lubrication properties of biological tissues and synthetic implants. Understanding how these environmental factors affect tribological behavior is critical for improving the performance and longevity of implants and for developing biomimetic materials that can adapt to these conditions [131, 132].

3.35.1 Temperature and its impact on tribological behavior

Human body temperature, typically around 37 °C [98.6 °F], remains stable but can fluctuate slightly with activity, infection, or localized inflammation. These variations in temperature can influence tribological performance.

3.35.1.1 Lubricant viscosity

Synovial fluid, the natural lubricant found in joints, changes viscosity with temperature. An increase in temperature typically reduces viscosity, leading to thinner fluid films and potentially higher friction and wear in joints. Conversely, lower temperatures may increase fluid viscosity, affecting movement and possibly hindering smooth joint function [133].

3.35.1.2 Material properties

Temperature fluctuations affect the mechanical properties of both biological tissues and synthetic materials. For instance, polymers used in implants may soften or harden with changes in temperature, impacting their wear resistance. Similarly, the viscoelastic properties of tissues like cartilage are temperature-sensitive, with stiffness often increasing at lower temperatures, which can exacerbate wear under load [134].

3.35.1.3 Inflammation and localized heat

Inflammatory responses, often triggered by injury or infection, lead to localized increases in temperature. This can accelerate wear in surrounding tissues and cause degradation of implant materials. Elevated temperatures may also influence the activity of enzymes that break down tissues, adding to the wear and degradation processes within the joint or implant site [135].

3.35.1.4 pH and biochemical environment

The pH levels within the body, typically maintained around 7.4, can vary significantly due to factors such as inflammation, disease, or local biochemical activity. These pH changes, along with other biochemical factors, impact tribological behavior:

3.35.1.5 Corrosion and material degradation

Implants, particularly those made from metals, are susceptible to corrosion in acidic environments. When pH levels drop – common in inflamed or infected areas – the risk of metal ion release from implants increases. This not only affects the integrity of the implant but can also introduce toxic ions into surrounding tissues, leading to further complications, such as inflammation or osteolysis, where bone around the implant weakens.

3.35.1.6 Wear particle reactivity

The acidity and biochemical composition of the environment can influence the reactivity and impact of wear particles generated from implants. In low pH conditions, metal wear particles may undergo accelerated degradation, increasing the release of metallic ions and leading to greater inflammatory responses. For polymer-based implants, acidic conditions can degrade polymer bonds, accelerating the breakdown of materials.

3.35.1.7 Enzymatic activity

The presence of enzymes in biological fluids can influence wear rates, particularly in cartilage. Enzymes such as matrix metalloproteinases are activated during inflammation and degrade the ECM of tissues, making them more susceptible to wear. Enzyme activity also affects the degradation rate of polymeric materials, underscoring the need for enzyme-resistant materials in implant design.

3.36 Mechanical loading and cyclic stress

Mechanical loads in biological systems vary with movement, posture, and external activities, influencing the tribological properties of tissues and implants:

3.36.1 Cyclic loading and fatigue wear

Biological systems are subject to continuous cycles of loading and unloading. Joints experience cyclical stresses from activities like walking or running, which can lead to fatigue wear in materials. Over time, cyclic loading can cause microcracks and degradation in both natural tissues and synthetic implants, especially in polymers and ceramics used in joint replacements.

3.36.2 Load distribution and contact pressure

In joints, uneven load distribution can lead to increased contact pressure on specific areas, resulting in localized wear. For example, poor alignment in joint implants can lead to "edge loading," where high stresses are concentrated at the edges, increasing wear. Natural joint surfaces are typically designed to distribute loads evenly, and replicating this property in implants is key to reducing wear.

3.36.3 Adaptation to variable loads

Biological tissues, such as bone and cartilage, can adapt to changes in load through remodeling, increasing or decreasing their density and structure in response to stress. Implants, however, lack this adaptive capability. As a result, surrounding tissues often bear the brunt of load changes, leading to potential issues such as bone resorption or implant loosening over time.

3.37 Presence of bodily fluids and tribocorrosion

The presence of bodily fluids, such as synovial fluid, blood, and interstitial fluids, introduces unique challenges in biological tribology, often leading to tribocorrosion – a combination of wear and corrosion due to both mechanical and chemical interactions:

3.37.1 Synovial fluid in joints

Synovial fluid provides lubrication in joint spaces, but its composition can change due to inflammation or injury. Increased protein levels, such as fibrinogen from inflammatory responses, can increase fluid viscosity and create thicker films that may alter lubrication behavior. This change can increase wear in joints and artificial implants, as the fluid's altered composition affects its flow and protective function.

3.37.2 Electrolyte content and corrosive reactions

Bodily fluids contain electrolytes, which can accelerate corrosion in metallic implants. Electrochemical reactions between metal surfaces and the electrolyte-rich environment lead to material degradation, particularly in stainless steel and titanium alloys. To mitigate this, surface treatments like coatings are used to create protective barriers that reduce exposure to bodily fluids.

3.37.3 Tribocorrosion effects

In joint replacements, for instance, the combination of mechanical wear and electrochemical corrosion can lead to a degradation process known as tribocorrosion. This results in the release of ions and particles into the surrounding tissues, potentially causing inflammatory responses or metal sensitivity in patients.

3.38 Implications for implant design and material selection

Given the diverse environmental factors that impact tribological behavior, designing implants that can withstand these conditions is challenging.

3.38.1 Biocompatibility and resistance to corrosion

Selecting materials that are biocompatible and resistant to the corrosive effects of bodily fluids is essential. Titanium alloys, ceramics, and coated metals are often used in implants for their corrosion resistance and ability to reduce ion release under varying pH and electrolyte conditions.

3.38.2 Surface treatments and coatings

Surface engineering techniques, such as applying DLC or hydroxyapatite coatings, help enhance wear resistance and protect against corrosion. These coatings create a barrier that shields the material from environmental effects, reducing tribocorrosion and improving implant longevity.

3.38.3 Dynamic adaptation to environmental factors

Bioinspired materials that can dynamically respond to environmental changes, like temperature or pH, hold promise for future implants. Research in hydrogels and polymers that mimic the adaptive properties of biological tissues is ongoing, aiming to create implants that respond to environmental changes with minimal wear.

Environmental factors – temperature, pH, biochemical conditions, loading variations, and the presence of bodily fluids – significantly influence the tribological performance of biological tissues and implants. Recognizing the impact of these factors is crucial for designing implants and materials that can withstand the demanding and variable biological environment. Advances in material science, surface coatings, and adaptive materials offer promising solutions for improving the durability, biocompatibility, and performance of implants, ultimately enhancing their integration with the body's natural tissues and reducing the need for revision surgeries [136, 137].

3.39 Conclusion

Biological systems exhibit extraordinary adaptations to mitigate friction and wear, ensuring longevity and functionality in dynamic and complex environments. The study of friction and wear in these systems reveals a remarkable interplay between structure, material composition, and biomechanics. These mechanisms are intricately optimized through evolutionary processes to sustain high efficiency under varying physiological and environmental conditions.

Key findings in biological tribology demonstrate that natural systems employ a diverse range of strategies to minimize wear and control friction. These include surface texturing, lubrication by biofluids, adaptive material properties, and self-healing capabilities. For example, the synovial joints in humans are a testament to evolutionary ingenuity, where a combination of soft cartilage and a highly efficient lubricant minimizes wear over decades of use. Similarly, the tough, wear-resistant enamel on teeth and the friction-modulating properties of plant cuticles illustrate nature's solutions to different functional challenges. Understanding these mechanisms provides valuable insights for designing bioinspired materials and systems. Applications in areas such as robotics, biomedical implants, and advanced materials engineering can significantly benefit from mimicking biological strategies to enhance performance and durability

However, challenges remain in fully decoding the complex interactions and multiscale phenomena that govern friction and wear in biological systems. Advancements in analytical techniques, computational modeling, and material synthesis will continue to bridge these knowledge gaps, enabling further innovation.

In conclusion, biological systems offer a blueprint for sustainable and efficient tribological solutions. By drawing inspiration from nature, researchers can develop innovative technologies that align with ecological and functional demands, paving the way for advancements in science, engineering, and medicine. The journey of exploring biological friction and wear mechanisms is not only a testament to nature's ingenuity but also a foundation for future breakthroughs in material science and biomechanics.

References

[1] Dowson D. (2007). Bio-tribology. In Bhushan B. (Ed.), *Springer Handbook of Nanotechnology* (pp. 1117–1138). Springer, https://doi.org/10.1007/978-3-540-29857-1_43.

[2] Briscoe BJ, Johnson SA. (2008). The relevance of tribology in biological systems. *Tribology International*, 41(9–10), 820–823. https://doi.org/10.1016/j.triboint.2008.03.002.

[3] Klein J. (2006). Molecular mechanisms of synovial joint lubrication. *Proceedings of the Institution of Mechanical Engineers, Part J: Journal of Engineering Tribology*, 220(8), 691–710. https://doi.org/10.1243/13506501JET153.

[4] Caligaris M, Ateshian GA. (2008). Effects of sustained interstitial fluid pressurization under migrating contact area, and boundary lubrication by synovial fluid constituents on cartilage friction. *Osteoarthritis and Cartilage*, 16(10), 1220–1227. https://doi.org/10.1016/j.joca.2008.03.009.

[5] Derler S, Gerhardt LC. (2012). Tribology of skin: Review and analysis of experimental results for the friction coefficient of human skin. *Tribology Letters*, 45(1), 1–27. https://doi.org/10.1007/s11249-011-9854-y.

[6] Forster H, Fisher J. (1999). The influence of continuous sliding and subsequent surface wear on the friction of articular cartilage. *Proceedings of the Institution of Mechanical Engineers, Part H: Journal of Engineering in Medicine*, 213(4), 329–345. https://doi.org/10.1243/0954411991534932.

[7] Wang A, Essner A, Schmidig G. (2008). The effects of lubricant composition on in vitro wear testing of polymeric acetabular components. *Journal of Biomedical Materials Research Part B: Applied Biomaterials*, 85B(1), 274–283. https://doi.org/10.1002/jbm.b.30867.

[8] Ateshian GA. (2009). The role of interstitial fluid pressurization in articular cartilage lubrication. *Journal of Biomechanics*, 42(9), 1163–1176. https://doi.org/10.1016/j.jbiomech.2009.04.040.

[9] Tadmor R, Chen N, Tang JX, Israelachvili JN. (2003). Thin film rheology and lubricity of hyaluronic acid solutions at a normal physiological concentration. *Biomacromolecules*, 4(3), 710–715. https://doi.org/10.1021/bm0257300.

[10] Hills BA, Butler BD. (2000). Surfactants identified in synovial fluid and their ability to act as boundary lubricants. *Annals of the Rheumatic Diseases*, 59(2), 128–133. https://doi.org/10.1136/ard.59.2.128.

[11] Jay GD. (2004). Lubricin and surfactant protein A as boundary lubricants in articular cartilage. *Biorheology*, 41(3–4), 277–280.

[12] Schmidt TA, Sah RL. (2007). Effect of synovial fluid on boundary lubrication of articular cartilage. *Osteoarthritis and Cartilage*, 15(1), 35–47. https://doi.org/10.1016/j.joca.2006.06.005.

[13] Celli JP, Turner BS, Afdhal NH, Keates S, Ghiran I, Kelly CP, Bansil R. (2007). Helicobacter pylori moves through mucus by reducing mucin viscoelasticity. *Proceedings of the National Academy of Sciences*, 106(34), 14321–14326. https://doi.org/10.1073/pnas.0903438106.

[14] Ranc H, Servais C, Chauvin B, Debaud S, Mischler S. (2009). Friction properties of human skin in contact with sebum and skin cream. *Colloids and Surfaces B: Biointerfaces*, 72(1), 173–178. https://doi.org/10.1016/j.colsurfb.2009.03.005.

[15] Ateshian GA. (2009). The role of interstitial fluid pressurization in articular cartilage lubrication. *Journal of Biomechanics*, 42(9), 1163–1176. https://doi.org/10.1016/j.jbiomech.2009.04.040.

[16] Hannig M, Joiner A. (2006). The structure, function and properties of the acquired pellicle. *Monographs in Oral Science*, 19, 29–64. https://doi.org/10.1159/000090582.

[17] Carré A, Vauthier C. (2009). Biotribology and nanomechanics of artificial surfaces for biomedical applications. *International Journal of Nanomedicine*, 4, 299–311. https://doi.org/10.2147/ijn.s3381.

[18] Bansil R, Turner BS. (2006). Mucin structure, aggregation, physiological functions and biomedical applications. *Current Opinion in Colloid & Interface Science*, 11(2–3), 164–170. https://doi.org/10.1016/j.cocis.2005.11.001.

[19] Lin W, Klein J. (2021). Recent progress in cartilage lubrication. *Advanced Materials*, 33(18), Article e2005513. https://doi.org/10.1002/adma.202005513.

[20] Mow VC, Huiskes R. (2005). *Basic Orthopaedic Biomechanics & Mechano-biology* (3rd ed.). Lippincott Williams & Wilkins.

[21] Lee YH, Bhattarai N, Zhang M, Gunn J. (2008). The influence of hyaluronic acid on the viscoelastic behavior of synthetic synovial fluid. *Journal of Biomedical Materials Research Part A*, 87A(2), 377–383. https://doi.org/10.1002/jbm.a.31724.

[22] Weston A, Vladescu S-C, Reddyhoff T, Griffiths A, Crouzier T, Fielden M, Garnett JA, Carpenter GH. (2023). The influence of ions on the lubricative abilities of mucin and the role of sialic acids. *Colloids and Surfaces B: Biointerfaces*, 227, Article 113327.

[23] Prekasan D, Saju KK. (2016). Review of the tribological characteristics of synovial fluid. *Procedia Technology*, 25, 1170–1174. https://doi.org/10.1016/j.protcy.2016.08.235.

[24] Rajankunte Mahadeshwara M, Al-Jawad M, Hall RM, Pandit H, El-Gendy R, Bryant M. (2024). How do cartilage lubrication mechanisms fail in osteoarthritis? A comprehensive review. *Bioengineering*, 11(6), 541. https://doi.org/10.3390/bioengineering11060541.

[25] Derler S, Gerhardt LC. (2012). Tribology of skin: Review and analysis of experimental results for the friction coefficient of human skin. *Tribology Letters*, 45(1), 1–27. https://doi.org/10.1007/s11249-011-9854-y.

[26] Tomlinson SE, Lewis R, Liu X, Carré MJ. (2009). Understanding the friction mechanisms between human skin and medical textiles. *Skin Research and Technology*, 15(3), 230–246. https://doi.org/10.1111/j.1600-0846.2009.00360.x.

[27] Pasumarty SM, Johnson SA, Watson SA, Adams MJ. (2010). Friction of the human finger pad: Influence of moisture, occlusion and velocity. *Tribology Letters*, 37, 283–294. https://doi.org/10.1007/s11249-009-9524-1.

[28] Adams MJ, Johnson SA, Lefèvre P, Lévesque V, Hayward V, André T, Thonnard JL. (2007). Finger pad friction and its role in grip and touch. *Journal of the Royal Society Interface*, 4(24), 255–269. https://doi.org/10.1098/rsif.2006.0193.

[29] Hendriks FM, Franklin SE. (2010). Influence of hydration and experimental setup on skin friction measurements. *Skin Research and Technology*, 16(4), 401–409. https://doi.org/10.1111/j.1600-0846.2010.00441.x.

[30] Sanders JE, Greve JM, Mitchell SB, Zachariah SG. (2004). Material properties of commonly used interface materials and their static coefficients of friction with skin and socks. *Journal of Rehabilitation Research and Development*, 41(6A), 821–832. https://doi.org/10.1682/JRRD.2003.09.0137.

[31] Gholizadeh H, Abu Osman NA, Eshraghi A. (2014). Interface pressure and pain perception in amputees with transfemoral sockets made of different materials. *Clinical Biomechanics*, 29(6), 630–634. https://doi.org/10.1016/j.clinbiomech.2014.04.008.

[32] Affatato S, Ruggiero A, Merola M. (2013). Materials and surface treatments for hip and knee prostheses. *Materials*, 6(12), 5460–5476. https://doi.org/10.3390/ma6125460.

[33] Wang A, Essner A, Polineni VK. (2015). Wear and damage of UHMWPE in total joint replacements. *Tribology International*, 95, 643–653. https://doi.org/10.1016/j.triboint.2015.01.004.

[34] Bergmann G, Graichen F, Rohlmann A. (2001). Hip joint loading during walking and running, measured in two patients. *Journal of Biomechanics*, 34(7), 859–869. https://doi.org/10.1016/S0021-9290(01)00033-3.

[35] Khoshdel A, Faezipour M, Elahinia M. (2018). Review of catheter friction and design for urological applications. *Biomedical Engineering Letters*, 8(1), 13–22. https://doi.org/10.1007/s13534-017-0049-z.

[36] Song JE, Koo H, Park K. (2020). Antifouling strategies for catheters and stents: From surface modification to endothelium mimicry. *ACS Applied Bio Materials*, 3(7), 3898–3914. https://doi.org/10.1021/acsabm.0c00335.

[37] Stickler DJ. (2008). Bacterial biofilms and the encrustation of urethral catheters. *Biofouling*, 24(5), 349–360. https://doi.org/10.1080/08927010802216126.

[38] Geerligs M, Peters GWM, Ackermans PAJ, Oomens CWJ, Baaijens FPT. (2008). Linear viscoelastic behavior of subcutaneous adipose tissue. *Biorheology*, 45(6), 677–688. https://doi.org/10.3233/BIR-2008-0507.

[39] Bhushan B. (2008). *Nanotribology and Nanomechanics: An Introduction* (2nd ed.). Springer, https://doi.org/10.1007/978-3-540-88546-5.

[40] Mate CM. (2008). *Tribology on the Small Scale: A Bottom up Approach to Friction, Lubrication, and Wear.* Oxford University Press.

[41] Urbakh M, Klafter J, Gourdon D, Israelachvili J. (2004). The nonlinear nature of friction. *Nature*, 430(6999), 525–528. https://doi.org/10.1038/nature02750.

[42] Bhushan B. (2008). *Nanotribology and Nanomechanics: An Introduction* (2nd ed.). Springer, https://doi.org/10.1007/978-3-540-88546-5.

[43] Schwarz UD, Zwörner O, Köster P, Wiesendanger R. (2003). Quantitative analysis of lateral force microscopy. *Physical Review B*, 65(24), 245402. https://doi.org/10.1103/PhysRevB.65.245402.

[44] Reitsma MG, Watanabe H. (2011). Quartz crystal microbalance with dissipation for studying tribology at the nanoscale. *Review of Scientific Instruments*, 82(10), 105109. https://doi.org/10.1063/1.3648135.

[45] Martin JM, Donnet C, Le Mogne T, Epicier T. (1993). Superlubricity of molybdenum disulphide. *Physical Review B*, 48(14), 10583–10586. https://doi.org/10.1103/PhysRevB.48.10583.

[46] Dienwiebel M, Verhoeven GS, Pradeep N, Frenken JWM, Heimberg JA, Zandbergen HW. (2004). Superlubricity of graphite. *Physical Review Letters*, 92(12), 126101. https://doi.org/10.1103/PhysRevLett.92.126101.

[47] Gao J, Müser MH. (2020). Molecular dynamics simulations of friction at the nanoscale. *Tribology Letters*, 68(2), 1–23. https://doi.org/10.1007/s11249-020-01291-1.

[48] Fisher J, Jin Z, Tipper J, Ingham E, Stone MH. (2004). Tribology of natural and artificial joints. In Bhushan B. (Ed.), *Modern Tribology Handbook* (Vol. 2, pp. 851–878). CRC Press.

[49] Wang H, Ateshian GA. (1997). The normal indentation stiffness of articular cartilage is determined by equilibrium tension. *Journal of Biomechanics*, 30(8), 771–775.

[50] Addy M. (2005). Tooth wear and sensitivity: Clinical advances in restorative dentistry. *British Dental Journal*, 198(7), 405–410.

[51] Liu F, Kozanek M, Hosseini A, Van de Velde SK, Gill TJ, Li G. (2015). In vivo tibiofemoral cartilage deformation during the stance phase of gait. *Journal of Biomechanics*, 43(4), 658–665.

[52] Neu CP, Komvopoulos K, Reddi AH. (2007). The interface of functional biomaterials with cells: Biotribological mechanisms of synovial joint lubrication. *Journal of Biomechanics*, 40(5), 1038–1046.

[53] Bragdon CR, Jasty M, Muratoglu OK, O'Connor DO, Harris WH. (2001). Third-body wear of highly cross-linked polyethylene in a hip simulator. *Journal of Arthroplasty*, 16(5), 658–665.

[54] Setton LA, Elliott DM, Mow VC. (1999). Altered mechanics of cartilage with osteoarthritis: Human osteoarthritis and an experimental model of joint degeneration. *Osteoarthritis and Cartilage*, 7(1), 2–14.

[55] Griffin TM, Guilak F. (2005). The role of biomechanics and inflammation in cartilage injury and repair. *Clinical Orthopaedics and Related Research*, 427, S17–S26.

[56] Burr DB. (2002). Bone microdamage and skeletal fragility in osteoporotic and stress fractures. *Journal of Bone and Mineral Research*, 17(12), 2094–2102.

[57] Buckwalter JA, Mankin HJ, Grodzinsky AJ. (2005). Articular cartilage and osteoarthritis. *Instructional Course Lectures*, 54, 465–480.

[58] Yan Y, Neville A, Dowson D. (2006). Biotribocorrosion – An appraisal of the time dependence of wear and corrosion interactions: Review. *Proceedings of the Institution of Mechanical Engineers, Part J: Journal of Engineering Tribology*, 220(2), 55–63. https://doi.org/10.1243/13506501JET88.

[59] Mischler S. (2008). Triboelectrochemical techniques and interpretation methods in tribocorrosion: A comparative evaluation. *Tribology International*, 41(7), 573–583. https://doi.org/10.1016/j.triboint.2007.11.003.

[60] Lussi A, Hellwig E, Zero D, Jaeggi T. (2011). Erosive tooth wear: Diagnosis, risk factors and prevention. *American Journal of Dentistry*, 19(6), 319–325.

[61] Shellis RP, Featherstone JDB, Lussi A. (2013). Understanding the chemistry of dental erosion. *Monographs in Oral Science*, 25, 163–179. https://doi.org/10.1159/000350478.

[62] Geetha M, Singh AK, Asokamani R, Gogia AK. (2009). Ti-based biomaterials, the ultimate choice for orthopaedic implants – A review. *Progress in Materials Science*, 54(3), 397–425. https://doi.org/10.1016/j.pmatsci.2008.06.004.

[63] Bartlett D, Ganss C, Lussi A. (2013). Basic erosive wear examination (BEWE): A new scoring system for scientific and clinical needs. *Clinical Oral Investigations*, 17(1), 21–26. https://doi.org/10.1007/s00784-012-0704-3.

[64] Gao A, Hang R, Huang X, Zhao L, Zhang X, Wang L, Chu PK. (2014). The effects of titania nanotubes with embedded silver oxide nanoparticles on bacteria and osteoblasts. *Biomaterials*, 35(13), 4223–4235. https://doi.org/10.1016/j.biomaterials.2014.01.067.

[65] Tsujimoto A, Barkmeier WW, Fischer NG, Nojiri K, Nagura Y, Takamizawa T, Latta MA, Miazaki M. (2018). Wear of resin composites: Current insights into underlying mechanisms, evaluation methods and influential factors. *Japanese Dental Science Review*, 54(2), 76–87. https://doi.org/10.1016/j.jdsr.2017.11.002.

[66] Buckwalter JA, Mankin HJ. (1998). Articular cartilage: Tissue design and chondrocyte-matrix interactions. *Instructional Course Lectures*, 47, 477–486.

[67] Mow VC, Holmes MH, Lai WM. (1992). Fluid transport and mechanical properties of articular cartilage: A review. *Journal of Biomechanics*, 25(12), 1371–1385. https://doi.org/10.1016/0021-9290(92)90058-9.

[68] Poole CA. (1997). Articular cartilage chondrons: Form, function and failure. *Journal of Anatomy*, 191 (Pt 1), 1–13.

[69] Mow VC, Ratcliffe A. (1997). Structure and function of articular cartilage and meniscus. In Nordin M, Frankel VH. (Eds.), *Basic Biomechanics of the Musculoskeletal System* (3rd ed., pp. 113–137). Lippincott Williams & Wilkins.

[70] Benninghoff A. (1925). Form und Bau der Gelenkknorpel in ihren Beziehungen zur Funktion. *Zeitschrift Für Zellforschung Und Mikroskopische Anatomie*, 2, 783–862.

[71] Athanasiou KA, Darling EM, Hu JC. (2013). *Articular Cartilage Tissue Engineering*. Morgan & Claypool Life Sciences.

[72] Burr DB, Gallant MA. (2012). Bone remodelling in osteoarthritis. *Nature Reviews Rheumatology*, 8(11), 665–673.

[73] Robinson WH, Lepus CM, Wang Q, Raghu H, Mao R, Lindstrom TM, Sokolove J. (2016). Low-grade inflammation as a key mediator of the pathogenesis of osteoarthritis. *Nature Reviews Rheumatology*, 12(10), 580–592.

[74] Zhang Y, Malfait AM, Rabago D. (2020). Innovations in intra-articular therapies for osteoarthritis. *Bone & Joint Research*, 9(10), 645–658.

[75] Chevalier X, Eymard F, Richette P. (2009). Biologic agents in osteoarthritis: Hopes and disappointments. *Nature Reviews Rheumatology*, 6(7), 393–398.

[76] Hunziker EB. (2002). Articular cartilage repair: Are the intrinsic biological constraints undermining this process insuperable?. *Osteoarthritis and Cartilage*, 10(6), 432–463.

[77] Felson DT, Zhang Y, Hannan MT, et al. (2000). Risk factors for incident radiographic knee osteoarthritis in the elderly: The Framingham study. *Arthritis & Rheumatism*, 43(5), 1049–1055.

[78] Shellis RP, Addy M. (2014). The interactions between attrition, abrasion and erosion in tooth wear. *Monographs in Oral Science*, 25, 32–45.

[79] Wetselaar P, Manfredini D, Ahlberg J, et al. (2019). Associations between tooth wear and dental sleep disorders: A narrative overview. *Journal of Oral Rehabilitation*, 46(6), 765–775.

[80] Grippo JO, Simring M, Schreiner S. (2004). Attrition, abrasion, corrosion and abfraction revisited: A new perspective on tooth surface lesions. *Journal of the American Dental Association*, 135(8), 1109–1118.

[81] Wetselaar P, Lobbezoo F. (2016). The tooth wear evaluation system: A modular clinical guideline for the diagnosis and management planning of worn dentitions. *Journal of Oral Rehabilitation*, 43(1), 69–80. https://doi.org/10.1111/joor.12336.

[82] Lobbezoo F, Van der Zaag J, Naeije M. (2006). Bruxism: Its multiple causes and its effects on dental implants – An updated review. *Journal of Oral Rehabilitation*, 33(4), 293–300. https://doi.org/10.1111/j.1365-2842.2006.01609.x.

[83] Rees JS. (2002). The effect of variation in occlusal loading on the development of abfraction lesions: A finite element study. *Journal of Oral Rehabilitation*, 29(2), 188–193. https://doi.org/10.1046/j.1365-2842.2002.00830.x.

[84] Grippo JO, Simring M, Coleman TA. (2012). Abfraction, abrasion, biocorrosion, and the enigma of noncarious cervical lesions: A 20-year perspective. *Journal of Esthetic and Restorative Dentistry*, 24(1), 10–23. https://doi.org/10.1111/j.1708-8240.2011.00487.x.

[85] Affatato S. Ed., (2012). *Wear of Orthopaedic Implants and Artificial Joints*. Woodhead Publishing, https://doi.org/10.1533/9780857096128.

[86] Bhushan B. Ed., (2013). *Encyclopedia of Tribology*. Springer, https://doi.org/10.1007/978-0-387-92897-5.

[87] Cicek H, Toptan F, Ozdemir I. (2021). Wear mechanisms in biomedical implants: A review. *Journal of the Mechanical Behavior of Biomedical Materials*, 115, 104282. https://doi.org/10.1016/j.jmbbm.2020.104282.

[88] Dumbleton JH, Manley MT. (2004). Metal-on-metal total hip replacement: What does the literature say? *Journal of Arthroplasty*, 19(8), 22–26. https://doi.org/10.1016/j.arth.2004.06.015.

[89] Hallab NJ, Jacobs JJ, Black J. (2000). Hypersensitivity to metallic biomaterials: A review of leukocyte migration inhibition assays. *Biomaterials*, 21(13), 1301–1314. https://doi.org/10.1016/S0142-9612(99)00233-2.

[90] Kurtz SM. (2009). *UHMWPE Biomaterials Handbook: Ultra High Molecular Weight Polyethylene in Total Joint Replacement and Medical Devices* (2nd ed.). Academic Press, https://doi.org/10.1016/B978-0-12-374721-1.X0001-1.

[91] Malchau H, Herberts P, Eisler T, Garellick G, Söderman P. (2002). The Swedish Total Hip Replacement Register. *Journal of Bone and Joint Surgery*, 84 B(Suppl(2), 2–20.

[92] Narayan RJ. Ed., (2010). *Biomedical Materials*. Springer, https://doi.org/10.1007/978-1-4419-9643-4.

[93] Patil N, Lee K, Goodman SB. (2008). Wear particles and periprosthetic osteolysis in hip and knee arthroplasty. *Journal of Bone and Joint Surgery*, 90(1), 99–102. https://doi.org/10.2106/JBJS.G.01264.

[94] Pruitt LA, Chakravartula AM. (2011). *Mechanics of Biomaterials: Fundamental Principles for Implant Design*. Cambridge University Press, https://doi.org/10.1017/CBO9780511976034.

[95] Ramakrishna S, Mayer J, Wintermantel E, Leong KW. (2001). Biomedical applications of polymer-composite materials: A review. *Composites Science and Technology*, 61(9), 1189–1224. https://doi.org/10.1016/S0266-3538(00)00241-4.

[96] Wang A. (2001). A unified theory of wear for ultra-high molecular weight polyethylene in multidirectional sliding. *Wear*, 248(1–2), 38–47. https://doi.org/10.1016/S0043-1648(00)00562-9.

[97] Willmann G. (2000). Ceramic femoral head retrieval data. *Clinical Orthopaedics and Related Research*, 379, 22–28. https://doi.org/10.1097/00003086-200010000-00004.

[98] Choudhury D, Paul S, Dhar A. (2015). Wear and corrosion properties of diamond like carbon (DLC) coating on different orthopedic implant materials. *Surface & Coatings Technology*, 276, 646–653. https://doi.org/10.1016/j.surfcoat.2015.06.020.

[99] Grill A. (2003). Diamond-like carbon: State of the art. *Diamond and Related Materials*, 12(2), 166–170. https://doi.org/10.1016/S0925-9635(03)00018-1.

[100] Rahmani R, Dini D, Rahnejat H. (2019). Modelling the tribological performance of modular orthopaedic implants: A review. *Tribology International*, 132, 158–172. https://doi.org/10.1016/j.triboint.2019.01.048.

[101] Müller F, Sonntag R, Kretzer JP, Bader R. (2020). Lubrication in total joint replacements: Natural synovial lubrication versus artificial lubricants. *Lubricants*, 8(5), 49. https://doi.org/10.3390/lubricants8050049.

[102] Dowson D. (2006). Tribological principles in orthopedic biomaterials. *Journal of Engineering in Medicine*, 220(8), 719–726. https://doi.org/10.1243/09544119JEIM103.

[103] Korhonen RK, Laasanen MS, Töyräs J, Rieppo J, Hirvonen J, Jurvelin JS, Helminen HJ. (2003). Comparison of the equilibrium response of articular cartilage in unconfined compression, confined compression and indentation. *Journal of Biomechanics*, 36(6), 907–914. https://doi.org/10.1016/S0021-9290(03)00084-8.

[104] Mak AF. (1986). The apparent viscoelastic behavior of articular cartilage – The contributions from the intrinsic matrix viscoelasticity and interstitial fluid flows. *Journal of Biomechanical Engineering*, 108(2), 123–130. https://doi.org/10.1115/1.3138603.

[105] Link J. M., Salinas , E. Y., Hu, J. C., & Athanasiou, K. A. (2020). The tribology of cartilage: Mechanisms, experimental techniques, and relevance to translational tissue engineering. *Clinical Biomechanics*, 79, 104880.

[106] Lin W, Klein J. (2023). Recent progress in cartilage lubrication: Insights into ultra-low friction mechanisms. *arXiv preprint*

[107] Mow VC, Lai WM (2023). Mixed lubrication mechanisms in synovial joints: Weeping, elastohydrodynamic, and squeeze film lubrication. MDPI Biotechnology Journal. Vol 3, Pg. 234–254.

[108] Karpiński R, Prus A, Baj J, Radej S, Prządka M, Krakowski P, Jonak K. (2025). Articular cartilage: Structure, biomechanics, and the potential of conventional and advanced diagnostics. *Applied Sciences*, 15(12), 6896. https://doi.org/10.3390/app15126896.

[109] Karpiński R, Prus A, Baj J, Radej S, Prządka M, Krakowski P, Jonak K. (2025). Articular cartilage: Structure, biomechanics, and the potential of conventional and advanced diagnostics. *Applied Sciences*, 15(12), 6896. https://doi.org/10.3390/app15126896.

[110] Vincent TL, et al. (2019). Mechanoadaptation: Articular cartilage through thick and thin. *The Journal of Physiology*, 597, 1271–1289.

[111] Catalano E. (2023). Biophysical and biomechanical properties of cartilage. *arXiv preprint.*

[112] Bioengineering. (2022). The role of the innate immune system in wear debris-induced osteolysis. *Bioengineering*, 9(12), 764. https://doi.org/10.3390/bioengineering9120764.

[113] Panez-Toro, I., Heymann, D., Gouin, F., Amiaud, J., Heymann, M. F., & Cordova, L. A. (2023). Roles of inflammatory cell infiltrate in periprosthetic osteolysis. *Frontiers in Immunology*, 14, 1310262.

[114] Lee C-Y, Kung P-C, Huang -C-C, Shih S-J, Huang E-W, Chen S-Y, Wu M-H, Tsou N-T. (2025). In vivo study of bone growth around additively manufactured implants with Ti-6Al-4V and bioactive glass powder composites. *arXiv preprint.*

[115] Ghosh, S., Bowen, J., Jiang, K., Espino, D. M., & Shepherd, D. E. (2013). Investigation of techniques for the measurement of articular cartilage surface roughness. Micron, 44, 179–184.

[116] Naka, M. H., Morita, Y., & Ikeuchi, K. (2005). Influence of proteoglycan contents and of tissue hydration on the frictional characteristics of articular cartilage. *Proceedings of the Institution of Mechanical Engineers, Part H: Journal of Engineering in Medicine*, 219(3), 175–182.

[117] Ateshian GA. (2009). The role of interstitial fluid pressurization in articular cartilage lubrication. *Journal of Biomechanics*, 42(9), 1163–1176. https://doi.org/10.1016/j.jbiomech.2009.04.040.

[118] Lawless BM, Sadeghi H, Temple DK, Dhaliwal H, Espino DM, Hukins DWL. (2017). Viscoelasticity of articular cartilage: Analysing the effect of induced stress and the restraint of bone in a dynamic environment. *Journal of the Mechanical Behavior of Biomedical Materials*, 75, 293–301. https://doi.org/10.1016/j.jmbbm.2017.07.040.

[119] Lin W, Klein J. (2023). Recent progress in cartilage lubrication: Insights into ultra-low friction mechanisms. *arXiv preprint*. https://doi.org/10.48550/arXiv.2311.02681

[120] Rawlings AV, Harding CR. (2004). Moisturization and skin barrier function. *Dermatologic Therapy*, 17 (S1), 43–48. https://doi.org/10.1111/j.1396-0296.2004.04S1005.x.

[121] Wang Y, Rudym DD, Walsh A, Abrahamsen L, Kim HJ, Kim HS, Kaplan DL. (2017). In vivo degradation of three-dimensional silk fibroin scaffolds. *Biomaterials*, 28(17), 2994–3004. https://doi.org/10.1016/j.biomaterials.2007.03.006.

[122] Brinckmann J, Notbohm H, Müller PK. (2005). *Collagen: Primer in Structure, Processing and Assembly*. Springer.

[123] Chien S. (2007). Mechanotransduction and endothelial cell homeostasis: The wisdom of the cell. *American Journal of Physiology-Heart and Circulatory Physiology*, 292(3), H1209–H1224. https://doi.org/10.1152/ajpheart.01047.2006.

[124] Reitsma S, Slaaf DW, Vink H, Van Zandvoort MA, Oude Egbrink MG. (2007). The endothelial glycocalyx: Composition, functions, and visualization. *Pflügers Archiv-European Journal of Physiology*, 454, 345–359. https://doi.org/10.1007/s00424-007-0212-8.

[125] Weinbaum S, Tarbell JM, Damiano ER. (2007). The structure and function of the endothelial glycocalyx layer. *Annual Review of Biomedical Engineering*, 9, 121–167. https://doi.org/10.1146/annurev.bioeng.9.060906.151959.

[126] Tarbell JM, Pahakis MY. (2006). Mechanotransduction and the glycocalyx. *Journal of Internal Medicine*, 259(4), 339–350. https://doi.org/10.1111/j.1365-2796.2006.01620.x.

[127] Berg JM, Tymoczko JL, Stryer L. (2006). *Biochemistry* (6th ed.). W.H. Freeman.

[128] Reitsma S, Slaaf DW, Vink H, van Zandvoort MA, Oude Egbrink MG. (2007). The endothelial glycocalyx: Composition, functions, and visualization. *Pflügers Archiv-European Journal of Physiology*, 454, 345–359. https://doi.org/10.1007/s00424-007-0212-8.

[129] Forero M, Yakovenko O, Sokurenko E, Thomas WE. (2006). Uncoiling mechanics of the P pilus rod of Escherichia coli. *Journal of Bacteriology*, 188(17), 5833–5839. https://doi.org/10.1128/JB.00204-06.

[130] Mischler S. (2008). Triboelectrochemical techniques and interpretation methods in tribocorrosion: A comparative evaluation. *Tribology International*, 41(7), 573–583. https://doi.org/10.1016/j.triboint.2007.11.003.

[131] Pawar P, Brünig A, Basu B. (2019). Environment-sensitive tribology in natural and biomimetic systems: An overview. *Journal of the Mechanical Behavior of Biomedical Materials*, 90, 515–528. https://doi.org/10.1016/j.jmbbm.2018.10.014.

[132] Bhushan B. (2012). *Encyclopedia of Tribology*. Springer, https://doi.org/10.1007/978-0-387-92897-5.

[133] Ratner BD, Hoffman AS, Schoen FJ, Lemons JE. (2020). *Biomaterials Science: An Introduction to Materials in Medicine* (4th ed.). Academic Press.

[134] Gold GE. (2001). Cartilage and joint imaging. *Radiologic Clinics of North America*, 39(1), 29–52. https://doi.org/10.1016/S0033-8389(05)70276-6.

[135] Kurtz SM, Devine JN. (2007). PEEK biomaterials in trauma, orthopedic, and spinal implants. *Biomaterials*, 28(32), 4845–4869. https://doi.org/10.1016/j.biomaterials.2007.07.013.

[136] Goldring MB, Otero M. (2011). Inflammation in osteoarthritis. *Current Opinion in Rheumatology*, 23(5), 471–478. https://doi.org/10.1097/BOR.0b013e328349c2b1.

[137] Caló E, Khutoryanskiy VV. (2015). Biomedical applications of hydrogels: A review of patents and commercial products. *European Polymer Journal*, 65, 252–267. https://doi.org/10.1016/j.eurpolymj.2014.11.024.

Sabra Banu and Sachin Kumar

4 Lubrication in biological systems

Abstract: In biological system, lubrication plays a fundamental role in biological systems by supporting smooth and efficient movement, from basic cellular interactions to the coordinated mechanics of the entire body. This chapter delves into the science of biotribology, which focuses on understanding friction, wear, and lubrication as they occur naturally in living organisms. One of the key objectives in biological lubrication is to reduce surface resistance, especially in movable joints, where synovial fluid acts as a vital biolubricant. In addition to joints, other body parts – such as the respiratory and gastrointestinal tracts – benefit from natural lubricants like mucus and interstitial fluids. These substances not only aid mobility but also protect tissues and facilitate physiological functions. The text explores how disruptions in lubrication can lead to pathological conditions like arthritis or mucosal injuries. Furthermore, the chapter discusses the growing field of biomimetics, which draws inspiration from these natural systems to develop artificial lubrication methods for medical devices and implants. Advances in smart materials, nanotechnology, and adaptive lubricants are also examined, pointing toward future innovations in both healthcare and mechanical systems. By understanding how biological systems manage movement and reduce internal strain, researchers and engineers can better design tools, therapies, and technologies that work in harmony with the human body.

Keywords: Biotribology, lubrication, synovial fluid, friction, wear, biomimetics, nanotechnology

4.1 Introduction

Biotribology, composed of two words of biology and tribology, is the study of wear, friction, and lubrication in biological system. It is a multidisciplinary branch of science, which also involved the principles of biology, engineering, and materials science [1]. Tribology is the science and technology of interacting surfaces in relative motion, encompassing the study of friction, wear, and lubrication [2]. Biotribology focuses on phenomena of tribology within living organisms, including humans and animals, and it is a collective approach where principles of tribology merge with principles of biology [3].

Acknowledgments: The authors are thankful to their parents and Dr Gagan Bansal for providing this wonderful opportunity to contribute to this prestigious book as authors and support and guidance.

Authors' contribution statement: Sabra Banu: conceptualization and writing – original draft preparation; Sachin Kumar: literature survey, study design, and editing.

https://doi.org/10.1515/9783111563176-004

Conclusively, biotribology is the study of friction, wear, and lubrications in biological systems [4]. Friction, which is broadly described in dictionaries as "the resistance that one object encounters when moving over another" is, in principle, easy to measure. It is obvious, however, that the resistance is not a function of the object in question alone [5]. It will depend, for example, on the nature of the surface over which it is being moved and, on the presence, the nature of any lubricant. In terms of biological system, every living organism faces friction, from our joints to cellular movement within tissues [6]. Although, wear is the displacement of material from one body to another body when subjected to contact and relative motion [7]. Both friction and wear are the property or ability of a system but not the property of material. Abrasive wear, adhesive wear, fretting corrosion, erosive wear, rolling contact fatigue, and tribo-corrosion are several distinct wear regimes and some of these can be operated sequentially or simultaneously [8]. The wear rate significantly changes according to the wear mode which is a function of the tribosystem. Figure 4.1 depicts the core component of biotribology which are considered as fundamental pillars.

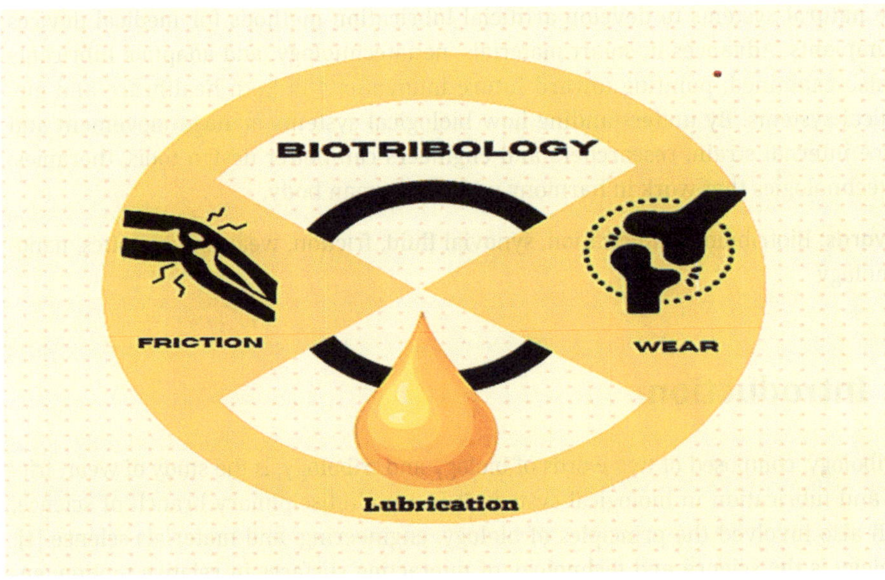

Fig. 4.1: Core components/three fundamental pillars of biotribology.

Lubrication in biological systems is a fundamental aspect of maintaining function and health. This natural process involves substances that reduce friction and facilitate smooth movement within the body like how lubricants function in mechanical systems [9]. For instance, synovial fluid in human joints acts as a lubricant, reducing friction between cartilage surfaces and enabling smooth articulation [10]. In addition to synovial fluid, other biological lubricants include mucus in the respiratory and gastrointestinal tracts, which reduce friction and protect epithelial surfaces. The effectiveness of these natural lubricants is often attributed to their unique composition and structure. For example, synovial fluid contains hyaluronic acid (HA) and lubricin, which contribute to its viscoelastic properties and ability to form a protective film over cartilage [11]. Figure 4.2 demonstrates the role of biotribology in joint lubrication replenishment therapy.

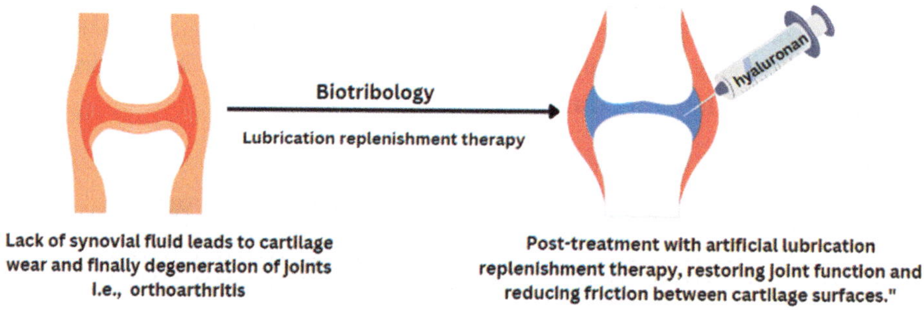

Fig. 4.2: Role of biotribology in joint lubrication replenishment therapy.

To accomplish systematic literature survey, we employed a selective and systematic approach to search, screen, and select relevant and scientific references for this chapter. Initially, we have conducted a comprehensive literature search using many reputed scientific electronic databases or search engines such as Web of Science (SCI), Scopus, and ScienceDirect, PMC, MDPI, and Google Scholar. In these databases, we were utilized relevant keywords such as lubrication, biological system, tribology, and biotribology to ensure a thorough search and get specific and relevant literatures. After that, we have screened each and every literature and thoroughly evaluated on behalf of inclusion criteria that involve a critical review of the identified references to ensure that only high-quality and pertinent studies were incorporated into this chapter. The aim of this chapter is to provide a comprehensive understanding of lubrication in biological systems, highlighting its crucial role in ensuring efficient and smooth movement across various scales, from cellular functions to the biomechanics of living organisms. The chapter seeks to explain the fundamental principles of lubrication, its significance in reducing friction and wear, and the role of biological lubricants such as synovial fluid, mucus, and interstitial fluids in promoting organ functions and joint mobility. Additionally, it explores the implications of lubrication failure in diseases, the potential of biomimetic approaches in artificial lubrication, and future advancements in materials and technologies to enhance lubrication performance in both natural and artificial environments.

4.2 Historical background

Biotribology deals with the study of lubrication in biological system which has a rich history. About four decades ago, Dowson and Wright coined the term "biotribology" in 1973. The field's principles have been applied to a range of engineering disciplines since ancient times, but the specialized branch of biotribology which focuses on biological systems began to develop in the mid-twentieth century. More specifically, the development of monolayer lubricants of new adhesives and the construction of advanced artificial joints can result from such studies [12, 14]. Figure 4.3 illustrates the historical journey of lubrication and biotribology effectively.

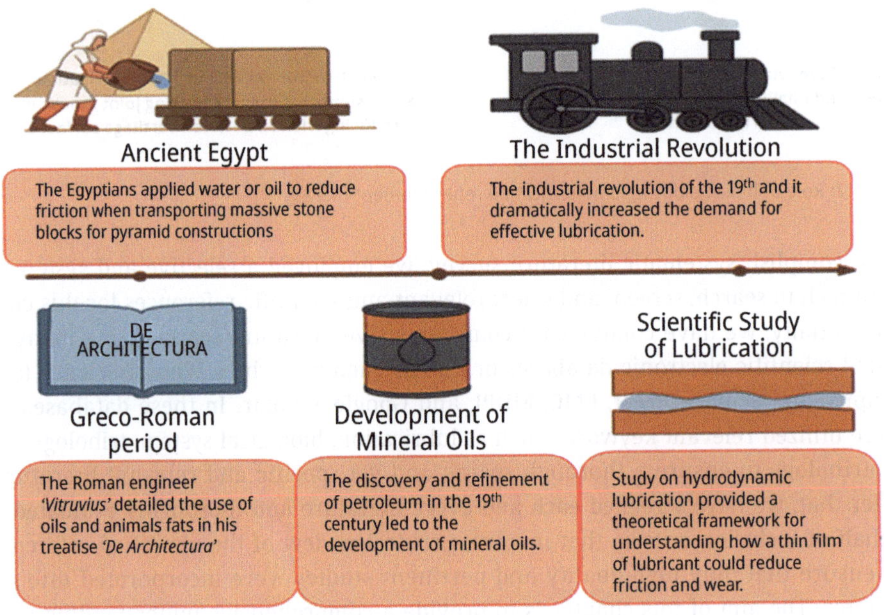

Fig. 4.3: Timeline depicting the historical evolution of lubrication technology.

4.2.1 Twentieth century: the emergence of biotribology

The formal study of biotribology began in the mid-twentieth century, primarily driven by the need to understand and improve artificial joints and prosthetic devices. The field began to develop as scientists and engineers realized that the principles of tribology could be applied to biological systems, particularly in understanding the mechanics of synovial joints like the knee and hip.

During 1950s–1960s, the first significant studies on the lubrication of human joints were conducted. Scientists like David Tabor and Frank Bowden laid the groundwork for tribology as a field, which later extended into biological systems in 1966. The term "tribology" was officially coined by Peter Jots in his report to the British government, which highlighted the economic impact of wear and friction. This led to a broader recognition of the field including its applications in biology [13, 14].

4.2.2 Twenty-first century: the modern era

Biotribology has evolved to address a wide range of biological and medical challenges beyond joint prosthetics. Modern biotribology research includes the study of dental tribology examining the wear and tear of teeth and dental implants skin tribology understanding the friction and wear properties of human skin, which is essential for developing medical devices and consumer products ocular tribology investigating the lubrication of the eye, which is crucial for contact lens development and the treatment of dry eye syndrome [14]. The history of lubrication is deeply intertwined with the development of tools, machines, and vehicles throughout human history. Some major milestones related to biological lubrication are listed in Tab. 4.1. The understanding and application of lubrication have evolved from ancient practices to sophisticated modern science [14, 15].

Tab. 4.1: Key milestones in the study of biological lubrication [12–15].

Ancient Egypt	The Egyptians, who applied water or oil to reduce friction when transporting massive stone blocks for pyramid construction.
Greco-Roman period	The Roman engineer Vitruvius detailed the use of oils and animal fats in his treatise *De Architectura*, emphasizing their role in reducing friction in rotating and sliding parts of machines.
The Industrial Revolution	The Industrial Revolution of the eighteenth and nineteenth centuries dramatically increased the demand for effective lubrication.
Development of mineral oils	The discovery and refinement of petroleum in the nineteenth century led to the development of mineral oils, the dominant type of lubricant. These oils were more stable and effective than the animal and vegetable fats previously used, particularly in high-temperature and high-pressure environments such as steam engines and early industrial machines.
Scientific study of lubrication	Study on hydrodynamic lubrication provided a theoretical framework for understanding how a thin film of lubricant could prevent direct contact between surfaces, reducing friction and wear.

4.3 Principle of lubrication

The primary purpose of a lubricant is to maintain the separation between two solids such that their asperities are shielded from one another's forces. For the lubricant to prevent being squeezed out of the contact, it must have a suitable viscosity [15]. It is usually sufficient to use a Newtonian fluid model to describe lubricating effects at the macroscale. Nonlinear effects must be taken into consideration as the dimensions and pressures decrease. Adsorbed liquids have a strong effect on friction and adhesion forces operating on contact regions in the micrometer range, with magnitudes less than around 1 mN [16]. The basic principle of lubrication involves reducing friction and wear between two surfaces in relative motion by introducing a substance, typically a lubricant that forms a thin film separating the surfaces. This thin film can be in the form of a liquid, solid, or gas and its primary functions (see Fig. 4.4) [17].

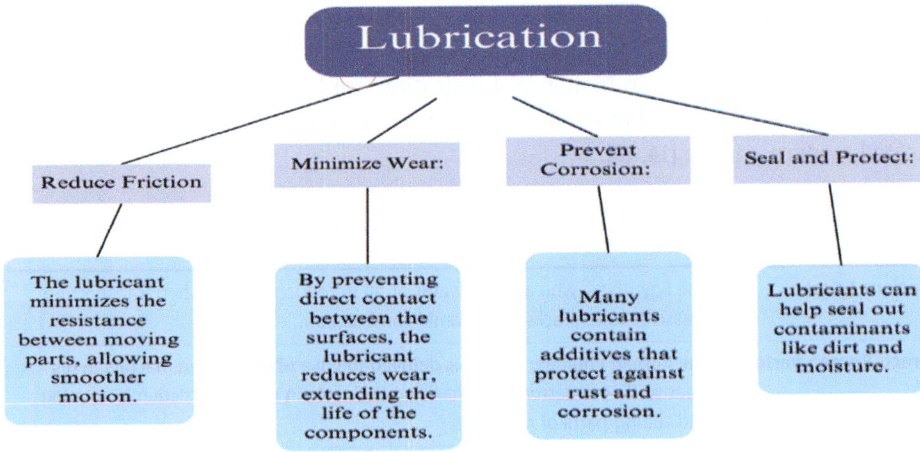

Fig. 4.4: Basic functions of lubricants or biolubricants.

On the one hand, friction is a desirable quality – it is really required for some things like an insect to start moving. Conversely, friction results in the loss of energy and, when combined with wear, also causes harm and devastation. Because the contacting surfaces are rough, mechanical interlock may result in macro-scale friction [18]. Work must be done to keep the body moving against the friction force. In addition to the friction force experienced by a moving body, inertia and static friction must also be overcome with force. Making the distinction between rolling, sliding, and static friction is essential [19]. Wear is a factor in sliding and rolling friction, and debris creates a third body. A minimal energy state that is reached when two solids come into close contact is known as adhesion [20]. This indicates that a specific force is required to break the solids apart. Adhesion exhibits a particular range of action as it increases with decreasing roughness, meaning that the smoother they are, two surfaces can adhere more effectively. Since their particles move against one another so easily, solid lubricants reduce friction. Examples are ceramic particles or powders (dry graphite, Teflon, molybdenum disulfide, aluminum, copper, etc.) [21, 22]. Solid lubricants are frequently employed in thermally stressed areas as grease additives. Since a closed lubrication cycle is frequently impractical, solid lubricants – which can be applied in a highly localized manner – offer significant advantages over fluid lubricants. Carefully choosing the molecular characteristics that result in a strong lubrication film is one potential method for creating novel lubricants that resemble solids. Research on natural lubricants may yield important hints on desirable molecular characteristics [21–23].

4.4 Biomimetics and artificial lubricants

Biomimetics and **artificial lubricants** are two related fields that leverage biological principles and synthetic technologies to solve problems involving friction and lubrication. The technique of creating new materials or technologies through mimicking natural biological systems, processes, and structures is known as biomimetics or biomimicry [24]. For usage in mechanical and biological applications, biomimetics aims to mimic the lubricating systems that naturally occur in living things. Biological systems with highly effective lubrication mechanisms that decrease friction and wear include human joints (synovial fluid), eyes (tear film), and reproductive systems (mucus) [25]. In this case, synovial fluid, a complex lubricant found in joints, reduces friction between cartilage and bones to promote fluid movement and less wear. Artificial lubricants are synthetic or engineered substances are created to reduce friction, wear, and heat between surfaces in contact or in motion. They can be liquids (oils), semisolids (greases), or solids (powders), and they are widely used in industrial and medical applications (see Fig. 4.5) [25, 26].

4.5 Applications of biomimetics in artificial lubricants

Biomimetics, which draws inspiration from nature to solve engineering challenges, has extensive applications in the development of artificial lubricants, particularly in joint replacements. By mimicking the natural lubrication mechanisms found in human joints, biomimetic approaches can improve the performance, longevity, and functionality of artificial joints. Here are some key applications of biomimetics in artificial lubricants.

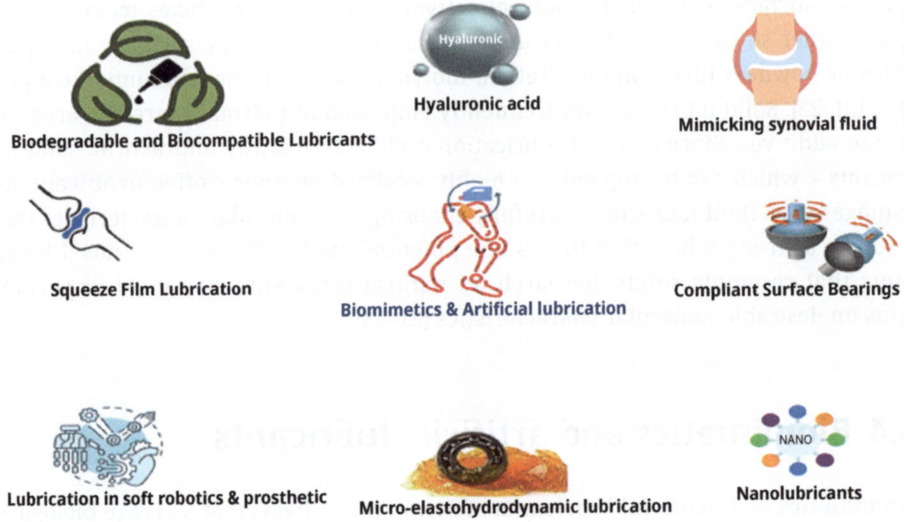

Fig. 4.5: Emerging strategies in biomimetics and artificial lubrication.

4.5.1 Mimicking synovial fluid lubrication

Human joints rely on synovial fluid, which provides lubrication through fluid-film mechanisms that reduce friction between cartilage surfaces. Artificial lubricants can be designed to replicate this mechanism, ensuring the proper separation of joint surfaces during movement, particularly under varying load and speed conditions. For instance, synovial fluid-inspired lubricants have been engineered to maintain a stable fluid film during joint motion, reducing friction and wear. These lubricants are often incorporated into artificial joints to enhance movement efficiency and minimize surface degradation [27, 28].

4.5.2 Compliant surface bearings

The compliant bearing surfaces are designed to have the soft, elastomeric properties of articular cartilage in human joint. Under the action of load, the surfaces deform in order to maintain a thin film of lubricant between the bearing surfaces. Such engineering enables bearings made of elastomeric or polymeric materials to withstand the kinds of changes in load and motion found in natural joints and provides them with the ability to act in ways similar to actual cartilage. Such surfaces provide for smoother and longer-lasting lubrication in the artificial joints than does a stiff material [29, 30].

4.5.3 Use of hyaluronic acid for viscosity control

HA is a glycosaminoglycan native to synovial fluid. One reason joint fluids are more viscous and elastic is due to HA contributions. It should be possible to reproduce the viscoelastic behavior of synovial fluid by supplementing synthetic lubricants with either HA or similar substances. For instance, artificial joint lubricants containing HA help in maintaining fluid-film thickness under varying load conditions, particularly when the person is walking or performing any low-velocity motion of joints. HA improves the resistance capability of artificial joints against wear by providing continuous lubrication under high pressure conditions [31, 32].

4.5.4 Squeeze film lubrication mechanism

Squeeze film lubrication occurs when a thin film of lubricant is maintained between joint surfaces during low-speed or compressive motions. This mechanism is crucial for preventing direct surface contact and reducing wear in artificial joints. In artificial joints, biomimetic designs use materials and structures that facilitate squeeze film lubrication, especially during slow or compressive movements such as walking or bending. This is particularly important in knee or hip replacements, where dynamic load bearing is frequent [33, 34].

4.5.5 Micro-EHL (micro-elastohydrodynamic lubrication)

Micro-elastohydrodynamic lubrication (micro-EHL) refers to the formation of high-pressure zones between rough surfaces, which can create a localized fluid film. This helps smooth out surface asperities (rough spots) and enhances lubrication. Biomimetic artificial lubricants are engineered to perform micro-EHL by generating local high pressures that prevent surface-to-surface contact in artificial joints. This results in less wear and better protection of joint surfaces over time, especially in complex motions such as those of the fingers or spine [35–37].

4.5.6 Adaptive lubricants with self-healing properties

Natural joint lubrication adapts to changing conditions including wear and tear. Biomimetic research aims to create artificial lubricants with similar self-healing properties, allowing them to repair surface damage and regenerate the lubricating film. Some advanced artificial lubricants are formulated with polymers that respond to mechanical stress by repairing damaged areas or replenishing depleted lubricating films, improving the longevity and durability of artificial joints [38, 39].

4.5.7 Integration of nanotechnology

Nanotechnology allows for the creation of nanoscale additives that can mimic the behavior of synovial fluid components, enhancing lubrication at the molecular level. These nanoparticles can reduce friction and improve load-bearing capacity in artificial joints. Nanoparticles that mimic the lubricating properties of proteins and lipids found in synovial fluid can be added to artificial lubricants. These nanoparticles can create a smoother interface between joint surfaces and enhance overall joint functionality [40–42].

4.5.8 Improved lubricants for soft robotics and prosthetics

Beyond such joint replacements, biomimetic lubricants have since been applied in softer robotics and prosthetics. Such application is primarily made with the aim of providing the robotic or prosthetic joints with friction-free lubrication to give them smooth motion without stickiness for long time. Biomimetic lubricants are installed between the joints of soft robots or prosthetic limbs for fluid motion without sticking. They are formulated to replicate mechanisms of human joint lubrication with extended durability and the least need to go for periodical maintenance [43, 44].

4.5.9 Biodegradable and biocompatible lubricants

Biomimetics can be instructive for the development of biodegradable and biocompatible lubricants designed to be used without adverse effects in the organism over protracted periods. Absorption or natural displacement can minimize the possibility of adverse reactions over a long period. Biodegradable lubricants, similar in composition to natural synovial fluid, could be applied to implants as temporary constructs or could serve as injectable solutions for joint repair. These lubricants reduce further operations and are safer for patients with permanent artificial implants of joints [45–47].

4.6 Current trends in biological lubrication

In biolubrication, the recent trends have witnessed vast advancements in the domains of biolubricants, hydration lubrication, and biomimetic materials [48]. Here, biolubricants, prepared through renewable resources, including vegetable oils as well as waste-derived oils have gained a lot of attention, mainly due to their environmental benefits and biodegradability. This volume describes recent work on their physicochemical properties, stability, and potential applications in machining and engine oils [49, 50]. The increasing focus on waste-derived oils is particularly significant, as it counters waste and assures sustainability. Water molecules play a significant role in hydration lubrication as low-friction environments are created by water molecules to improve lubrication between two nonconforming surfaces. This paradigm is essential for understanding and designing efficient boundary lubrication systems in biological media [51–53]. Such hydrated layers, supported by the charges they surround, can be stretched to extreme pressure without squeezing out, thus maintaining fluidity and reducing friction. The mechanism is very strongly relevant to synovial joints, contributing to the ultra-low friction present in healthy cartilage [54, 55]. Advances in biomimetic materials are also changing the landscape. Synthetic cartilage and other tissues capturing the key lubrication properties of natural biological systems are being developed. Such biomimetic materials are supposed to mimic the low friction and wear resistance associated with natural cartilage, and their promising applications thus lie in medical implants and tissue engineering [56]. Integration of the principles of hydration lubrication into these materials makes them more effective for applications in different biomedical applications. Sustainability is still the key, and the scientists work toward more environmentally friendly lubricants [57–59]. This includes lubricants that could be more biodegradable and use oils obtained from waste products. An important area of study is the lubrication in medical implants such as prosthetic joints. The better one can improve the lubrication of such medical implants, the better they perform, and their lifespan could be. This avoids many needs for replacements. The present study also deals with the tribological properties of biodegradable lubricants with high viscosity indexes, flash points, and a high resistance to shear. In that sense, they are applicable to different fields such as emulsifier, plasticizer, surfactant, or resins [60, 61]. One promising direction in the context of biolubricants is through the use of the following oils: nonedible oils and waste-derived oils, such as the bio-oil coming from pyrolysis and bio-crude arising from hydrothermal liquefaction. These can not only offer a somewhat sustainable substitute for conventional oils but also be applied toward waste management. The inherently interdisciplinary nature of biological lubrication – from biology, chemistry, materials science, and engineering – all push toward the development of innovative solutions that promise human health and environmental sustainability benefits. Further improvements in biolubricants properties, understanding hydration lubrication mechanisms, and development of biomimetic materials with more capabilities for better replication of natural tissue properties are highly probable areas of future research [49–51, 62–64].

4.7 Conclusion and future directions

In the context of disease development, particularly conditions like arthritis, a key factor is the disruption of lubrication in biological systems including joints. Healthy joints rely on efficient lubrication from synovial fluid to function optimally. This fluid's primary role is to reduce wear and friction, ensuring smooth movement within the joint. For biological joints to operate effectively and prevent damage, sufficient lubrication is essential. By understanding the fundamental principles of biotribology, which deals with the study of wear, lubrication, and friction in biological systems, we can devise innovative solutions for both engineering and medicinal applications. These principles not only help in extending the lifespan and enhancing the functionality of artificial joints but also offer insights into improving various mechanical systems by mimicking the efficiency found in nature. Studies in biotribology shed light on the microscopic interactions between the interfacing surfaces in biological joints, demonstrating that synovial fluid acts as a naturally occurring lubricant, minimizing wear and friction. This insight highlights the complexity of joint mechanics and its vital role in enabling smooth movement. Moreover, synovial fluid plays a crucial role in shock absorption and nutrient distribution to cartilage, which is key to maintaining joint health. These natural mechanisms serve as inspiration for designing more effective biomimetic systems including medical implants and artificial joints that replicate natural motion. The application of biomimetics has already led to significant advancements in the development of materials and coatings that enhance the mechanical performance of artificial joints. By closely replicating the properties of synovial fluid, engineers are creating devices that can move more naturally and withstand greater wear. The connection between lubrication and joint health further emphasizes the potential of incorporating these biotribological principles into treatments for conditions like arthritis, thus improving the overall quality of life.

Moreover, as we enhance the durability and functionality of artificial joints with solutions inspired by nature, we open the door to further innovations in materials science and biomaterials. This chapter examines the critical role of lubrication in biological systems, specifically focusing on the importance of synovial fluid in joint lubrication. By understanding how synovial fluid reduces wear and lowers friction in the joints, we are also exploring the potential of biomimetic lubrication techniques that can be applied to human-made devices. Such advancements will likely lead to improved designs for medical implants and equipment, ensuring better performance and longevity. The functions of natural lubricants such as synovial fluid, mucus, and interstitial fluids have been discussed, with an emphasis on how they reduce friction and protect biological structures. Biotribological concepts show that beyond their friction-reducing properties, these natural lubricants possess unique biochemical characteristics that enhance their effectiveness. For instance, the presence of HA and lubricin in synovial fluid contributes to its viscoelasticity and provides a protective coating on cartilage surfaces. As we look to the future, advances in materials science, nanotechnology, and smart materials will deepen our understanding of lubrication processes at the molecular level, leading to further improvements in lubricant performance. In particular, adaptive self-healing lubricants may address issues of wear and deterioration in both biological and nonbiological systems, offering potential solutions for the enhancement of artificial joints. The evolving science of lubrication in biological systems promises innovative breakthroughs that will improve health outcomes for individuals requiring medical interventions. Furthermore, these advancements will enhance the performance of engineered systems in diverse fields, making them more efficient and effective.

References

[1] Zhou ZR, Jin ZM. (2015). Biotribology: Recent progresses and future perspectives. *Biosurface and Biotribology*, 1(1), 3–24. https://doi.org/10.1016/j.bsbt.2015.03.001.

[2] Dowson D. (2012). Bio-tribology. *Faraday Discussions*, 156(1), 9–30. https://doi.org/10.1039/C2FD20103H.

[3] Abdelbary A. (2020). *Extreme Tribology: Fundamentals and Challenges*. CRC press, https://doi.org/10.1201/9780429448867.

[4] Furey MJ, Burkhardt BM. (1997). Biotribology: Friction, wear, and lubrication of natural synovial joints. *Lubrication Science*, 9(3), 255–271. https://doi.org/10.1002/ls.3010090303.

[5] Bansal, G., Gautam, R. K., Misra, J. P., & Mishra, A. (2024). Tribological behavior of silver-doped eggshell-derived hydroxyapatite reinforcement in PMMA-based composite. *Proceedings of the Institution of Mechanical Engineers, Part L: Journal of Materials: Design and Applications*, 238(11), 2142–2157.

[6] Ludema KC, Ajayi L. (2018). *Friction, Wear, Lubrication: A Textbook in Tribology*. CRC press, https://doi.org/10.1201/9780429444715.

[7] Jin Z, Dowson D. (2013). Bio-friction. *Friction*, 1, 100–113. https://doi.org/10.1007/s40544-013-0004-4.

[8] Wagner RM, Maiti R, Carré MJ, Perrault CM, Evans PC, Lewis R. (2021). Bio-tribology of vascular devices: A review of tissue/device friction research. *Biotribology*, 25, 100169. https://doi.org/10.1016/j.biotri.2021.100169.

[9] Gebeshuber IC. (2007). Biotribology inspires new technologies. *Nano Today*, 2(5), 30–37. https://doi.org/10.1016/S1748-0132(07)70141-X.

[10] Alexander RM. (2006). Introduction to biotribology: Animal locomotion. *Proceedings of the Institution of Mechanical Engineers, Part J: Journal of Engineering Tribology*, 220(8), 649–656. https://doi.org/10.1243/13506501JET1.

[11] Ghosh S, Choudhury D, Das NS, Pingguan-Murphy B. (2014). Tribological role of synovial fluid compositions on artificial joints – A systematic review of the last 10 years. *Lubrication Science*, 26(6), 387–410. https://doi.org/10.1002/ls.1266.

[12] Affatato S, Traina F. (2011). Bio and medical tribology. In *Tribology for Engineers* (pp. 243–286). Woodhead Publishing, https://doi.org/10.1533/9780857091444.243.

[13] Singh H, Kaur K. (2024). Biotribology: Recent advancements, applications. *Metal Matrix Composites: A Modern Approach to Manufacturing*, 147. https://doi.org/10.2174/9789815223439124010008.

[14] Akhaia S, Wadhwab AS. (2024). Recent advances in bio-tribology from joint lubrication to medical implants: A review. *Journal of Materials*, 2(2), 125–135. 10.61552/JME.2024.02.004.

[15] Meng Y, Xu J, Jin Z, Prakash B, Hu Y. (2020). A review of recent advances in tribology. *Friction*, 8, 221–300. https://doi.org/10.1007/s40544-020-0367-2.

[16] Sharma SK, Grewal HS. (2023). Tribological behavior of bioinspired surfaces. *Biomimetics*, 8(1), 62. https://doi.org/10.3390/biomimetics8010062.

[17] Khot S, Guttal R. (2021, February). Mechanics of bio-tribology in bionic joints-analytical approach. In *IOP Conference Series: Materials Science and Engineering* (Vol. 1070, No. 1, pp. 012113). IOP Publishing, 10.1088/1757-899X/1070/1/012113.

[18] Kolm R, Jogl C, Kleiner R, Gebeshuber IC, Werner WSM, Störi H. (2004, March). Characterisation of monomolecular lubricant films. In *Proceedings 14th International Colloquium Tribology* (Vol. 3, pp. 1663–1666). http://hdl.handle.net/20.500.12708/44144.

[19] Cecilia JA, Ballesteros Plata D, Alves Saboya RM, Tavares de Luna FM, Cavalcante CL Jr, Rodríguez-Castellon E. (2020). An overview of the biolubricant production process: Challenges and future perspectives. *Processes*, 8(3), 257. https://doi.org/10.3390/pr8030257.

[20] Mohan S, Mohan A. (2015). Wear, friction and prevention of tribo-surfaces by coatings/nanocoatings. In *Anti-Abrasive Nanocoatings* (pp. 3–22). Woodhead Publishing, https://doi.org/10.1016/B978-0-85709-211-3.00001-7.

[21] Fuller KNG, Tabor D. (1975). The effect of surface roughness on the adhesion of elastic solids. *Proceedings of the Royal Society of London. A. Mathematical and Physical Sciences*, 345(1642), 327–342. https://doi.org/10.1098/rspa.1975.0138.

[22] Jallo LJ, Chen Y, Bowen J, Etzler F, Dave R. (2011). Prediction of inter-particle adhesion force from surface energy and surface roughness. *Journal of Adhesion Science and Technology*, 25(4–5), 367–384. https://doi.org/10.1163/016942410X525623.

[23] Bhushan B. (2003). Adhesion and stiction: Mechanisms, measurement techniques, and methods for reduction. *Journal of Vacuum Science & Technology B: Microelectronics and Nanometer Structures Processing, Measurement, and Phenomena*, 21(6), 2262–2296. https://doi.org/10.1116/1.1627336.

[24] Singh B, Basri AA, Yidris N, Pai R, Ahmad KA. (2024). Introduction to biomimetics, modelling and analysis. In *High Performance Computing in Biomimetics: Modeling, Architecture and Applications* (pp. 1–20). Singapore: Springer Nature Singapore, https://doi.org/10.1007/978-981-97-1017-1_1.

[25] Dėdinaitė A. (2012). Biomimetic lubrication. *Soft Matter*, 8(2), 273–284. https://doi.org/10.1039/C1SM06335A.

[26] Yang L, Zhao X, Ma Z, Ma S, Zhou F. (2023). An overview of functional biolubricants. *Friction*, 11(1), 23–47. https://doi.org/10.1007/s40544-022-0607-8.

[27] Krishna DV, Sankar MR. (2023). Bioinspired artificial synovial fluid for in vitro frictional behavior of bovine articular cartilage and auxiliary biomaterials. *Journal of Molecular Liquids*, 388, 122836. https://doi.org/10.1016/j.molliq.2023.122836.

[28] Hu L, Li J, Yang X, Ruan M. (2024). A common hydration lubrication based on hydrophobically modified sulfobetaine polyurethane and synovial components in the application of artificial meniscus. *Chemistry of Materials*, 36(15), 7388–7396. https://doi.org/10.1021/acs.chemmater.4c01331.

[29] Jones E, Scholes SC, Burgess IC, Ash HE, Unsworth A. (2009). Compliant layer bearings in artificial joints. Part 2: Simulator and fatigue testing to assess the durability of the interface between an elastomeric layer and a rigid substrate. *Proceedings of the Institution of Mechanical Engineers, Part H: Journal of Engineering in Medicine*, 223(1), 1–13. https://doi.org/10.1243/09544119JEIM.

[30] Unsworth A. (1991). Tribology of human and artificial joints. *Proceedings of the Institution of Mechanical Engineers, Part H: Journal of Engineering in Medicine*, 205(3), 163–172. https://doi.org/10.1243/PIME_PROC_1991_205_287.

[31] Cardona-Muñoz JI. (2012). La viscosuplementacióncomotratamiento alternativo en la osteoartritis. *Orthotips AMOT*, 8(2), 87–92.

[32] Kho HS, Chang JY, Kim YY. (2020). Viscosities of mixtures of hyaluronic acids with different molecular weights and their effects on enzymatic activities of lysozyme and peroxidase. *Clinical Oral Investigations*, 24, 3961–3970. https://doi.org/10.1007/s00784-020-03262-z.

[33] Hou JS, Mow VC, Lai WM, Holmes MH. (1992). An analysis of the squeeze-film lubrication mechanism for articular cartilage. *Journal of Biomechanics*, 25(3), 247–259. https://doi.org/10.1016/0021-9290(92)90024-U.

[34] Hori Y. (2006). Squeeze film. *Hydrodynamic Lubrication*, 137–160. https://doi.org/10.1007/4-431-27901-6_7.

[35] Dowson D, Jin ZM. (1986). Micro-elastohydrodynamic lubrication of synovial joints. *Engineering in Medicine*, 15(2), 63–65. https://doi.org/10.1243/EMED_JOUR_1986_015_019_0.

[36] Dowson D. (1995). Elastohydrodynamic and micro-elastohydrodynamic lubrication. *Wear*, 190(2), 125–138. https://doi.org/10.1016/0043-1648(95)06660-8.

[37] Snidle RW, Evans HP. (1995). Elastohydrodynamic lubrication of real surfaces. In *Tribology Series* (Vol. 30, pp. 219–227). Elsevier, https://doi.org/10.1016/S0167-8922(08)70632-1.

[38] Yu M, Liu M, Fu S. (2021). Slippery antifouling polysiloxane–polyurea surfaces with matrix self-healing and lubricant self-replenishing. *ACS Applied Materials & Interfaces*, 13(27), 32149–32160. https://doi.org/10.1021/acsami.1c07132.

[39] Gong H, Song Y, Li GL, Zhang L, Guo D, Xie G. (2023). An intelligent polymer composite with self-lubricating and self-healing functionalities. *Composites Part B: Engineering*, 260, 110776. https://doi.org/10.1016/j.compositesb.2023.110776.

[40] Goswami R, Kama B, Mishra A. (2024). 12 Nanotechnology for enhancement of bio-based lubricant production. *Biolubricants: Feedstocks, Catalysts, and Nanotechnology*, 1, 193. https://doi.org/10.1515/9783111145136.

[41] Garcia Tobar M, Contreras Urgiles RW, Jimenez Cordero B, Guillen Matute J. (2024). Nanotechnology in lubricants: A systematic review of the use of nanoparticles to reduce the friction coefficient. *Lubricants*, 12(5), 166. https://doi.org/10.3390/lubricants12050166.

[42] Ali ZAAA, Takhakh AM, Al-Waily M. (2022). A review of use of nanoparticle additives in lubricants to improve its tribological properties. *Materials Today: Proceedings*, 52, 1442–1450. https://doi.org/10.1016/j.matpr.2021.11.193.

[43] Majidi C. (2014). Soft robotics: A perspective – Current trends and prospects for the future. *Soft Robotics*, 1(1), 5–11. https://doi.org/10.1089/soro.2013.00.

[44] Powell SK, Cruz RL, Ross MT, Woodruff MA. (2020). Past, present, and future of soft-tissue prosthetics: Advanced polymers and advanced manufacturing. *Advanced Materials*, 32(42), 2001122. https://doi.org/10.1002/adma.202001122.

[45] Khemchandani B, Somers A, Howlett P, Jaiswal AK, Sayanna E, Forsyth M. (2014). A biocompatible ionic liquid as an antiwear additive for biodegradable lubricants. *Tribology International*, 77, 171–177. https://doi.org/10.1016/j.triboint.2014.04.016.

[46] Ghosh P, Das M. (2016). Biocompatible multifunctional lubricating oil additives. *Petroleum Science and Technology*, 34(15), 1367–1373. https://doi.org/10.1080/10916466.2016.1202967.

[47] Khadem M, Kang WB, Kim DE. (2024). Green Tribology: A review of biodegradable lubricants – Properties, current status, and future improvement trends. *International Journal of Precision Engineering and Manufacturing-Green Technology*, 11(2), 565–583. https://doi.org/10.1007/s40684-023-00556-x.

[48] Rawat SS, Harsha AP. (2019). Current and future trends in grease lubrication. *Automotive Tribology*, 147–182. https://doi.org/10.1007/978-981-15-0434-1_9.

[49] Hörner D. (2002). Recent trends in environmentally friendly lubricants. *Journal of Synthetic Lubrication*, 18(4), 327–347. https://doi.org/10.1002/jsl.3000180407.

[50] Cecilia JA, Ballesteros Plata D, Alves Saboya RM, Tavares de Luna FM, Cavalcante CL Jr, Rodríguez-Castellón E. (2020). An overview of the biolubricant production process: Challenges and future perspectives. *Processes*, 8(3), 257. https://doi.org/10.3390/pr8030257.

[51] Akhaia S, Wadhwab AS. (2024). Recent advances in bio-tribology from joint lubrication to medical implants: A review. *Journal of Materials*, 2(2), 125–135.

[52] Johns-Rahnejat PM, Rahmani R, Rahnejat H. (2023). Current and future trends in tribological research. *Lubricants*, 11(9), 391. https://doi.org/10.3390/lubricants11090391.
[53] Ojaomo KE, Samion S, Yusop MZM. (2024). *Nano Bio-Lubricant as a Sustainable Trend in Tribology Towards Environmental Stability: Opportunities and Challenges*. https://doi.org/10.5109/7172279.
[54] Salih N, Salimon J. (2021). A review on new trends, challenges and prospects of ecofriendly friendly green food-grade biolubricants. *Biointerface Res. Appl. Chem*, 12(1), 1185–1207.
[55] Reeves CJ, Menezes PL. (2016). Advancements in eco-friendly lubricants for tribological applications: Past, present, and future. *Ecotribology: Research Developments*, 41–61. https://doi.org/10.1007/978-3-319-24007-7_2.
[56] Syahir AZ, Zulkifli NWM, Masjuki HH, Kalam MA, Alabdulkarem A, Gulzar M, . . . Harith MH. (2017). A review on bio-based lubricants and their applications. *Journal of Cleaner Production*, 168, 997–1016. https://doi.org/10.1016/j.jclepro.2017.09.106.
[57] Siddaiah A, Menezes PL. (2016). Advances in bio-inspired tribology for engineering applications. *Journal of Bio-and Tribo-corrosion*, 2, 1–19. https://doi.org/10.1007/s40735-016-0053-0.
[58] Zhou ZR, Jin ZM. (2015). Biotribology: Recent progresses and future perspectives. *Biosurface and Biotribology*, 1(1), 3–24. https://doi.org/10.1016/j.bsbt.2015.03.001.
[59] Khadem M, Kang WB, Kim DE. (2024). Green Tribology: A review of biodegradable lubricants – Properties, current status, and future improvement trends. *International Journal of Precision Engineering and Manufacturing-Green Technology*, 11(2), 565–583. https://doi.org/10.1007/s40684-023-00556-x.
[60] Almasi S, Ghobadian B, Najafi G, Soufi MD. (2021). A review on bio-lubricant production from non-edible oil-bearing biomass resources in Iran: Recent progress and perspectives. *Journal of Cleaner Production*, 290, 125830. https://doi.org/10.1016/j.jclepro.2021.125830.
[61] Darminesh SP, Sidik NAC, Najafi G, Mamat R, Ken TL, Asako Y. (2017). Recent development on biodegradable nanolubricant: A review. *International Communications in Heat and Mass Transfer*, 86, 159–165. https://doi.org/10.1016/j.icheatmasstransfer.2017.05.022.
[62] Wang M, He H, Fang X, Li H. (2024). The development status and future trends of lubricant additives technology: Based on patents analysis. *Plos One*, 19(6), e0304888. https://doi.org/10.1371/journal.pone.0304888.
[63] Gao L, Lu X, Zhang X, Meng Q, Jin Z. (2022). Lubrication modelling of artificial joint replacements: Current status and future challenges. *Lubricants*, 10(10), 238. https://doi.org/10.3390/lubricants10100238.
[64] Johns-Rahnejat PM, Rahmani R, Rahnejat H. (2023). Current and future trends in tribological research. *Lubricants*, 11, 391. https://doi.org/10.3390/lubricants11090391.

Sarla Yadav, Gunjan Varshney, Pooja Yadav, Riya Malik, and Raminder Kaur

5 Biotribology of articular cartilage

Abstract: Articular cartilage (AC), a specialized connective tissue, is crucial for the optimal functioning of diarthrodial joints. The distinctive composition of AC, including collagen fibers, proteoglycans, and chondrocytes, provides the structural and functional integrity required to withstand repetitive mechanical loads and enable near-frictionless joint movement. The organized configuration of its extracellular matrix and significant water content provides effective load distribution and shock absorption, maintaining joint integrity during dynamic activities. The tribological properties of AC, including its low coefficient of friction and wear resistance, are crucial to its functionality. These characteristics emerge from complex interactions between the solid and fluid phases of the tissue, with mechanisms including fluid pressurization and border lubrication being significant. The synergistic impact of components like hyaluronan and superficial zone protein in synovial fluid ensures extended lubrication under various movement and stress circumstances. Pathological disorders, such as osteoarthritis, can compromise the integrity of AC by disrupting its structural and mechanical balance. This deterioration leads to increased friction, surface damage, and joint impairment, significantly affecting mobility and quality of life. This chapter analyses the structure and function of AC, the critical tribological mechanisms that regulate its performance, and the pathological conditions that affect its tribology. This chapter aims to improve therapeutic methods and develop tissue-engineered solutions that replicate the intrinsic properties of AC through understanding these components.

Keywords: Articular cartilage (AC), tribology, osteoarthritis (OA), lubrication, mechanical behavior, tissue engineering

5.1 Introduction

Articular cartilage (AC) is a specialized connective tissue positioned on opposing bone surfaces in diarthrodial joints, such as the knee, hip, and shoulder. It is crucial for enabling nearly frictionless and fluid joint movement while withstanding lifetimes of repetitive mechanical stress [1]. Its unique physiological and anatomical characteristics make it a highly specialized tissue, exhibiting tribological function that exceeds engineering standards in spite of the lack of vascular, lymphatic, and innervational structures. This allows for load distribution, shock absorption, and joint lubrication. It is important to understand its biotribological characteristics because of its complex

https://doi.org/10.1515/9783111563176-005

mechanical qualities and susceptibility to degradation, particularly in conditions like osteoarthritis (OA).

The exceptional mechanical characteristics of AC originate from its hierarchical structure. This anisotropic and heterogeneous tissue mostly consists of type II collagen, high molecular weight proteoglycans (particularly aggrecans), and interstitial fluid abundant in physiological electrolytes, exhibiting remarkable load-bearing and lubricating properties [2, 3]. The collagen network, organized in a depth-dependent manner, transitions from a perpendicular orientation at the bone interface to a parallel orientation at the articular surface. The structural arrangement of the "Benninghoff arcades" enhances tissue resilience under mechanical stress and facilitates fluid pressurization during loading. Proteoglycans, distinguished by their bottlebrush-like structure, interact with collagen to create a low-permeability matrix that retains interstitial fluid and provides compressive stiffness, while their fixed negative charges produce ionic gradients vital for cartilage hydration and swelling [4, 5].

The complex tribological mechanics of cartilage include mixed lubrication modes, border lubrication, and fluid pressurization. According to weeping lubrication theory, fluid exudation during loading encourages low-friction motion, while boundary lubrication suggests that synovial fluid components are used to create an extremely thin lubricating covering [6]. Recent studies underscore the superiority of fluid pressurization in reducing friction under dynamic loading, influenced by the interplay of various lubrication regimes and mechanical and biological variables. However, prolonged loading decreases fluid pressurization, increasing reliance on the solid matrix and potentially causing tissue injury and friction. The degradation of AC, particularly in OA, disrupts the intricate equilibrium of its structural, mechanical, and biochemical properties [6]. This syndrome hampers mechanical function and compromises tribological performance, leading to discomfort, inflammation, and reduced joint mobility. The increase in OA cases is likely to affect over 130 million individuals worldwide by 2050, requiring the formulation of effective treatment strategies and regenerative approaches [6]. The replication of the tribological properties of natural cartilage in tissue-engineered constructs poses a significant challenge, characterized by substantial gaps in the understanding and standardization of quantifying these parameters.

This chapter provides an in-depth examination of the architecture and functionality of AC, exploring its unique zonal organization and critical contributions to joint biomechanics. It analyzes the tribological properties and mechanical behavior of cartilage, emphasizing the frictional, lubrication, and wear characteristics that influence its function in both healthy and pathological conditions. Furthermore, it investigates the impact of conditions such as OA on cartilage tribology, clarifying the biomechanical and biochemical factors contributing to joint degeneration. This chapter seeks to enhance the understanding of AC and facilitate the development of innovative strategies for maintaining joint health and preventing degenerative conditions by integrating experimental and computational research discoveries.

5.2 Articular cartilage (AC)

AC is a complex, specialized connective tissue that encases and protects the terminal ends of bones in synovial joints. The knee joint comprises the smooth surfaces that envelop the femoral and tibial condyles, as well as the inferior surface of the patella. The surface has a soft and slippery texture, characterized by an exceptionally low coefficient of friction (CoF), while the underlying layer integrates with a calcified stratum (the tidemark) that interlocks with the subchondral bone [4].

5.2.1 Structure and composition of AC

AC is a type of hyaline cartilage measuring 2–4 mm in thickness. In contrast to most tissues, AC lacks blood vessels, nerves, and lymphatics. It is a flexible, robust tissue that functions as a stress absorber to safeguard the underlying bone. The characteristics of AC are contingent upon the composition and architecture of the extracellular matrix (ECM) [6, 7]. In addition, the synthesis and preservation of this matrix rely on a sparse arrangement of specialized cells known as chondrocytes. Throughout skeletal development, AC originates from tightly packed mesenchymal cells that differentiate into chondrocytes, which proliferate rapidly and produce substantial quantities of ECM. The ECM mostly consists of water (up to 80%), collagen, proteoglycans, and smaller quantities of other non-collagenous proteins and glycoproteins, which are synthesized and maintained by the relatively few chondrocytes (Fig. 5.1) [8].

Collagen establishes a network of fibrils that provides the structural framework and shape of AC, creating compartments filled with water-binding proteoglycan complexes that regulate compressibility. The capacity of AC to withstand compressive deformation is mainly attributed to the trapping of elevated concentrations of the polyanionic proteoglycan aggrecan inside the collagen fibrillar matrix. The osmotic pressure exerted by the glycosaminoglycan (GAG) chains on the aggregated aggrecan molecules is countered and confined inside the insoluble collagenous matrix [9]. The deterioration of cartilage is a fundamental pathological characteristic of arthropathies, including OA and rheumatoid arthritis (RA), and entails the proteolytic cleavage of its primary structural components, specifically aggrecan and type II collagen.

Proteolysis and the subsequent depletion of the GAG-rich area of aggrecan from cartilage constitute an initial event in cartilage degeneration, but substantial catabolism of the collagen fibrillar structure transpires later and may signify the onset of irreversible cartilage degradation [10]. The functional integrity of AC in a healthy joint relies on chondrocytes producing various matrix components in appropriate quantities and correct sequences. Due to the absence of blood and lymphatic arteries in cartilage, chondrocytes' viability and synthetic function rely on the diffusion and movement of nutrients and metabolites within the matrix. Consequently, a delicate equilibrium is maintained between anabolism (synthesis) and catabolism (tissue degradation), with

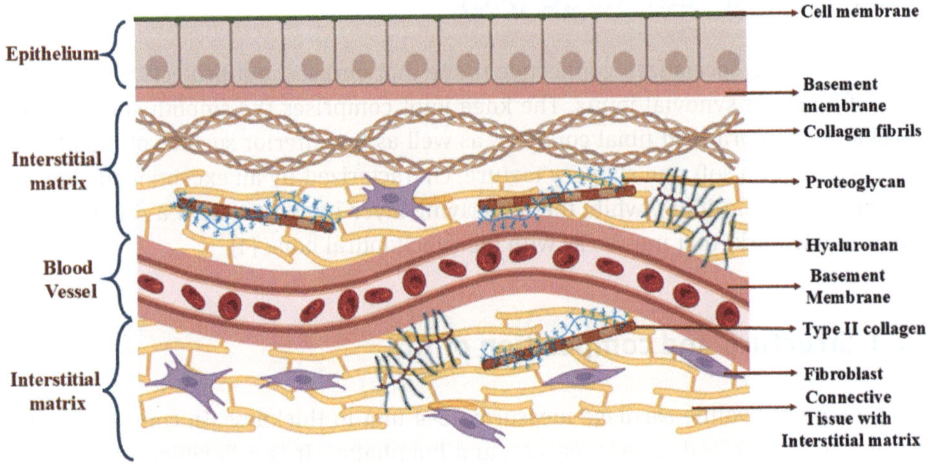

Fig. 5.1: A detailed overview of extracellular matrix (ECM) of AC.

continuous tissue remodeling being integral to the typical healthy process. In OA, the equilibrium is disrupted, favoring degenerative changes, whereas chondrocytes are unable to maintain their function, despite an initial rise in their numbers and synthetic activities.

5.2.1.1 Chondrocyte

Chondrocytes uniquely regulate both synthetic and catabolic processes in health and illness. They constitute less than 5% of the overall volume of the AC. Each chondrocyte creates a particular microenvironment and is accountable for the turnover of the ECM in its surrounding area [11]. Cells in various regions of the tissue exhibit metabolic disparities due to the necessity of distinct matrix components for structural and functional maintenance.

Chondrocytes exhibit variability in morphology, dimensions, and cellular density per unit area, contingent upon their distinct anatomical locations, including various parts of the typical knee joint. Near the surface, they are more flattened, smaller, and often exhibit a higher density than the cells located deeper into the matrix. Each chondrocyte resides within a lacuna, while collagen fibers extend to the periphery of the lacunae, which are filled with fine fibrillar material (Fig. 5.2). In contrast to osteocytes in bone, chondrocytes do not establish contact between their cytoplasmic processes, resulting in minimal or absent cell-cell interaction [12]. They possess a limited quantity of mitochondria, indicative of their reduced oxygen consumption rates. In mature cartilage, the rate of cellular division is minimal; however, division occurs in response to damage or disease.

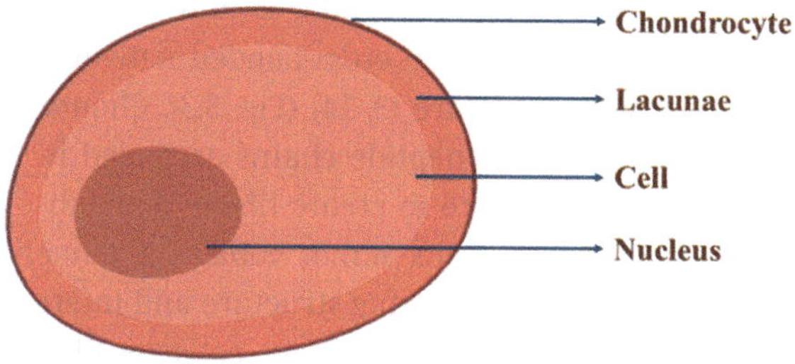

Fig. 5.2: Structure of chondrocytes.

Chondrocytes, especially in the deeper uncalcified zone, possess a well-developed endoplasmic reticulum and Golgi apparatus, essential for protein synthesis and the sulfation of mucopolysaccharides that constitute the proteoglycan side chains. Lipid is present intracellularly and as a diffuse layer around the cells, perhaps contributing to cartilage lubrication [7].

5.2.1.2 Water

Water is the primary constituent of AC, comprising up to 80% of its wet weight. Approximately 30% of this water is associated with the intrafibrillar region of the collagen, whereas a smaller portion is located in the intracellular space. The remainder resides within the pore space of the matrix. Inorganic ions such as sodium, calcium, chloride, and potassium are dissolved in the tissue fluid [13, 14]. The relative water content decreases from approximately 80% in the surface zone to 65% in the deep zone. Water flow through cartilage and over the articular surface promotes the transport and allocation of nutrients to chondrocytes while providing lubrication. A substantial fraction of the interfibrillar water appears to exist in a gel form, and most may be conveyed across the ECM by creating a pressure gradient across the tissue or by compressing the solid matrix. The resistance to flow through the matrix is significant; hence, the tissue's permeability is poor. The capacity of AC to withstand significant loads, often surpassing several times an individual's body weight, is due to the interaction between frictional resistance to water movement and the pressurization of water within the matrix.

5.2.1.3 Collagen framework

A minimum of 13 distinct forms of collagen are present in the human body's connective tissues. Articular hyaline cartilage mostly comprises types II, IX, and XI collagen, along with minor quantities of other types (Tab. 5.1). Type II collagen constitutes the principal fibrous component, comprising 90–95% of the total collagen and 40–70% of the overall dry weight of AC. It creates a three-dimensional cross-banded network of fibrils that preserves the "shape" of the cartilage structure [3, 14] (Fig. 5.3). Chondrocytes secrete procollagen molecules, composed of polypeptide chains arranged in a triple-helix configuration. These proliferate and coalesce to create fibrils that subsequently establish robust covalent inter-fibrillar cross-links. Minor collagens, namely kinds IX and XI, are believed to play a crucial role in regulating structure and interactions. Type II collagen serves as the primary component of cartilage collagen fibrils, with type XI collagen present in the core and type IX collagen located on the surface of the fibril. Collagen XI is predominantly found within the fibrils and is covalently cross-linked to type II collagen. It is present in higher concentrations within smaller fibrils. It has been proposed that its role in collagen fibrillogenesis is the nucleus for type II fibril formation [2].

Type IX collagen constitutes merely 1% of the total collagen in mature AC, despite its significantly higher concentration in fetal tissues. The covalent cross-linking occurs with type II collagen on the fibril surface in an antiparallel orientation. The covalent links resemble the intramolecular and intermolecular covalent cross-links found in type II collagen, specifically the trifunctional hydroxy pyridinium cross-link [15]. Type IX extends into the matrix from the surface of type II fibrils and may engage in covalent interactions with other type IX molecules. This interaction may contribute to the formation and stabilization of the collagen fibril network [14]. Type IX collagen plays a significant role in stabilizing the three-dimensional organization of the collagen network, thereby contributing to collagen's capacity to withstand the swelling pressure exerted by proteoglycans and the tensile stresses generated within the tissue during in situ loading. Type IX collagen plays a crucial role in collagen fibril formation, as evidenced by the occurrence of chondrodysplasia due to defects in this collagen. Moreover, a decrease in the function of type IX collagen may contribute to the degeneration of AC, which results in OA [16].

In OA, the swelling of the collagen network has been observed as one of the initial characteristics of the fibrillation process. When activated by inflammatory mediators like the cytokine interleukin-1 (IL-1), the chondrocyte produces enzymes known as metalloproteinases (including stromelysin and collagenases) that fragment collagen molecules. This facilitates the degradation of the collagen matrix and permits the elevated water content characteristic of early OA [6].

5.2.1.3.1 Structure of collagen

Collagen is a rod-like molecule characterized by a significant level of structural organization. A right-handed triple helical collagenous domain is present, composed of three polypeptide alpha (α) chains. Each chain consists of repeating triplets (Gly-X-Y) that create a natural left-handed helix, with X frequently being proline and Y commonly hydroxyproline, an uncommon amino acid that is abundant in collagen [7].

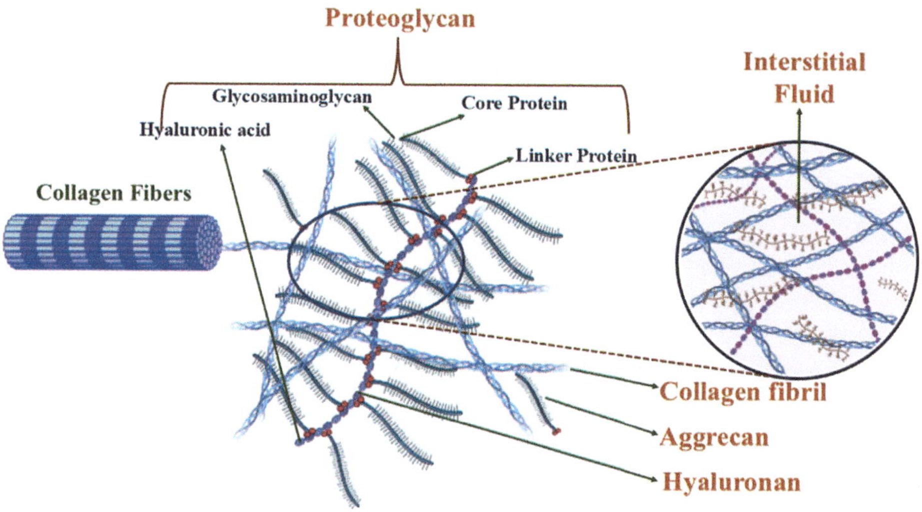

Fig. 5.3: Network of collagen fibrils in ECM.

The frequency of repeats differs across collagen types [5]. A continuous helical (Gly-X-Y) structure is typically present throughout the approximately 1,000-amino-acid length in fibril-forming collagen. In contrast, nonfibrillar collagens often exhibit interruptions in these repeats due to the presence of non-collagenous domains. The occurrence of glycine at every third residue, along with proline, hydroxyproline, and hydroxylysine, contributes to the tightly packed, rigid, triple helical structure that imparts collagen's distinctive tensile strength. The proline and hydroxyproline ring structures maintain the α chain in an extended configuration. The center of the helix exhibits a spatial restriction that permits the accommodation of glycine, the smallest amino acid (Fig. 5.4). Each of the α chains contains a short sequence of approximately 25 amino acid residues at each terminus (telopeptides) that lacks a triple helix structure [16]. Telopeptides have a significant but inadequately characterized function in assembling proteins into fibrils. This organized collagen structure seems to be specifically engineered to withstand tension.

The triple-helical collagenous domains in fibrillar collagens can polymerize extracellularly through staggered lateral associations, forming collagen fibrils. The fibrils exhibit a periodicity of 680 Å, which can be observed using transmission electron microscopy [17, 18]. Collagen fibrils in AC serve two primary functions. First, the structure imparts tensile stiffness and strength to the tissue. Second, the collagen network also limits the swelling pressure of the embedded proteoglycans, thereby contributing to compression stiffness. The structure of collagen is highly appropriate for performing these functions.

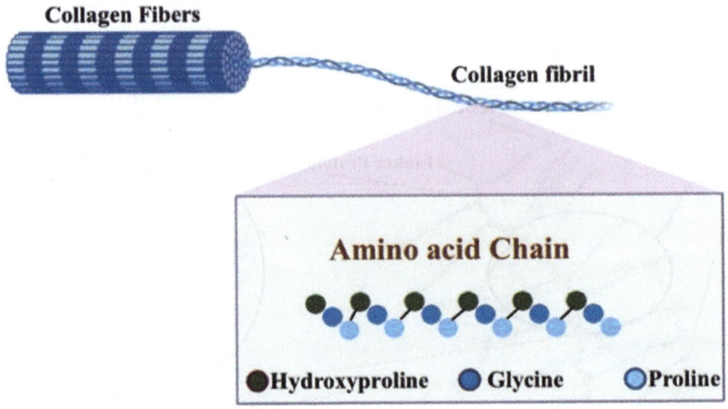

Fig. 5.4: Structure of collagen in AC.

5.2.1.3.2 Synthesis of collagen

The processes of collagen biosynthesis, secretion, and aggregation into fibrils are intricate. Chondrocyte cell nuclei exhibit extensive splicing, facilitating tissue-specific and well-regulated alternative splicing of transcripts for collagens, notably collagen types II and IX. In fibril-forming collagens, collagen α chains are synthesized and transported to the endoplasmic reticulum as prepro-α chains. Following the cleavage of signal peptides, procollagen chains undergo extensive and complex posttranslational modifications before the assembly and folding of the triple helix. The processes involve the hydroxylation of proline and lysine to form hydroxyproline and hydroxylysine, as well as glycosylation. Chain assembly initiates at the non-collagenous COOH terminus and progresses toward the NH_2 terminus, where one chain interacts with two others in the lumen of the endoplasmic reticulum to create the triple-helical molecule. This process also involves the formation of inter- and intrachain disulfide bridges. Upon formation of the triple helix, procollagen proceeds into the secretory pathway [7].

Following secretion into the extracellular space, the COOH and NH_2-terminal propeptides undergo cleavage. The cleaved amino and carboxyl propeptides may play a role in the feedback regulation of procollagen synthesis. Collagen molecules aggregate and self-assemble into fibrils. The fibrils are reinforced by covalent cross-links both within and between collagen molecules, facilitated by lysol oxidase [2]. The processing and assembly of non-fibril-forming collagens occur through analogous steps, albeit with some variations. Types VI, IX, and XI collagens maintain their post-secretion of amino and carboxyl groups in non-collagenous domains. Various types of collagen are described and characterized in Tab. 5.1. Certain collagens undergo further processing, including the addition of chondroitin sulfate (CS) chains, likely occurring intracellularly within the Golgi apparatus [14].

Tab. 5.1: Different types of collagen present in AC [14].

Collagen	Definition	Type	Length	Function
Type II collagen	A rope-like fibrillar collagen is the primary collagen found in cartilage. It is an intervertebral disc and vitreous body marker for cartilage differentiation.	Homotrimer	300 nm	It provides tensile strength and stiffness to the cartilage.
Type IX collagen	Fibril-associated collagen with disrupted triple helices (FACIT). It is reduced collagen, covalently bonded to type II fibrils via intrafibrillar hydroxypridinium crosslinks. Single chondroitin sulfate chain linked to α_2 (IX). It is a marker for cartilage differentiation.	Heterotrimer	200 nm	It gives lateral strength with flexibility to the fibrillar network.
Type X collagen	It is a short-chain collagen restricted to hypertrophic mineralizing cartilage.	Homotrimer	150 nm	Not defined
Type XI collagen	It is a lesser fibril-forming collagen in mixed fibrils of types II and IX collagens. The cartilage of this type is fragile and lacks cohesion, with abnormally thick fibrils.	Heterotrimer	300 nm	Not defined

5.2.1.4 Proteoglycans

Proteoglycans are the primary molecules embedded in collagen fibrils, constituting 15–40% of the dry weight of AC. They possess several negative charges, making them very hydrophilic, which facilitates water retention and thus enhances shock-absorbing ability. Proteoglycans consist of a core protein adorned with GAG side chains; GAGs are linear polysaccharides composed of several repeating disaccharide units, including keratan sulfate (KS) and CS, along with various other oligosaccharides. The majority of proteoglycans are present in the cartilage matrix as aggregates. Up to 100 proteoglycan monomers will attach to a central hyaluronic acid (HA) filament, stabilized by link protein (Fig. 5.5). Their total molecular weights are substantial, ranging from 1 to 2 million daltons (Da) [3].

Proteoglycans are essential for the reversible absorption of loading forces. They are accountable for regulating water flow and, hence, impede deformation under compressive force. When the tissue is compressed, such as during weight-bearing, some water is expelled, although a significant amount is retained due to the attraction forces of the proteoglycans. The proteoglycans come into closer proximity but are repelled by their strong charges, which prevents deformation [14]. The tissue swiftly restores its shape upon the removal of stress by reabsorbing water. The elevated charge of proteoglycans inhibits the passage of big molecules through the tissue while permitting the diffusion of smaller molecules, a crucial mechanism for food delivery to avascular cartilage.

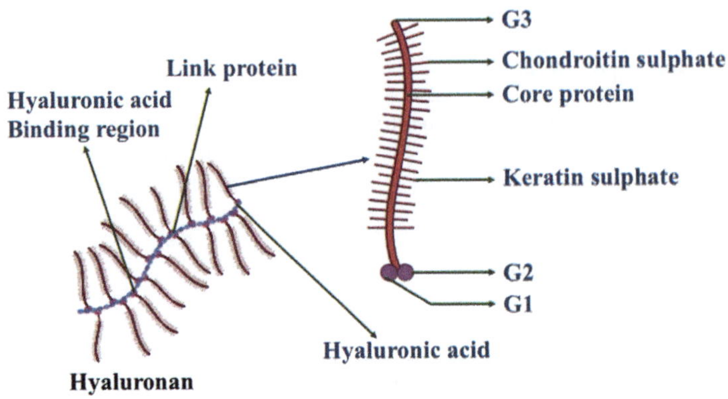

Fig. 5.5: Structure of proteoglycans in the ECM of AC.

AC consists of several essential proteoglycans for proper function, including aggrecan, decorin, biglycan, lumican, and fibromodulin [10]. The nonaggregating proteoglycans are characterized by their ability to bond with collagen. Although decorin, biglycan, and fibromodulin are far smaller than aggrecan, even if they may be present in similar molar ratios. These molecules demonstrate a close association in protein structure; nevertheless, they differ in GAG composition and function. Decorin possesses one dermatan sulfate chain, biglycan comprises two, and fibromodulin contains many KS chains. Decorin and fibromodulin interact with type II collagen fibrils in the matrix, facilitating fibrillogenesis and inter-fibril contacts. Biglycan is primarily situated near chondrocytes, where it may engage with collagen [4, 7].

5.2.1.5 Non-collagenous proteins and glycoproteins

Non-collagenous proteins contribute to the construction and stability of the ECM. They establish connections between chondrocytes and the ECM. Non-collagenous proteins encompass sticky glycoproteins like fibronectin, thrombospondin, and chondroadherin, as well as additional matrix proteins, including link protein, cartilage matrix oligomeric protein, cartilage matrix protein, and proline/arginine-rich and leucine-rich repeat proteins. AC tissue fluid comprises water, dissolved gases, tiny proteins, and metabolites. The loading of AC induces fluid pressurization and displacement. The tissue fluid is intimately linked with the joint's synovial fluid, which is crucial for nearly frictionless movement of the articulating surfaces [2, 3].

5.2.2 Zones of AC

AC is heterogeneous, with distinct structural and functional properties across its depth. AC is often categorized into four distinct zones (Fig. 5.6):

- Superficial layer
- Intermediate (or transitional) zone
- Deep (or radial) zone
- Calcified layer

5.2.2.1 Superficial zone

The superficial zone is the thinnest and most external layer, serving as the gliding surface for joints, making up 10–20% of the AC volume. It comprises a superficial layer (lamina splendens) that is approximately 2 µm thick, consisting of randomly oriented collagen fibrils, and a deeper layer of collagen fibrils aligned parallel to the cartilage surface, adhering to the "split line" pattern, which corresponds to the direction of normal joint movement. The collagen fibrils in the superficial zone have a wave-like configuration known as crimp. The deep layer of the superficial zone comprises flat, metabolically dormant AC cells (chondrocytes), contains minimal proteoglycan, and has the highest water concentration among all zones (about 80%) [9].

5.2.2.2 Transitional zone

The middle (or transitional) zone is generally more substantial than the superficial zone. It represents 40–60% of the total cartilage volume and provides an anatomic and functional bridge between the superficial and radial zones. In this zone, collagen fibrils have a larger width compared to the superficial zone and are oriented in a random manner. The proteoglycan concentration is higher, and the aggregate complexes are larger than those in the superficial zone. The chondrocytes are roughly spherical and are believed to be metabolically more active than those in the superficial zone [19].

5.2.2.3 Radial zone

The deep (or radial) zone comprises the biggest diameter collagen fibrils. They are oriented perpendicularly to the subchondral bone and the cartilage surface. The chondrocytes are often arranged in radial columns and are considered to be metabolically active. The radial zone comprises 30% of the total AC volume. The tide mark delineates the deep zone from the calcified cartilage [7, 10]. Due to its elevated proteogly-

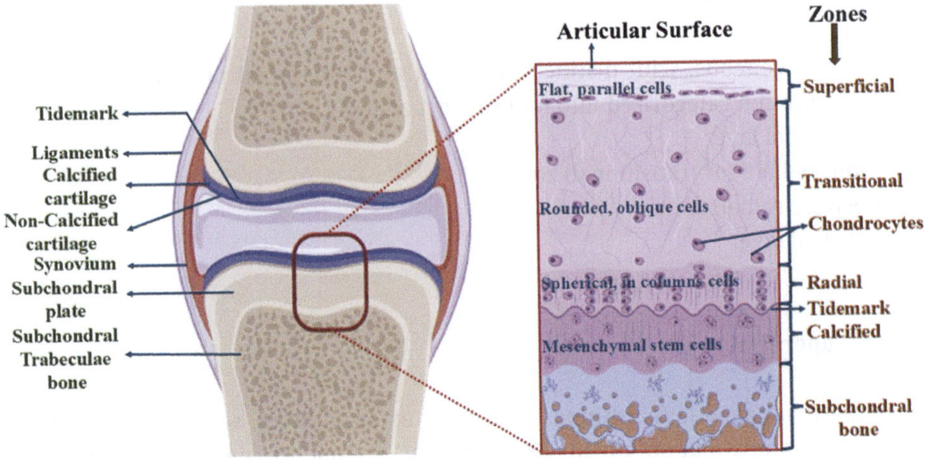

Fig. 5.6: Distinct zones of AC.

can concentration, the deep zone provides the most resistance to compressive stresses. The collagen fibrils are oriented perpendicularly to the AC. The tidemark serves to attach the more flexible and vertically oriented collagen fibrils located in the deepest layer of non-calcified AC. This anchoring mechanism is believed to protect these fibrils from disruption or rupture at their junction with the calcified zone.

5.2.2.4 Calcified zone

The calcified zone/layer is essential for attaching the cartilage to the bone by locking the collagen fibrils of the deep zone to the subchondral bone. The calcified zone consists of hydroxyapatite, acting as a barrier between the inflexible bone and the flexible cartilage, thereby diminishing a mechanical gradient [9, 20]. The calcified zone contains metabolically active chondrocytes, serves structural integration, and is considered important for nutrition and cartilage repair arising from the underlying bone. In this region, the cellular population is limited, and the chondrocytes are hypertrophic. Figure 5.1 illustrates the stratified architecture of cartilage.

5.2.3 Ultrastructure of AC

The matrix comprises various regions characterized by proximity to the chondrocytes, composition, and the diameter and organization of collagen fibrils, in addition to zonal variations in structure and composition. The ECM can be categorized into pericellular, territorial, and interterritorial regions (Fig. 5.7) [2].

5.2.3.1 Pericellular matrix

The pericellular matrix is a thin layer that encircles the cell membrane and fully envelops the chondrocyte. It mainly comprises proteoglycans, along with glycoproteins and other non-collagenous proteins. This matrix area may serve a functional role in initiating signal transduction inside load-bearing cartilage [12].

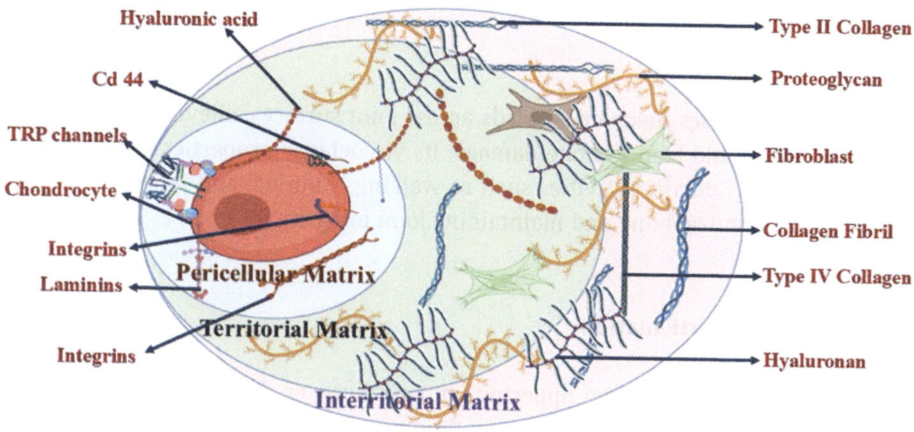

Fig. 5.7: Regions of ECM of AC.

5.2.3.2 Territorial matrix

The territorial matrix encases the pericellular matrix and primarily consists of delicate collagen fibrils, forming a basket-like structure around the cells. This region is denser than the pericellular matrix, and it has been suggested that the territorial matrix may shield the cartilage cells from mechanical stresses and enhance the resilience of the AC structure, enabling it to endure significant pressures [2].

5.2.3.3 Interterritorial matrix

The interterritorial region is the largest of the three matrix regions and significantly contributes to the biomechanical properties of AC. This region is distinguished by randomly oriented bundles of large collagen fibrils aligned parallel to the surface in the superficial zone, obliquely in the intermediate zone, and perpendicular to the joint surface in the deep zone. Proteoglycans are prevalent in the interterritorial region [11, 19].

5.2.4 Functions of AC

The principal functions of AC are to transfer forces between articulating bones, to distribute forces within joints, and to provide a relatively frictionless surface for joint movement. AC is crucial for optimal joint function, improving biomechanical efficiency and longevity. The primary functions of AC include:

5.2.4.1 Load distribution

AC uniformly disperses mechanical loads across joint surfaces, alleviating stress concentrations that could lead to bone damage. Its viscoelastic properties allow it to absorb impact forces during activities such as walking, running, and jumping, thereby protecting subchondral bone and maintaining joint integrity [4].

5.2.4.2 Frictionless articulation

AC diminishes friction between opposing joint surfaces, enabling smooth and nearly frictionless motion. The interplay between fluid pressurization and border lubrication mechanisms ensures little friction despite varying mechanical forces. This function is crucial for maintaining joint health and preventing degradation over time [3].

5.2.4.3 Articular lubrication

AC aids in joint lubrication through its interaction with synovial fluid, which contains essential macromolecules like hyaluronan and superficial zone protein (SZP). These molecules facilitate boundary lubrication and maintain the fluid-film barrier between articulating surfaces, thereby reducing friction and wear [6, 21].

5.2.4.4 Load bearing

During joint activities, AC experiences substantial mechanical loads, often exceeding several times the body weight. The collagen matrix imparts significant tensile strength, and the proteoglycan network offers compressive stiffness, enabling it to withstand recurrent stress without structural failure [22].

5.2.4.5 Articular stability

Despite not being directly accountable for ligamentous or capsular stability, the congruence and smoothness of the AC augment the mechanical stability of the joint. The cartilage surface promotes proper articulation between bones, allowing for precise joint movement [23].

5.2.4.6 Preservation of subchondral bone

The AC serves as a protective shield for the subchondral bone by absorbing and distributing compressive forces. This function reduces the risk of microdamage and pathological changes in the bone beneath the cartilage [19].

5.3 Tribological performance and mechanical behavior of AC

AC provides low-friction surfaces and exceptional load-bearing capability; therefore, it plays an essential role in joint functioning. Its mechanical behavior and tribological performance, having a special emphasis on the relationships between wear, friction, and structural integrity. Understanding these characteristics clarifies cartilage biomechanics and opens the door to the development of cutting-edge materials and methods for cartilage regeneration and repair [24].

5.3.1 Tribological properties

Surface roughness and CoF are the two quantitative tribological properties that are assessed in both native and engineered AC. This section defines the CoF and surface roughness and displays the native AC values [25]. For additional context and perspective, the tribological characteristics of native cartilage are juxtaposed to the CoF and surface roughness of synthetic materials.

5.3.1.1 Surface roughness

Asperities on the articulating surface are quantified by R_a, a widely used metric. The average height deviation from the surface midline is used to determine the surface roughness, which is commonly expressed in nanometers [26]. In natural hyaline AC, surface roughness varies from 1 to 150 nm across the body. In contrast, the surface roughness of the femoral head components of complete hip replacements usually ranges from 40 to 200 nm [27, 28].

5.3.1.2 Coefficient of friction (CoF)

The horizontal force required to move two surfaces across one another in relation to the normal force is known as the CoF, or μ. The most researched tribological characteristic in the field of AC is CoF. CoF is influenced by lubrication mode, normal load, articular surface roughness, and experimental parameters, including testing modality, in both native and experimental contexts. CoF can be calculated in both kinetic and static conditions.

Additionally, it is possible to measure the initial and equilibrium CoF. It has been found that the CoF of native AC falls between 0.001 and 0.45 [29, 30]. Under certain conditions, AC can be more frictionless than a man-made bearing, as evidenced by the CoF of 0.005 offered by standard new and cleaned rolling bearings [31].

5.3.2 Lubrication mechanisms

The mechanics of AC lubrication have been of interest and significance, and many researchers have dedicated their lives to studying these mechanisms. Numerous cartilage lubrication models have been put forth since the 1860s. Researchers proposed the existence of a low-friction layer between the AC surfaces in the first one, that is, fluid film lubrication. Several other lubrication models, including the squeeze-film mechanism, boundary lubrication, weeping lubrication, hydration lubrication, etc., were proposed as related research progressed. Among them, current research focuses on hydration lubrication and boundary lubrication. A thorough examination of cartilage lubrication mechanisms is necessary because an increasing number of studies now demonstrate that various systems work together to lubricate cartilage [32].

Lubrication can be broadly divided into two categories: fluid-film lubrication and boundary lubrication. At rest, boundary lubrication is most crucial. When loaded, direct articular contact is prevented by a thin layer of lubricating fluid. A layer of lubricant molecules that are chemically bonded to the surfaces forms a boundary layer when heavy loads are sustained for extended periods of time; otherwise, the fluid film is exhausted. Compared to unlubricated surfaces, this boundary layer offers less friction because of its low shear strength. Lubrican is the lubricating glycoprotein found in synovial joints. On each articulating surface, the lubrican molecules form a monolayer that lowers surface friction and stops articular wear [33].

There are three different kinds of fluid-film lubrication: elastohydrodynamic, squeeze-film, and hydrodynamic.

When two surfaces are angled and one glides over the other, with a fluid wedge separating the bearing surfaces, hydrodynamic lubrication takes place. A layer of fluid is dragged through a narrowing wedge-shaped gap as the surfaces slide over each other. The action generates a lift force that maintains the sliding surfaces apart by increasing the fluid's hydrodynamic pressure. The fluid layer prevents the two sliding surfaces from coming into contact, and the amount of friction is determined by the fluid's viscosity and the form of the space between the two surfaces [34].

Squeeze-film lubrication aligns the joint surfaces in parallel and places the two surfaces opposite each other without any sliding movement. The fluid's high viscosity prevents it from being forced out of the space between the two surfaces, causing a pressure buildup that can sustain comparatively hefty loads. However, the lubricating fluid runs out, and the surfaces come into contact if the heavy loads are maintained over extended periods of time [35].

Elastohydrodynamic lubrication is analogous to squeezing film lubrication. However, in synovial joints, the relative velocity of the two opposing surfaces is typically too low to maintain the minimal fluid thickness to assure absolute separation. Due to reduced fluid-film dissipation, the articular surface experiences transitory elastic deformation and displays a greater surface area, improving load tolerance over time (lowering the requisite relative velocity) [36].

Synovial joints exhibit characteristics of both fluid-film and boundary lubrication, and in practice, none of the lubricating mechanisms can adequately explain joint lubrication on its own. The dominant mechanism at any single moment depends on the circumstances. Fluid-film lubrication is observed across the rest of the bearing surface during the range of motion, while boundary lubrication is typical at the areas of intimate contact. Two types of lubrication – weeping/self-lubrication and boosted lubrication – are believed to occur during movement, in addition to boundary and elastohydrodynamic lubrication. In weeping lubrication, fluid is pushed beneath and in front of the loading surface until the optimum compression load is achieved. When the load is removed, the cartilage comes back together because the water is reabsorbed [34]. In boosted lubrication, the lubricating fluid is separated, allowing concentrated HA to stay on the joint surface while water is forced into the superficial cartilage layers. These puddles are defined as fluid pools that are trapped by the bearing surfaces' contact points [37].

5.3.3 Wear mechanisms

Wear is a complicated process that occurs between articulating surfaces and can cause the substrate to lose material. In most instances, the countersurface uses mechanical techniques to remove material from one surface. During this procedure, certain materials are eliminated as tiny particles from the contact surfaces [6]. The primary cause of artificial joints' brief lifespan is the aseptic loosening brought on by wear particles formed from polyethylene. Therefore, it is crucial for bioengineers to have a thorough grasp of wear mechanisms in order to obtain some insight into how to reduce bearing surface wear loss. Various techniques for removing materials from the bearing surfaces can be used to describe various wear mechanisms. In other words, the mechanical methods of material loss from bearing surfaces can be used to describe various wear mechanisms [38]. Five wear mechanisms may be identified in this situation: fatigue, corrosion, erosion, adhesion, and abrasion. Although hip or knee prostheses may have distinct wear modes over the course of their service life, different wear mechanisms can be identified by evaluating the wear on the artificial joints' worn surfaces [39, 40].

The loss of material that happens when a hard, rough material glides over a softer material is known as abrasive wear. The most frequent cause of wear is adhesion, which results from surface asperities coming into contact with one another under shear. It is the displacement of material compounds from one surface to another and is characterized as the transfer of material between two surfaces. Impacting particles or molecules cause erosion, which gradually erodes the material's surface through repetitive deformations and material loss. Fatigue occurs when cyclic loading and the development of surface microcracks damage a material's surface. Fretting takes place when two surfaces scrape against each other repeatedly in a cycle, removing material from both opposing surfaces. The only kind of wear that is independent of relative material motion is corrosion, which happens when a substance reacts chemically with its surroundings [1].

5.3.4 Biomechanics of AC

In addition to its exceptional low-friction qualities, cartilage is frequently referred to be a "spongy shock absorber" that shields the joint from heavy impact loads. During regular activities, cartilage in the human body, such as the hip joint, can be subjected to contact pressures of up to 10 MPa [41]. Different frequencies of compressive, tensile, and shear stresses are applied to cartilage under physiological conditions. It's critical to understand that the mechanical properties of AC differ based on age, species, joint, joint position, and tissue zone.

During unloading, interstitial fluid is soaked back into the tissue after being forced out of the cartilage with every joint movement. This action is continuous as the contact point moves across the surface of the cartilage. Interstitial fluid pressurization is the primary load support mechanism that primarily affects cartilage's compressive qualities. The interstitial fluid trapped in the aggrecan network experiences frictional drag, which dissipates the compressive load. For cartilage, compressive Young's moduli ranging from 0.24 to 0.85 MPa have been found [42, 43]. When the surrounding areas are drawn toward the loading point, cartilage tension develops as a by-product of compression. Collagen fibers dominate the tensile properties, which are nonlinear and particularly influenced by their arcadic alignment: Higher deformations cause the crosslinked fibers to stretch, increasing tissue stiffness at greater strains, while smaller deformations cause the collagen fibers to realign in the loading direction. Human AC has a tensile modulus that ranges from 5 to 25 MPa [41]. The interaction between the collagen fibers and the proteoglycan network determines the shear properties of cartilage. The superficial zone's collagen fibers, which are arranged horizontally and closely together, maximize the tissue's resistance to shear forces. Cartilage's shear storage modulus ranges from 0.2 to 2.5 MPa [30, 44].

5.3.5 Failure mechanism of AC

Numerous mechanical and biological factors can cause damage to AC [45]. Acute trauma is one of the main causes, and it arises from high-energy hits, like those that occur during vehicle crashes or sports activities. These forces may result in injuries that are classified as osteochondral (involving both subchondral bone and cartilage) or chondral (affecting only cartilage). This type of stress can result in flaps, fissures, or full-thickness abnormalities, which reduce the cartilage's capacity to absorb shocks and distribute loads. The underlying matrix injury frequently remains unnoticed, allowing gradual degeneration to occur even when the articular surface seems intact [46].

Another important factor contributing to cartilage degradation is repetitive loading, which is frequently connected to sports or physically taxing jobs. Prolonged joint stress causes the cartilage matrix to gradually deteriorate, which results in increased tissue permeability, a decrease in compressive stiffness, and the loss of proteoglycans [45]. The cartilage structure becomes weaker as a result of this cumulative damage, increasing its vulnerability to cracks and tears. Surface fibrillation, matrix swelling, and subchondral bone thickening are examples of apparent degenerative alterations that can develop from these microdamages over time, further destabilizing the joint.

Despite the surface appearing visibly intact, blunt trauma (such as falls or direct impacts) can also cause cartilage damage at the cellular and structural levels. Blunt forces frequently cause chondrocyte death, which hinders the cartilage's ability to heal itself. Furthermore, swelling and softness result from the compromise of the macromolecular structure of collagen and proteoglycans, making the tissue more susceptible to additional harm. Although they might not show up right away, these hidden injuries frequently lead to the deterioration of cartilage and joint function over time.

Another threat to the integrity of cartilage is physiological stress that surpasses typical levels. Despite cartilage's ability to tolerate normal joint loads, abrupt high-impact forces – like those from jumping or sports collisions – can alter its mechanical parameters. Similarly, the cartilage may be subjected to excessive stress by sudden joint motions that do not involve stabilizing muscle contractions. The cartilage's susceptibility to severe loading conditions is highlighted by experimental data showing that impact pressures greater than 25 MPa can result in chondrocyte death and surface fissures, even in the absence of subchondral bone fracture [47].

Damage to cartilage is also influenced by age and genetics. The mechanical characteristics of cartilage alter with age; its tensile strength decreases and it becomes stiff, which makes it more vulnerable to tears and fissures. After skeletal maturity, fully mineralized subchondral bone also raises the possibility of mechanical failure. Genetic predispositions can also cause structural flaws in cartilage, making certain people more vulnerable to harm from mechanical stress that others may be able to withstand without suffering any harm.

Another frequent cause of cartilage deterioration is chronic overuse injuries, which arise from prolonged, repetitive joint motions without adequate recuperation. Prolonged overuse undermines cartilage resilience and function by gradually reducing vital matrix components like proteoglycans. Surface fibrillation, capillary invasion into the calcified cartilage zone, and thickening of the subchondral bone are examples of the accumulation of structural damage. Long-term usage eventually results in structural collapse and decreased joint function, even though the appearance of damage may not be noticeable promptly.

Frequently, a combination of acute trauma and repetitive stress leads to cartilage deterioration. Post-traumatic OA, for example, might result from an acute injury that weakens the cartilage and increases its vulnerability to further abuse. Likewise, damage to the ligaments or menisci that surround the joint might result in aberrant loading patterns that eventually worsen cartilage deterioration. In order to prevent long-term joint dysfunction and maintain cartilage integrity, the interaction of these factors emphasizes the significance of early diagnosis, intervention, and efficient load control [45].

5.4 Pathological conditions affecting AC tribology

The tribological properties of AC can be greatly impacted by pathological diseases, which can result in increased friction, wear, and tissue degeneration. RA, OA, and joint traumas are among the conditions that change the mechanical characteristics of cartilage, making it less susceptible to bear load and lubricate. This causes joint pain, stiffness, and rapid cartilage deterioration. Under pathological circumstances, AC undergoes biochemical and structural alterations that reduce its ability to support loads and minimize friction.

5.4.1 Osteoarthritis (OA)

Osteoarthritis (OA), the most common musculoskeletal disorder in the world, is a joint disease that causes pathological AC degeneration. It has been proposed that chondrotoxic proteins are released when the synovial membrane becomes inflamed, which causes cartilage tissue to gradually deteriorate. Meniscal lesions, malalignments, joint instability, and hereditary factors can all be secondary causes of OA. The most prevalent signs of OA are stiffness and discomfort in the joints. Furthermore, the functional impairment and reduced perfusion brought on by the pain may accelerate the disease's progression [41, 48].

The fluid content of cartilage rises in the early (typically subclinical) stages of OA. Here, the aggregation ability and the concentration of proteoglycans decrease, which leads, in combination with the loosening of the collagen network, to a higher water content and tissue swelling. Therefore, the physiological load support mechanism fails, the fluid phase can no longer dissipate energy through interstitial fluid pressurization, and the collagen matrix must support the applied loads [49]. This minimizes the resilience of the tissue and leads to an accelerated degradation of the collagen fibers during the onset of the disease. As side effects, the underlying subchondral bone changes as the fluid content increases and the degree of mineralization decreases. Cartilage degeneration can lead to inflammatory reactions of the synovium and the joint capsule, as well as to fibrosis of the ligaments. Moreover, the impaired tribological function of cartilage often leads to damage of the menisci.

5.4.1.1 Pathophysiology of OA

When the delicate balance of cartilage matrix synthesis and destruction is upset, the net result is AC loss, and the degenerative process that ensues is OA. Despite significant advances in our understanding of cartilage structure, there is still a lot to learn about the mechanisms of cartilage destruction. While it is felt that the initial pathology in OA involves the AC, it is evident from very early stages that there are changes also in the synovial lining, the synovial fluid, and the underlying subchondral bone.

Macroscopically, the cartilage surface undergoes three phases of degeneration, starting with fibrillation and progressing through erosion and cracking to expose the underlying subchondral bone. Eventually, the bone surface may be denuded of its cartilage cover, a process termed eburnation. In the knee joint, the areas of cartilage damage are most apparent on the surfaces exposed to excessive load-bearing and may be quite focal in the early phases.

At the biochemical level, the main changes in OA involve an increase in the water content of the AC, a decrease in the proteoglycan concentration, and loss of the collagen network. In the early phases, the AC will actually thicken and swell because of the increased water and early increased synthesis of proteoglycans. However, these changes leave the cartilage less compressible, more permeable to tissue breakdown products, and thus more prone to damage from impact loading. The collagen network also breaks down as enzymes are released from stressed chondrocytes and synovial lining cells. The enzymes belong to a family called metalloproteinases, of which collagenase is a member. The cartilage softens (termed chondromalacia) and progresses to fibrillation of the surface layers.

Microscopically, there is evidence of chondrocyte necrosis as well as focal clumps or clones of increased proliferation. The proliferating cells are attempting repair and have been shown to increase proteoglycan production locally. The "repaired" cartilage tends to be more cellular, and at points where erosions have occurred and bone has been denuded, it exhibits the properties of fibrocartilage rather than the resilient hyaline cartilage of a normal, healthy joint. The main collagen type in fibrocartilage is type I rather than type II, and the increased fibrous content offers less protection to the underlying bone. Pathology occurs in all joint tissues in OA, including bones, menisci, ligaments, and synovium, but the central feature is the progressive degeneration of the AC. AC has a very poor reparative capacity, and ultimately, it is the breakdown and erosion of this tissue that signifies end-stage arthritis, necessitating joint replacement surgery.

5.4.2 Rheumatoid arthritis (RA)

The National Cancer Institute accurately describes RA as "an autoimmune disease that causes pain, swelling, and stiffness in the joints, may cause severe joint damage, loss of function and disability. The disease may last from months to a lifetime, and symptoms may improve and worsen over time" [50]. RA can appear in the upper extremities, lower extremities, spine, and axial joints, with characteristics including:

- inflammation (joint swelling with characteristic soft tissue involvement),
- symmetry (same joint regions of both the upper and lower extremities),
- polyarthritis (involvement of >5 joints),
- antibodies (RF, anti-CCP, anti-RA33),
- chronicity (duration of >6 weeks),
- erosions (bony destruction as seen on conventional X-ray),
- absence of recent infections or comorbid conditions associated with arthritis,
- painful metacarpophalangeal (MCP) or metatarsophalangeal (MTP) compression,
- genetics, and
- morning stiffness (lasting more than a few minutes).

The disease process may be characterized as having four distinct phases: An initial phase in which there is no clinical evidence of the disease, but some patients may have markers in the blood that denote autoimmunity; an early inflammatory phase that includes clinical manifestations that may or may not be accompanied by a confirmed diagnosis of RA; a destructive phase that includes erosions and disease progression; and an ongoing phase accompanied by irreversible joint destruction [51]. Simultaneously, there are two overlapping subpopulations of patients with RA: individuals who are positive for the presence of rheumatoid factor (RF) and individuals who are positive for the presence of antibodies that can bind cyclic citrullinated peptides (CCPs). Patients with either of these biomarkers tend to have a more severe course of RA, with anti-CCP antibodies having a greater prognostic value than RF. Patients with neither of these biomarkers tend to have a more benign course and are referred to as having "seronegative" RA [52].

RA progression and the pace of joint destruction are quite variable. SE + DR4 alleles, RF production, gender, and the presence of erosive disease at presentation are among the factors that may influence the pace of joint destruction in RA, independent of inflammatory disease activity.

5.4.2.1 Pathophysiology of RA

The pathophysiology of RA involves chronic inflammation of the synovial membrane, which can destroy AC and juxta-articular bone. Recent discoveries regarding biologic pathways have improved understanding of the phenomena associated with rheumatoid inflammation and their consequences. New molecules and cells in the biological pathway have been identified and are targets for therapeutic intervention. RA is characterized by infiltration of the synovial membrane in multiple joints with T cells, B cells, and monocytes. This process is preceded by activation of endothelial cells; neovascularization (growth of new blood vessels) is another hallmark of RA synovitis. Expansion of synovial fibroblast-like and macrophage-like cells leads to a hyperplastic synovial lining layer. This expanded synovial membrane, often termed "pannus," invades the periarticular bone at the cartilage-bone junction and leads to bony erosions and cartilage degradation. Molecules such as receptor activator of nuclear factor κB ligand (RANKL), prostaglandins, and matrix metalloproteinases are induced by proinflammatory cytokines, including tumor necrosis factor (TNF) and interleukin (IL)-6, and mediate signs and symptoms of the disease, including pain, swelling, and degradation of cartilage and bone. Stimulation by RANKL, TNF, and IL-6 generates osteoclasts within the synovial membrane and promotes bony damage. These molecular and cellular events result in the clinical disease expression. Progression of joint damage is intrinsically associated with joint swelling. The cause of RA is unknown. However, genetic and environmental factors both contribute to RA. Many gene loci are associated with RA (Box). However, certain HLA class II antigens, such as HLA-DRB1*01 and HLA-DRB1*04, contain the "shared" epitope – a stretch of five amino acids in the region responsible for antigen presentation to T lymphocytes – and are most closely associated with RA. Genes with weaker associations (Box) may also contribute, especially by gene-gene and gene-environment interactions. Environmental risk factors for RA include smoking, periodontitis, and characteristics of the microbiome of the gut, mouth, and lungs, as well as viral infections [53–55].

5.5 Conclusion

The biotribology of AC is crucial for comprehending joint function and enhancing therapies for degenerative diseases such as OA. Its peculiar composition and structure allow for exceptional wear resistance and lubrication, which are essential for maintaining low-friction, load-bearing joint action. Research into cartilage repair methods and biomaterials, along with a more thorough investigation of lubrication mechanisms, could lead to the development of novel treatments that replicate or restore the biomechanical function of native cartilage. To address the issues of cartilage degeneration and repair and enhance long-term joint health outcomes, ongoing interdisciplinary efforts are essential.

References

[1] Both K, Universit T. Biotribology and biomechanics of articular cartilage.

[2] Poole AR, Kojima T, Yasuda T, Mwale F, Kobayashi M, Laverty S. (2001). Composition and structure of articular cartilage: A template for tissue repair. *Clinical Orthopaedics and Related Research*, 391 (SUPPL.), 26–33.

[3] Mow VC, Ratcliffe A, Poole AR. (1992). Cartilage and diarthrodial joints as paradigms for hierarchical materials and structures. *Biomaterials*, 13(2), 67–97.

[4] Lai YY, Li D, Chang SW. (2021). Computational insights into the substrate recognition mechanism of cartilage extracellular matrix degradation. *Computational and Structural Biotechnology Journal*, 19, 5535–5545.

[5] Roemer EJ, West KL, Northrup JB, Iverson JM. (2016). Cho. 乳鼠心肌提取 HHS public access. *Physiol Behav*, 176(1), 100–106.

[6] Link JM, Salinas EY, Hu JC, Athanasiou KA. (2020). The tribology of cartilage: Mechanisms, experimental techniques, and relevance to translational tissue engineering. *Clinical Biomechanics*, 79 (October), 0–1.

[7] Sophia Fox AJ, Bedi A, Rodeo SA. (2009). The basic science of articular cartilage: Structure, composition, and function. *Sports Health*, 1(6), 461–468.

[8] Xiang Y, Bunpetch V, Zhou W, Ouyang H. (2019). Optimization strategies for ACI: A step-chronicle review. *Journal of Orthopaedic Translation*, 17, 3–14.

[9] Krakowski P, Rejniak A, Sobczyk J, Karpiński R. (2024). Cartilage integrity: A review of mechanical and frictional properties and repair approaches in osteoarthritis. *Healthcare MDPI*, (Vol. 12, No. 16, p. 1648).

[10] Klika V, Gaffney EA, Chen YC, Brown CP. (2016). An overview of multiphase cartilage mechanical modelling and its role in understanding function and pathology. *Journal of the Mechanical Behavior of Biomedical Materials*, 62, 139–157.

[11] Pettenuzzo S, Arduino A, Belluzzi E, Pozzuoli A, Fontanella CG, Ruggieri P, et al. (2023). Biomechanics of chondrocytes and chondrons in healthy conditions and osteoarthritis: A review of the mechanical characterisations at the microscale. *Biomedicines*, 11(7), 1942.

[12] Poole CA. (1997). Articular cartilage chondrons: Form, function and failure. *The Journal of Anatomy*, 191(1), 1–13.

[13] Zhao X, Zhao W, Zhang Y, Zhang X, Ma Z, Wang R, et al. (2022). Recent progress of bioinspired cartilage hydrogel lubrication materials. *Biosurface and Biotribology*, 8(3), 225–243.

[14] Muir H. (1995). The chondrocyte, architect of cartilage. Biomechanics, structure, function and molecular biology of cartilage matrix macromolecules. *BioEssays*, 17(12), 1039–1048.

[15] Gulsen A, Merve G, Meltem P. (2018). Biotribology of cartilage wear in knee and hip joints review of recent developments. *IOP Conference Series: Materials Science and Engineering*, (Vol. 295, No. 1, p. 012040).

[16] Berrien SLJ. (1999). Biotribology: Studies of the Effects of Biochemical Environments on the Wear and Damage of Articular Cartilage.

[17] Eyre DR. (1991). The collagens of articular cartilage. *Seminars in Arthritis and Rheumatism*, 21(3 SUPPL. 2), 2–11.

[18] Cremer MA, Rosloniec EF, Kang AH. (1998). The cartilage collagens: A review of their structure, organization, and role in the pathogenesis of experimental arthritis in animals and in human rheumatic disease. *Journal of Molecular Medicine*, 76(3–4), 275–288.

[19] March L, Little C. (2010). Articular cartilage in health and disease. In *The Musculoskeletal System: Systems of the Body Series* (2nd ed., pp. 77–95). Elsevier Limited.

[20] Čípek P, Vrbka M, Rebenda D, Nečas D, Křupka I. (2020). Biotribology of synovial cartilage: A new method for visualization of lubricating film and simultaneous measurement of the friction coefficient. *Materials*, 13(9), 2075.

[21] McNary SM, Athanasiou KA, Reddi AH. (2012). Engineering lubrication in articular cartilage. *Tissue Engineering Part B: Reviews*, 18(2), 88–100.

[22] Ostermeyer G-P, Popov VL, Shilko EV, Vasiljeva OS. (2021). *Multiscale Biomechanics and Tribology of Inorganic and Organic Systems*. Springer Tracts in Mechanical Engineering.

[23] Lloyd LL. (1991). Rigid macroporous copolymers as stationary phases in high-performance liquid chromatography. *Journal of Chromatography A*, 544(C), 201–217.

[24] Adetunla A, Afolalu A, Jen T-C, Ogundana A. (2023 Jun 5). The advances of tribology in materials and energy conservation and engineering innovation. *E3S Web of Conferences*, EDP Sciences. (Vol. 391, p. 01014).

[25] Link JM, Salinas EY, Hu JC, Athanasiou KA. (2020 Oct). The tribology of cartilage: Mechanisms, experimental techniques, and relevance to translational tissue engineering. *Clinical Biomechanics (Bristol, Avon)*, 79, 104880.

[26] Zappone B, Greene GW, Oroudjev E, Jay GD, Israelachvili JN. (2008 Feb 1). Molecular aspects of boundary lubrication by human lubricin: Effect of disulfide bonds and enzymatic digestion. *Langmuir*, 24(4), 1495–1508.

[27] Ghosh S, Abanteriba S. (2016 Jan 1). Status of surface modification techniques for artificial hip implants. *Science and Technology of Advanced Materials*, 17(1), 715–735.

[28] Ghosh S, Bowen J, Jiang K, Espino DM, Shepherd DET. (2013). Investigation of techniques for the measurement of articular cartilage surface roughness. *Micron*, 44, 179–184.

[29] Athanasiou KA, Darling EM, Hu JC, DuRaine GD, AHR. (2016). editor *Articular Cartilage* (2nd ed., p. 696). CRC Press.

[30] Middendorf JM, Griffin DJ, Shortkroff S, Dugopolski C, Kennedy S, Siemiatkoski J, et al. (2017 Oct 1). Mechanical properties and structure-function relationships of human chondrocyte-seeded cartilage constructs after in vitro culture. *Journal of Orthopaedic Research*, 35(10), 2298–2306.

[31] Woydt M, Wäsche R. (2010). The history of the Stribeck curve and ball bearing steels: The role of adolf martens. *Wear*, 268(11), 1542–1546.

[32] An H, Liu Y, Yi J, Xie H, Li C, Wang X, et al. (2022). Research progress of cartilage lubrication and biomimetic cartilage lubrication materials. *Frontiers in Bioengineering and Biotechnology*, 10, 1012653.

[33] Lees D, Partington P. (2016). Articular cartilage. *Journal of Orthopaedic Trauma*, 30(3), 265–272.

[34] McNary SM, Athanasiou KA, Reddi AH. (2012 Apr). Engineering lubrication in articular cartilage. *Tissue Engineering Part B: Reviews*, 18(2), 88–100.

[35] Sadique M, Shah SR, Sharma SK, Islam SMN. (2023). Effect of significant parameters on squeeze film characteristics in pathological synovial joints. *Mathematics*, 11(6), 1468.

[36] Jin ZM. (2013). *Elastohydrodynamic Lubrication of Natural Synovial Joints BT – Encyclopedia of Tribology* (pp. 889–892). Wang QJ, Chung Y-W. editors., Boston, MA: Springer US.

[37] Walker PS, Dowson D, Longfield MD, Wright V. (1968 Nov). "Boosted lubrication" in synovial joints by fluid entrapment and enrichment. *Annals of the Rheumatic Diseases*, 27(6), 512–520.

[38] Shen G, Fang F, Kang C. (2018 Jun 1). Tribological performance of bioimplants: A comprehensive review. *Nanotechnology and Precision Engineering*, 1(2), 107–122.

[39] Niemczewska-Wójcik M. (2017). Wear mechanisms and surface topography of artificial hip joint components at the subsequent stages of tribological tests. *Measurement*, 107, 89–98.

[40] Niemczewska-Wójcik M, Piekoszewski W. (2017). The surface topography of a metallic femoral head and its influence on the wear mechanism of a polymeric acetabulum. *Archives of Civil and Mechanical Engineering*, 17(2), 307–317.

[41] Athanasiou KA, Darling EM, Hu JC, DuRaine GD, AHR. (2013). *Articular Cartilage* (1st ed., p. 439). Taylor & Francis.

[42] Little CJ, Bawolin NK, Chen X. (2011 Aug). Mechanical properties of natural cartilage and tissue-engineered constructs. *Tissue Engineering Part B: Reviews* , 17(4), 213–227.

[43] Korhonen RK, Laasanen MS, Töyräs J, Rieppo J, Hirvonen J, Helminen HJ, et al. (2002 Jul). Comparison of the equilibrium response of articular cartilage in unconfined compression, confined compression and indentation. *Journal of Biomechanics*, 35(7), 903–909.

[44] Jin M, Grodzinsky AJ. (2001 Nov 1). Effect of electrostatic interactions between glycosaminoglycans on the shear stiffness of cartilage: A molecular model and experiments. *Macromolecules*, 34(23), 8330–8339.

[45] Buckwalter JA. (2002). Articular cartilage injuries. *Clinical Orthopaedics and Related Research*, 402(402), 21–37.

[46] Pathria MN, Chung CB, Resnick DL. (2016 Jul). Acute and stress-related injuries of bone and cartilage: Pertinent anatomy, basic biomechanics, and imaging perspective. *Radiology*, 280(1), 21–38.

[47] Haut RC. (1989). Contact pressures in the patellofemoral joint during impact loading on the human flexed knee. *Journal of Orthopaedic Research Is the Official Publication of the Orthopaedic Research Society*, 7(2), 272–280.

[48] Erggelet C, Mandelbaum BR, Mrosek EH, Scopp JM. (2008). Principles of cartilage repair. *Principals of Cartilage Repair*, Vol 1, 1–121.

[49] Eckstein F, Wirth W. (2011). Quantitative cartilage imaging in knee osteoarthritis. *Arthritis*, 2011, 475684.

[50] Jahid M, Khan KU, Rehan-Ul-Haq, Ahmed RS. (2023 Sep). Overview of rheumatoid arthritis and scientific understanding of the disease. *Mediterranean Journal of Rheumatology*, 34(3), 284–291.

[51] O'Neil LJ, Alpízar-Rodríguez D, Deane KD. (2024 Apr 1). Rheumatoid arthritis: The continuum of disease and strategies for prediction, early intervention, and prevention. *The Journal of Rheumatology*, 51(4), 337 LP – 349.

[52] Paroli M, Sirinian MI. (2023 Oct). When autoantibodies are missing: The challenge of seronegative rheumatoid arthritis. *Antibodies*, 12(4), 69.

[53] Gao Y, Zhang Y, Liu X. (2024 Mar 1). Rheumatoid arthritis: Pathogenesis and therapeutic advances. *MedComm*, 5(3), e509.

[54] Guo Q, Wang Y, Xu D, Nossent J, Pavlos NJ, Xu J. (2018). Rheumatoid arthritis: Pathological mechanisms and modern pharmacologic therapies. *Bone Research*, 6(1), 15.

[55] Firestein GS. (2018). Pathogenesis of rheumatoid arthritis: The intersection of genetics and epigenetics. *Transactions of the American Clinical and Climatological Association*, 129, 171–182.

Neetu Sharama, Ved Prakash, Bhawana, Diksha, Tanuj Kumar Bisht, and Lazreg Hadji

6 Lubrication in biological systems

Abstract: Lubrication in biological systems is crucial for smooth, pain-free movement and the protection of joints and tissues. Synovial fluid, found in joints like the knees and hips, not only reduces friction but also absorbs shocks to prevent damage. Beyond joints, lubrication is vital in areas like the digestive and reproductive systems, reducing friction and facilitating smooth functioning. Replicating this natural lubrication in artificial joints is challenging, but advances in materials and technology - such as self-lubricating materials and nanotechnology - improving implant durability and effectiveness, enhancing the quality of life for patients.

Keywords: Lubrication, biological systems, movement, artificial joints, nanotechnology

6.1 Introduction

Lubrication, a concept often associated with machines and engineering, is just as critical in the biological world. The human body, like a finely tuned machine, has many moving parts that need lubrication to function efficiently. This process is essential for reducing friction, preventing damage, and ensuring the smooth operation of various tissues and organs. The scientific study of lubrication in living organisms, called *biotribology*, delves into how these natural lubrication systems work and how we can apply these principles in medical and engineering fields to solve real-world problems.

6.1.1 Understanding biological lubrication

Our bodies have developed highly sophisticated lubrication mechanisms to navigate the complexities of our daily movements and environments. Unlike machines, which rely on external lubricants, biological systems generate their own. These lubricants, such as synovial fluid (in joints), mucus, and saliva, are unique in their composition and ability to adapt to specific conditions [1]. Think of synovial fluid as a natural cushion that makes your knee or shoulder movements seamless, or consider the protective tear film in our eyes that prevents discomfort when we blink. Even inside our digestive system, mucus serves as a vital lubricant, enabling the smooth passage of food while shielding the sensitive lining from digestive enzymes. Each lubricant in our body is tailored to the specific environment it serves, ensuring that every part operates optimally [2–5].

https://doi.org/10.1515/9783111563176-006

6.1.2 Evolution of lubrication mechanisms

Over time, biological lubrication systems have evolved to meet the demands of the environments organisms live in. For humans, this evolution is especially evident in the joints, which accommodate a range of movements, from simple bending to more complex rotations. Synovial fluid, with its ability to change viscosity depending on the pressure placed on the joint, is a prime example of how our bodies have adapted to handle mechanical stress efficiently [6]. Additionally, these lubrication systems help conserve energy. By reducing friction and wear, they minimize the energy required for movement, making biological systems incredibly efficient. This energy conservation is especially crucial during endurance activities, where efficiency is key to survival [2].

6.1.3 Clinical significance and applications

Biological lubrication systems have important implications in medicine, particularly in orthopedics. By studying *biotribology*, scientists and engineers have been able to design better artificial joints and implants that mimic the body's natural lubrication mechanisms. For instance, ultra-high-molecular weight polyethylene (UHMWPE), a material used in joint replacements, was inspired by the low-friction properties of cartilage. Hydrogels, which replicate the elasticity of synovial fluid, are another innovation improving the durability and comfort of artificial joints.

These advancements aim to reduce friction and wear in implants, improving their longevity and enhancing the quality of life for patients.

6.1.4 Future directions in biotribology

The field of biotribology is constantly evolving. As we learn more about how the body's natural lubrication works, we open doors to new innovations. Scientists are exploring bio-inspired lubricants and materials that behave more like their natural counterparts. This includes the use of nanotechnology and smart materials that could revolutionize the design of medical implants, making them more efficient and durable [4].

Beyond medicine, biotribology has the potential to impact other fields, such as robotics and bio-mimicry. By studying how animals, like insects, lubricate their joints, researchers can develop more agile, efficient robots. The lessons learned from nature continue to inspire and shape technological advancements in many unexpected areas [2].

6.2 Types and functions of lubrication in living organisms

Lubrication is a vital process in our bodies that helps things move smoothly without causing damage. Just like machines need oil to keep their parts running, our bodies have natural lubricants that reduce friction between moving tissues and organs. Whether it's our joints bending, our eyes blinking, or food moving through our digestive system, these processes rely on lubrication to prevent wear and tear. Over time, our bodies have evolved specific lubrication systems to meet the unique needs of different organs and tissues, ensuring everything works together efficiently and smoothly. These systems are incredibly complex, involving specialized fluids and structures that keep us functioning in balance.

6.2.1 Types of lubrication in living organisms

6.2.1.1 Boundary lubrication

Boundary lubrication happens when two surfaces in the body come into direct contact but are cushioned by a thin layer of lubricant. This is crucial in areas where tissues are very close together and experience a lot of pressure, like joints that bear heavy loads. In our bodies, molecules like **lubricin** and **hyaluronic acid** are the key players in this type of lubrication. These molecules help joints, such as the knees or hips, move smoothly and prevent damage. For example, lubricin creates a protective film on the surfaces of cartilage, reducing friction and allowing for easy, pain-free movement.

Lubricin is a type of glycoprotein found in synovial fluid, the liquid that lubricates our joints. It works by sticking to the surfaces of cartilage, forming a protective layer that reduces friction between the moving parts of the joint. This is especially important when joints are under heavy pressure, like during running or lifting [7].

Hyaluronic acid is another key player in joint health. It gives synovial fluid its thick, gel-like consistency, helping to maintain a smooth, protective layer over cartilage. Together, lubricin and hyaluronic acid create a system that ensures joint movement is smooth and pain-free, while protecting cartilage from wear and tear [8].

6.2.1.1.1 Examples in biological systems

- **Synovial joints:**
  In synovial joints like the knee, boundary lubrication plays a crucial role in protecting cartilage from wear and tear, especially during activities that put a lot of pressure on the joints, such as running or jumping. A thin layer of synovial fluid,

rich in lubricin and hyaluronic acid, ensures that the cartilage surfaces slide smoothly against each other with minimal friction [9–12].

– **Anatomy of a synovial joint:**

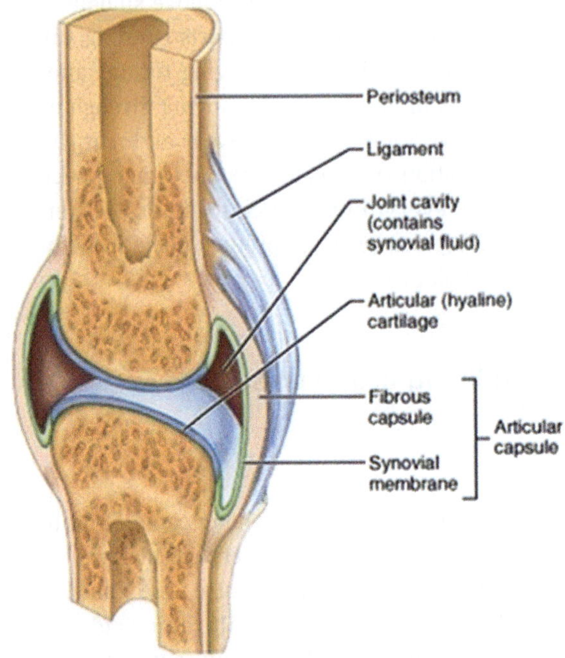

Fig. 6.1: Anatomy of a synovial joint. Reproduced with permission [13].

The joint is surrounded by an articular capsule, which consists of an outer fibrous layer for stability and an inner synovial membrane that produces synovial fluid, as shown in Fig. 6.1. This fluid fills the space between the bones, reducing friction and nourishing the cartilage that covers the ends of the bones, known as hyaline cartilage. Ligaments connect the bones, adding further stability, while the periosteum, a protective layer, covers the bone surfaces outside the joint. All these components work in harmony to allow the joint to move flexibly and with cushioning.

– **Skin:**
The skin also relies on boundary lubrication, particularly in areas like the eyelids, where smooth movement is essential. As the eyelids move over the eyes, minimal friction is needed to prevent irritation. This is made possible by the lipids secreted by sebaceous glands, which create a thin, oily layer on the skin's surface, reducing friction during blinking [4].

6.2.1.2 Fluid film lubrication

Fluid film lubrication happens when a thick layer of lubricant fully separates two moving surfaces, preventing them from making direct contact. This type of lubrication is commonly seen in joints and other biological areas where smooth, continuous movement is vital.

- **Synovial fluid:** In synovial joints, fluid film lubrication is made possible by synovial fluid, which creates a thick, protective layer between the moving surfaces. This not only reduces friction but also helps to evenly spread the load across the joint, preventing stress from being concentrated on any one spot. The viscosity of synovial fluid plays a key role in maintaining this layer, especially when the joint is moving quickly or bearing a light load [6].
- **Articulating joints:** In large, load-bearing joints like the hip and knee, fluid film lubrication is essential for smooth movement and protection. Synovial fluid in these joints is constantly replenished and maintained, ensuring that the protective layer remains intact and effective, even as the joint moves under different conditions and pressures. This allows the joint to function smoothly without damage [4].
- **Ocular system:** The cornea of the eye is protected by a thin layer of tear film that acts as a natural lubricant during blinking. This tear film helps reduce friction between the eyelid and the surface of the cornea, preventing irritation or potential damage. In addition to its lubricating function, the tear film also helps keep the eye clean by removing debris and providing essential nutrients to nourish the cornea, ensuring overall eye health [14].

6.2.1.3 Mixed lubrication

Mixed lubrication is a blend of both boundary and fluid film lubrication. It happens when part of the load is supported by a fluid layer, while the rest relies on direct contact between rough surface areas. This type of lubrication is quite common in biological systems, where the mechanical needs fluctuate; calling for a mix of lubrication techniques to ensure everything functions smoothly.

- **Synovial joints:** In synovial joints, mixed lubrication helps manage different types of movements and loads effectively. For example, during slow movements or when the load on a joint is light, the fluid film can adequately keep the cartilage surfaces apart. However, under higher stress or faster movements, boundary lubrication takes over, making sure the joint surfaces don't grind against each other and cause damage [15].

6.2.1.3.1 Examples in biological systems

- **Spine:** In the spine, the intervertebral discs serve as cushions between the vertebrae, helping to distribute the load and provide flexibility. The fluid inside the discs offers fluid film lubrication, allowing smooth movement, while the structure of the discs ensures that boundary lubrication kicks in at areas experiencing the most stress. This dual action helps protect the spine during movement and load-bearing activities [4].

6.2.1.4 Squeeze film lubrication

Squeeze film lubrication occurs when a fluid film is compressed between two surfaces, creating a high-pressure area that supports the load. This type of lubrication is particularly important in scenarios where surfaces move closer together under load, requiring a lubricant that can withstand the pressure without being expelled from the contact area.

- **Synovial fluid in joints**: In joints, squeeze film lubrication helps absorb shocks and evenly distribute loads across cartilage surfaces. For example, during activities like running or jumping, the pressure on joints increases significantly. Synovial fluid is squeezed into the contact area, generating enough pressure to cushion the impact and protect the cartilage from damage [16].

6.2.1.4.1 Examples in biological systems

- **Intervertebral discs:** Squeeze film lubrication is also found in the intervertebral discs of the spine. When the spine is compressed, such as during heavy lifting, the fluid inside the discs is pushed into the spaces between the vertebrae, providing cushioning that protects the spinal column from injury [6].

6.2.2 Functions of lubricants in living organisms

Lubrication is a vital physiological process in living organisms that enables the smooth and efficient functioning of various body systems. Biological lubricants, such as bodily fluids, primarily serve to reduce friction, minimize wear, aid smooth movement, and protect tissues. Here's a breakdown of their functions:

6.2.2.1 Reduction of friction

A key function of biological lubricants is reducing friction between moving parts of the body. For instance, in synovial joints, synovial fluid forms a thin layer between the cartilage surfaces, allowing them to glide over one another smoothly. This reduction in friction prevents the wear and tear of joint cartilage and helps in maintaining joint health by avoiding pain and degeneration [17].

6.2.2.2 Minimization of wear and tear

Lubricants help minimize tissue and joint wear. In synovial joints, the lubrication provided by synovial fluid helps distribute the load and absorb shocks, reducing stress on cartilage and other joint components. This protective effect is vital for preserving the long-term health of joint tissues [18].

6.2.2.3 Facilitation of smooth movement

By reducing resistance between moving surfaces, lubricants ensure smooth movement. Aside from synovial fluid in joints, lubrication is essential in other systems too. For example, mucus in the gastrointestinal tract (GIT) allows food to pass smoothly, and cervical mucus aids in sperm transport in the reproductive system [18].

6.2.2.4 Shock absorption

Biological lubricants also act as shock absorbers. Synovial fluid, in addition to reducing friction, helps cushion joints during activities that involve impact, such as running or jumping. This prevents damage to joint surfaces and lowers the risk of injury [16].

6.2.2.5 Protection and nutrient supply

Beyond mechanical roles, biological lubricants offer protection and supply nutrients. For example, synovial fluid contains glycoproteins like lubricin, which shield cartilage from damage and provide essential nutrients. Since cartilage lacks a direct blood supply, this nutrient distribution is crucial for its health and repair [19].

6.2.2.6 Facilitation of cellular functions

Lubrication also aids cellular functions and interactions. In the circulatory system, the viscosity of blood and the lining of blood vessels help reduce friction on red blood cells, ensuring smooth blood flow and protecting against damage to blood vessels [20].

6.2.2.7 Enhancement of tissue repair and regeneration

Lubricants contribute to tissue repair and regeneration. For instance, synovial fluid's lubricating properties create an environment conducive to joint healing, while mucus in the respiratory and digestive systems protects and repairs tissues by maintaining moisture and clearing debris [21].

6.2.2.8 Lubrication in the gastrointestinal tract

In the GIT, boundary lubrication is vital for the smooth passage of food. The GIT is lined with mucus, a viscoelastic fluid that facilitates food movement and protects epithelial cells from mechanical damage and digestive enzymes. This mucus consists mostly of water, mucins (glycoproteins), lipids, and electrolytes. The mucins form a gel that gives mucus its adhesive and lubricating properties [4].

- **Protective barrier:** Mucus protects epithelial cells from abrasive food particles and corrosive digestive enzymes, particularly in the stomach, where it prevents acidic gastric juices from damaging the stomach lining [22].
- **Facilitation of food movement:** Mucus allows food to pass smoothly through the digestive system, aiding nutrient absorption by maintaining close contact between the food and the absorptive surfaces of the intestines [4].

6.2.2.8.1 Examples in biological systems

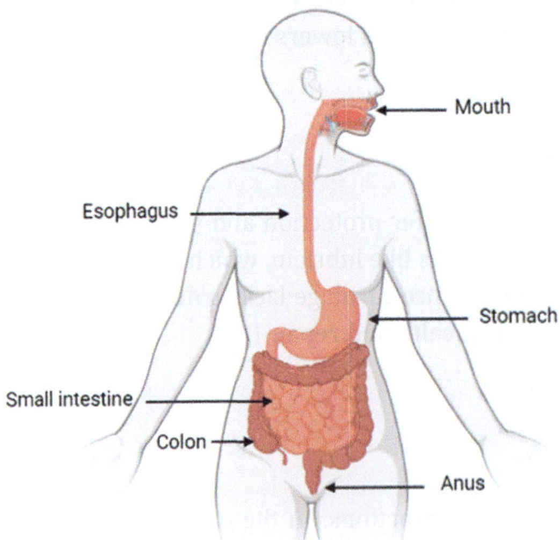

Fig. 6.2: Parts of the human digestive system, from mouth to anus. Reproduced with permission [23].

- **Esophagus:** Mucus in the esophagus helps food travel smoothly from the mouth to the stomach, reducing friction and protecting the esophageal lining from damage [22].
- **Stomach and intestines:** Mucus not only protects the stomach and intestines from acids and enzymes but also plays a critical role in digestion and absorption by ensuring smooth food movement [4], as shown in Fig. 6.2.

Biological lubricants perform essential roles beyond reducing friction. They absorb shocks, supply nutrients, and facilitate smooth movement, highlighting their importance in preserving the function and health of various body systems.

6.2.3 Composition of synovial fluid

Synovial fluid is a complex, non-Newtonian substance that contains several key components, each contributing to its lubricating properties.

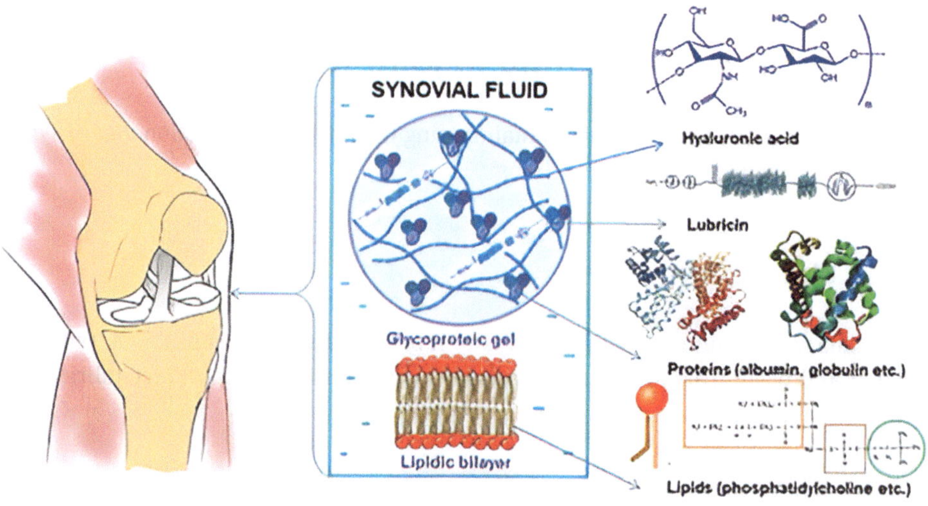

Fig. 6.3: Components of synovial fluid. Reproduced with permission [24].

- **Hyaluronic acid:** This high-molecular-weight glycosaminoglycan is the main factor behind the fluid's viscosity and elasticity. Hyaluronic acid and lubricin are present in synovial fluid, as shown in Fig. 6.3. Produced by synoviocytes, the cells lining the synovial membrane, hyaluronic acid gives synovial fluid its viscoelastic characteristics, allowing it to adapt to different mechanical stresses. The concentration of hyaluronic acid in synovial fluid can vary from 0.1 to 4 mg/mL, depending on factors like joint health and activity levels [19].

- **Lubricin:** Also known as proteoglycan 4 (PRG4), lubricin is a mucinous glycoprotein crucial for boundary lubrication. It creates a protective layer over the cartilage, minimizing friction and wear between the joint surfaces. Lubricin also has anti-adhesive properties, which prevent cells and proteins from sticking to the cartilage, helping to maintain joint health and function [7].
- **Phospholipids:** These lipids contribute to synovial fluid's lubricating abilities by forming a layer on the cartilage surface. This lubricating layer complements the actions of lubricin, further reducing friction between joint surfaces [4].
- **Water:** Synovial fluid is about 94% water, which acts as a medium for dissolved molecules and ensures the fluidity needed for effective joint lubrication. The water content also helps spread compressive forces evenly across the joint, protecting the cartilage from damage during movement [4–6].
- **Proteins and electrolytes:** Synovial fluid contains proteins like albumin and globulin, along with electrolytes such as sodium, potassium, and chloride. These components help maintain osmotic balance and create a biochemical environment that supports overall joint health [15].

6.3 Role of synovial fluid in joint lubrication

Synovial fluid plays a crucial role in maintaining the health and functionality of synovial joints through several key functions:

6.3.1 Lubrication

The primary function of synovial fluid is to lubricate the surfaces of joints, reducing friction and wear. This lubrication occurs in two main forms: boundary lubrication and fluid film lubrication.

6.3.1.1 Boundary lubrication

In this process, lubricin forms a thin, protective layer over the cartilage, preventing direct contact between the joint surfaces during movement. This type of lubrication is particularly important in joints that bear heavy loads or move slowly, where other forms of lubrication might not be enough to prevent friction [7].

6.3.1.2 Fluid film lubrication

Here, synovial fluid creates a thicker layer between cartilage surfaces, effectively preventing contact during rapid movements or under moderate loads. The viscoelastic nature of hyaluronic acid ensures the integrity of this film, allowing it to absorb and redistribute forces across the joint [19].

6.3.2 Shock absorption

Synovial fluid also helps absorb shocks during activities such as running or jumping. Thanks to the viscosity provided by hyaluronic acid, it can resist compressive forces, distributing them evenly across the joint surfaces. This function protects cartilage from damage caused by nutrient distribution. Since cartilage lacks its own blood supply, it relies on synovial fluid for nutrition. The fluid circulates through the joint, delivering nutrients and removing waste. This is crucial for the health and repair of cartilage tissue [15].

6.3.3 Maintenance of joint homeostasis

Synovial fluid helps maintain the joint's biochemical environment, regulating factors like pH and ionic composition. This creates a stable environment that supports the joint's overall function [4].

6.3.4 Mechanisms of joint lubrication by synovial fluid

Several biomechanical and biochemical mechanisms make synovial fluid effective in lubricating joints:

6.3.4.1 Squeeze film lubrication

This occurs when synovial fluid is compressed between cartilage surfaces during joint loading. The resulting high-pressure region prevents the surfaces from directly contacting one another. This is especially helpful in high-impact activities, where the fluid is "squeezed" into the contact area, cushioning the joint and protecting the cartilage from damage [6].

6.3.4.2 Hydrodynamic lubrication

In this mechanism, a wedge-shaped film of synovial fluid forms between moving joint surfaces. As they move, the fluid is drawn into the wedge, creating pressure that separates the surfaces. This is particularly effective in joints with curved surfaces, like the hip and shoulder, where the joint shape helps form this protective fluid layer [4–6].

6.3.4.3 Elastohydrodynamic lubrication

Here, the elasticity of cartilage comes into play. When the cartilage deforms under load, it increases the fluid film thickness, keeping surfaces apart and distributing the load more evenly. This mechanism is crucial in high-load joints like the knee and hip, where elasticity helps protect the joint from wear [5, 6].

6.3.4.4 Boundary layer lubrication

This mechanism involves a thin layer of lubricin forming on cartilage surfaces, creating a low-friction interface that prevents direct contact and wear. It is especially effective in joints under heavy loads or slow movements, where the fluid film alone might not be sufficient [7].

6.3.5 Pathological conditions affecting synovial fluid and joint lubrication

Several medical conditions can affect synovial fluid, impairing its ability to lubricate joints and leading to joint disorders:

6.3.5.1 Osteoarthritis

Osteoarthritis, a degenerative joint disease, involves the breakdown of cartilage and changes in synovial fluid composition. A decrease in hyaluronic acid levels leads to lower viscosity and poor lubrication, increasing friction and wear on the joint, which contributes to cartilage damage, pain, and stiffness [25].

6.3.5.2 Rheumatoid arthritis

In this autoimmune disorder, inflammation of the synovium leads to the release of enzymes that break down cartilage and synovial fluid components. This impairs the production of lubricin and hyaluronic acid, reducing lubrication and leading to joint deformities and dysfunction [26].

6.3.5.3 Synovitis

Synovitis is the inflammation of the synovial membrane, caused by injury, infection, or autoimmune diseases. It can result in an overproduction of synovial fluid, altering its quality and reducing its ability to lubricate the joint [27, 28].

6.3.5.4 Gout

Gout is a metabolic disorder in which uric acid crystals accumulate in the joints, causing inflammation and pain. These crystals interfere with normal lubrication, causing damage to cartilage and the synovial membrane, which can lead to joint degeneration over time [29].

6.3.6 Therapeutic interventions to enhance synovial fluid function

To improve synovial fluid function in people with joint disorders, several therapeutic interventions have been developed; a detailed discussion is given in Fig. 6.4.

6.3.6.1 Visco supplementation

This treatment involves injecting hyaluronic acid directly into the joint to restore the synovial fluid's viscosity and elasticity. Commonly used for osteoarthritis, this therapy enhances joint lubrication and reduces pain by replenishing hyaluronic acid levels [25–28].

6.3.6.2 Lubricin augmentation

Experimental treatments are exploring ways to increase lubricin levels in synovial fluid, such as gene therapy and direct injections of recombinant lubricin. This could

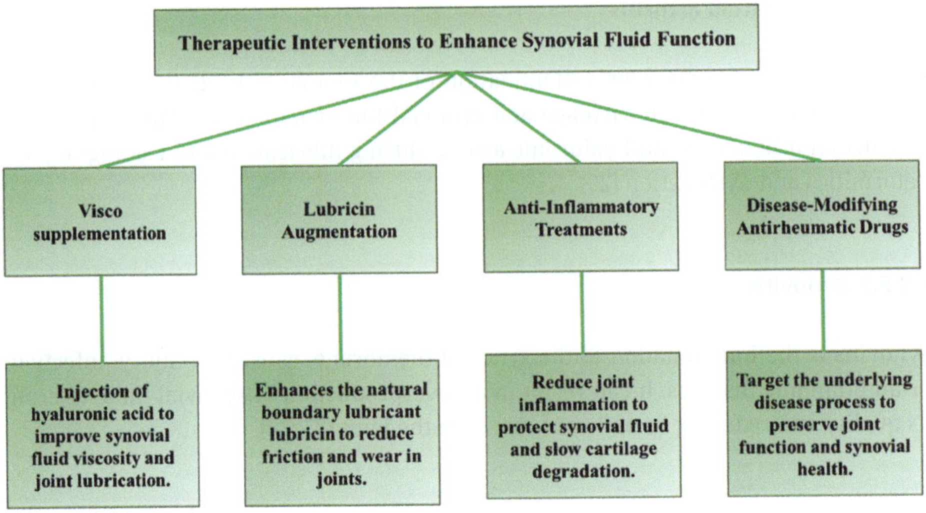

Fig. 6.4: Therapeutic interventions to enhance synovial fluid function.

help protect cartilage by improving boundary lubrication, particularly in conditions like osteoarthritis.

6.3.6.3 Anti-inflammatory treatments

Medications like NSAIDs and corticosteroids help reduce inflammation in joint disorders like rheumatoid arthritis and gout. By preserving the synovial fluid's integrity, these treatments maintain proper joint lubrication and prevent further damage.

6.3.6.4 Disease-modifying antirheumatic drugs (DMARDs)

DMARDs slow the progression of rheumatoid arthritis by targeting the underlying autoimmune response. This helps maintain the normal production of synovial fluid and preserves joint lubrication and function [4, 26].

6.3.6.5 Joint replacement surgery

In severe cases, joint replacement surgery might be necessary. Although it does not directly involve synovial fluid, replacing damaged joints with artificial ones can restore mobility and reduce pain, addressing the consequences of poor lubrication [25, 26].

6.3.7 Emerging research and future directions

Research on synovial fluid's role in joint lubrication continues, with promising developments on the horizon.

6.3.7.1 Biomechanical modeling of synovial fluid

Advances in biomechanical modeling are shedding light on how synovial fluid behaves under various conditions. These models can help predict how diseases and treatments affect joint lubrication, contributing to the development of better therapies and artificial joints [30].

6.3.7.2 Nanotechnology in joint lubrication

Nanotechnology holds the potential for enhancing joint lubrication by developing nanoparticles that act as lubricants or deliver therapeutic agents directly to the joint. This technology could help improve both natural and artificial joint function while reducing wear and tear [31–33].

6.4 Lubrication strategies for artificial joints and implants

In recent years, the development and optimization of artificial joints and implants have seen significant advancements, particularly in efforts to replicate the natural lubricating mechanisms of synovial joints. Effective lubrication plays a vital role in reducing wear and extending the lifespan of these implants. This section will discuss various lubrication strategies used in artificial joints and implants, while also addressing the challenges and innovations in this field.

6.4.1 Overview of artificial joints and their lubrication challenges

Artificial joints, commonly known as prostheses, are designed to replace damaged or diseased joints, with popular types including hip, knee, and shoulder replacements. These implants are generally constructed from a mix of metals, ceramics, and polymers, which are chosen for their mechanical strength, durability, and compatibility with the body. However, even with advancements in materials, replicating the intricate lubrication provided by natural synovial fluid continues to be a challenge.

In natural joints, synovial fluid serves to reduce friction and wear while providing shock absorption. For artificial joints, lubrication strategies must either mimic or replace these functions using engineered solutions or alternative lubricants. The effectiveness of an artificial joint is dependent not just on the materials used but also on how well it is lubricated [34, 35].

6.4.2 Types of lubrication in artificial joints

Types of lubrication in artificial joints include boundary lubrication, fluid film lubrication, mixed lubrication, and electrohydrodynamic lubrication, as shown in Fig. 6.5.

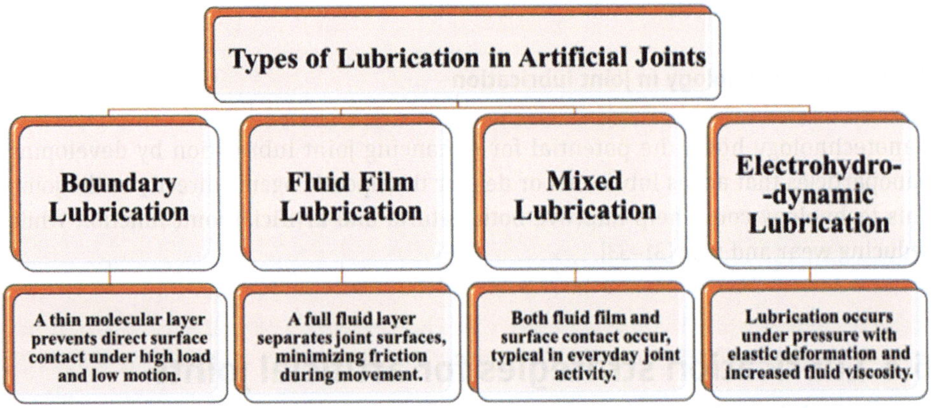

Fig. 6.5: Types of lubrication in artificial joints.

6.4.2.1 Boundary lubrication

Boundary lubrication occurs when the surfaces of an implant are separated by a thin molecular layer of lubricant. This type of lubrication is especially crucial in situations of high load or limited movement, where fluid film lubrication may not suffice. In artificial joints, boundary lubrication is achieved using self-lubricating materials or specialized surface coatings.

For instance, polyethylene and cross-linked polyethylene (XLPE) coatings are commonly applied to artificial joints to minimize friction and wear. These materials form a lubricating layer at the contact points, which is essential for maintaining joint function and longevity [36].

6.4.2.2 Fluid film lubrication

Fluid film lubrication occurs when a continuous layer of lubricant keeps the joint surfaces apart, preventing direct contact. In natural joints, this layer is provided by synovial fluid, while in artificial joints, it can be achieved using body fluids, synthetic lubricants, or engineered solutions.

The design of artificial joints often includes features to enhance fluid film lubrication by optimizing component geometry and alignment. For example, the shapes of articulating surfaces in hip and knee replacements are specifically designed to facilitate fluid film formation, thus reducing friction and wear during movement [37].

6.4.2.3 Mixed lubrication

Mixed lubrication refers to a combination of boundary and fluid film lubrication. This approach is particularly relevant in high-load situations or when motion speeds vary, as both lubrication mechanisms are beneficial.

To support mixed lubrication, designs need to focus on optimizing material properties and surface roughness. Recent advancements in hybrid materials and composite coatings are improving performance in mixed lubrication scenarios by effectively combining the advantages of both lubrication types [4–6].

6.4.2.4 Elastohydrodynamic lubrication

Elastohydrodynamic lubrication (EHL) occurs when the elastic deformation of joint surfaces under load allows for the formation of a lubricating film. This lubrication type is crucial in high-load joints, like the hip or knee, where slight surface deformation is necessary to maintain a continuous lubricating film.

Materials with favorable elastic properties, such as certain metals and ceramics, are utilized in implants to support EHL. For example, cobalt-chromium alloys and zirconia are selected for their hardness and elasticity, which contribute to effective lubrication and reduced wear [38].

6.4.3 Material selection for artificial joint lubrication

Choosing the right materials for artificial joints is vital for achieving effective lubrication and minimizing wear. These materials must be biocompatible, durable, and capable of supporting various lubrication strategies.

6.4.3.1 Metals

Metals like cobalt-chromium alloys, titanium alloys, and stainless steel are frequently used due to their strength, durability, and resistance to corrosion. Cobalt-chromium alloys, in particular, are preferred for their hardness and wear resistance but require careful surface polishing to lower friction [39].

6.4.3.2 Ceramics

Ceramics, such as alumina and zirconia, are favored for their excellent wear resistance and low friction coefficients. Ceramic-on-ceramic articulations work exceptionally well in hip replacements due to their smooth surfaces and hydrophilic properties, which enhance fluid film lubrication. However, ceramics tend to be more brittle than metals, making careful design necessary to prevent fractures [4, 40].

6.4.3.3 Polymers

Polymers like UHMWPE are commonly used in artificial joints, particularly for bearing surfaces. UHMWPE provides low friction and high wear resistance, while XLPE is used to boost wear resistance and increase the longevity of implants [41].

6.4.3.4 Composites and surface coatings

Composite materials and advanced surface coatings play a significant role in improving the performance of artificial joints. For instance, ceramic-coated metal implants leverage the strength of metals along with the wear resistance of ceramics. Additionally, diamond-like carbon coatings are used to minimize friction and wear [42–45].

6.4.4 Lubrication enhancements and innovations

A number of innovative strategies are currently being explored to enhance lubrication in artificial joints.

6.4.4.1 Lubricant-infused surfaces

Lubricant-infused surfaces are designed to retain lubricants within their material structure, providing ongoing lubrication. An example of this is hydrogel coatings, which absorb and release lubricants based on mechanical stress. These coatings mimic the natural lubrication mechanisms of synovial fluid, helping to reduce friction and wear [46].

6.4.4.2 Nanotechnology in lubrication

Nanotechnology presents new opportunities for enhancing lubrication through the application of nanoparticles. Materials such as graphene, molybdenum disulfide (MoS_2), and titanium dioxide (TiO_2) can improve lubricant characteristics by lowering friction and wear. Nanostructured coatings, including nanocomposite coatings, can also enhance boundary lubrication by interacting with both biological and synthetic lubricants at the nanoscale [47].

6.4.4.3 Smart materials and adaptive lubrication

Developments in smart materials that react to changes in joint conditions are on the rise. These materials can modify their properties, such as viscosity or surface roughness, in response to load, temperature, or motion. This adaptive functionality helps ensure optimal lubrication and performance in artificial joints [47].

6.4.4.4 Challenges in artificial joints and implants

Replicating the natural lubrication mechanisms of synovial fluid in artificial joints and implants presents significant challenges. The effectiveness of these implants hinges on lubrication strategies that can mimic or replace natural functions. Recent advancements in materials science and engineering have led to the creation of self-lubricating materials, fluid film lubrication techniques, and innovative coatings designed to improve the performance and longevity of artificial joints. Emerging technologies, like nanotechnology and smart materials, are paving the way for more effective and adaptable lubrication solutions in the future.

6.5 Conclusions

Lubrication in living organisms is essential for the smooth functioning and protection of various systems, especially in joints, where synovial fluid reduces friction and absorbs shock. This natural process not only minimizes wear and tear but also supports tissue health and repair. Insights into biological lubrication have inspired innovations in artificial implants, improving their performance and longevity. Ongoing research is vital for enhancing synthetic systems and improving the quality of life for those with joint replacements.

References

[1] Lee S, Spencer ND. (2008). Sweet, hairy, soft, and slippery. *Science*, 319, 575–576.

[2] Dowson D. (2009). *History of Tribology*. Wiley.

[3] Celli J, Gregor B, Turner B, Afdhal NH, Bansil R, Erramilli S. (2005). Viscoelastic properties and dynamics of porcine gastric mucin. *Biomacromolecules*, 6, 1329–1333.

[4] Hills BA. (2000). Boundary lubrication *in vivo*. *Journal of Engineering in Medicine*, 214, 83–94.

[5] Szeri AZ. (2001). Hydrodynamic and elastohydrodynamic lubrication. In Bhushan B. (Ed.), *Modern Tribology Handbook* (pp. 383–453). New York, NY, USA: CRC Press.

[6] Mow VC, Holmes MH, Lai WM. (1984). Fluid transport and mechanical properties of articular cartilage: A review. *Journal of Biomechanics*, 17(5), 377–394. DOI: 10.1016/0021-9290(84)90031-9.

[7] Jay GD, Waller KA. (2014). The biology of lubricin: Near frictionless joint motion. *Matrix Biology*, 39, 17–24.

[8] Wang M, Deng Z, Guo Y, Xu P. (2021). Designing functional hyaluronic acid-based hydrogels for cartilage tissue engineering. *Frontiers in Bioengineering and Biotechnology*, 9, 1–12.

[9] Hodge WA, Fijan RS, Carlson KL, Burgess RG, Harris WH, Mann RW. (1986). Contact pressures in the human hip joint measured *in vivo*. *Proceedings of the National Academy of Sciences of the United States of America*, 83, 2879–2883.

[10] Klein J. (2012). Polymers in living systems: From biological lubrication to tissue engineering and biomedical devices. *Polymers for Advanced Technologies*, 23, 729–735.

[11] Linn FC. (1968). Lubrication of animal joints: II the mechanism. *Journal of Biomechanics*, 1, 193–205.

[12] Wright V, Dowson D. (1976). Lubrication and cartilage. *Journal of Anatomy*, 121, 107–118.

[13] Bayer IS. (2018). Advances in tribology of lubricin and lubricin-like synthetic polymer nanostructures. *Lubricants*, 6(2), 30. https://doi.org/10.3390/lubricants6020030.

[14] Bron AJ, Tiffany JM, Gouveia SM, Yokoi N, Voon LW. (2004). Functional aspects of the tear film lipid layer. *Experimental Eye Research*, 78(3), 347–360. DOI: 10.1016/j.exer.2003.09.01.

[15] Lewis PR, McCutchen CW. (1959). Experimental evidence for weeping lubrication in mammaliancjoints. *Nature*, 184, 1285.

[16] Hlavácek M. (1993). The role of synovial fluid filtration by cartilage in lubrication of synovial joints–II. Squeeze-film lubrication: Homogeneous filtration. *Journal of Biomechanics*, 26(10), 1151–1160. DOI: 10.1016/0021-9290(93)90063-k, PMID: 8253820.

[17] Lu XL, Mow VC. (2008). Biomechanics of articular cartilage and determination of material properties. *Medicine & Science in Sports & Exercise*, 40(2), 193–199. DOI: 10.1249/mss.0b013e31815cb1fc.

[18] Tamer TM. (2013). Hyaluronan and synovial joint: Function, distribution and healing. *Interdisciplinary Toxicology*, 6(3), 111–125. DOI: 10.2478/intox-2013-0019.

[19] Swann DA, Silver FH, Slayter HS, Stafford W, Shore E. (1985). The molecular structure and lubricating activity of lubricin isolated from bovine and human synovial fluids. *Biochemical Journal*, 225(1), 195–201.

[20] Popel AS, Johnson PC. (2005). Microcirculation and hemorheology. *Annual Review of Fluid Mechanics*, 37, 43–69.

[21] Cone RA. (2009). Barrier properties of mucus. *Advanced Drug Delivery Reviews*, 61(2), 75–85.

[22] Allen A, Flemström G, Garner A, Kivilaakso E. (1993). Gastroduodenal mucosal protection. *Physiological Reviews*, 73(4), 823–857.

[23] Julián-Flores A, Aguilar-Zárate P, Michel MR, Sepúlveda-Torre L, Torres-León C, Aguilar CN, Chávez-González ML. (2025). Exploring the therapeutic potential of medicinal plants in the context of gastrointestinal health: A review. *Plants*, 14(5), 642. https://doi.org/10.3390/plants14050642.

[24] Onu I, Gherghel R, Nacu I, Cojocaru FD, Verestiuc L, Matei DV, . . . Galaction AI. (2024). Can combining hyaluronic acid and physiotherapy in knee osteoarthritis improve the physicochemical properties of synovial fluid?. *Biomedicines*, 12(2), 449. https://doi.org/10.3390/biomedicines12020449.

[25] Hinton R, Moody RL, Davis AW, Thomas SF. (2002). Osteoarthritis: Diagnosis and therapeutic considerations. *American Family Physician*, 1, 65(5), 841–848.

[26] Firestein GS, McInnes IB. (2017). Immunopathogenesis of rheumatoid arthritis. *Immunity*, 46(2), 183–196.

[27] Smith MD. (2011). The normal synovium. *Open Rheumatology Journal*, 5, 100–106.

[28] Scanzello CR, Goldring SR. (2012). The role of synovitis in osteoarthritis pathogenesis. *Bone*, 51(2), 249–257.

[29] Choi HK, Mount DB, Reginato AM. (2005). Pathogenesis of gout. *Annals of Internal Medicine*, 143(7), 499–516.

[30] Holmes MH, Mow VC. (1990). The nonlinear characteristics of soft gels and hydrated connective tissues in ultrafiltration. *Journal of Biomechanics*, 23(12), 1145–1156.

[31] Liu Y, Shah KM, Luo J. (2021). Strategies for articular cartilage repair and regeneration. *Frontiers in Bioengineering and Biotechnology*, 9, 770655. DOI: 10.3389/fbioe.2021.770655.

[32] Sotres J, Arnebrant T. (2013). Experimental investigations of biological lubrication at the nanoscale: The cases of synovial joints and the oral cavity. *Lubricants*, 1(4), 102–131. https://doi.org/10.3390/lubricants1040102.

[33] Anjum S, Ishaque S, Fatima H, Farooq W, Hano C, Abbasi BH, Anjum I. (2021). Emerging applications of nanotechnology in healthcare systems: Grand challenges and perspectives. *Pharmaceuticals (Basel)* 21, 14(8), 707. DOI: 10.3390/ph14080707, PMID: 34451803.

[34] Martell JM, Verner JJ, Incavo SJ. (2003). Clinical performance of a highly cross-linked polyethylene at two years in total hip arthroplasty: A randomized prospective trial. *Journal of Arthroplasty*, 18(7 Suppl 1), 55–59. DOI: 10.1016/s0883-5403(03)00341.

[35] Di Puccio F, Mattei L. (2015). Biotribology of artificial hip joints. *World Journal of Orthopedics*, 18, 6(1), 77–94. DOI: 10.5312/wjo.v6.i1.77.

[36] Link JM, Salinas EY, Hu JC, Athanasiou KA. (2020). The tribology of cartilage: Mechanisms, experimental techniques, and relevance to translational tissue engineering. *Clinical Biomechanics (Bristol, Avon)*, 79, 104880. DOI: 10.1016/j.clinbiomech.2019.10.016.

[37] Gao L, Lu X, Zhang X, Meng Q, Jin Z. (2022). Lubrication modelling of artificial joint replacements: Current status and future challenges. *Lubricants*, 10, 238. https://doi.org/10.3390/lubricants10100238.

[38] FC W, Liu F, Jin ZM. (2004). A general elastohydrodynamic lubrication analysis of artificial hip joints employing a compliant layered socket under steady state rotation. *Proceedings of the Institution of Mechanical Engineers, Part H*, 218(5), 283–291. DOI: 10.1243/0954411041932836.

[39] Tapscott DC, Wottowa C. (2024). *Orthopedic Implant Materials*. In StatPearls Treasure Island (FL): StatPearlsPublishing. https://www.ncbi.nlm.nih.gov/books/NBK560505.

[40] Bal BS, Garino J, Ries M, Rahaman MN. (2007 Jan-Mar). A review of ceramic bearing materials in total joint arthroplasty. *HIP International*, 17(1), 21–30. DOI: 10.1177/112070000701700105, PMID: 19197839.

[41] Goretti C, Polidoro F, Paderni S, Belluati A. (2019). Ceramic on ceramic total hip arthroplasty and liner fracture. Two case reports and review of literature. *Acta Biomed* 5, 90(12-S), 192–195. DOI: 10.23750/abm.v90i12-S.8961, PMID: 31821308; PMCID: PMC7233715.

[42] Gomes Souza F Jr, Bhansali S, Pal K, Silveira MFD, Santos Oliveira M, Valladão VS, Brandão e Silva DS Silva GB. (2024). A 30-year review on nanocomposites: Comprehensive bibliometric insights into

microstructural, electrical, and mechanical properties assisted by artificial intelligence. *Materials (Basel)* 27, 17(5), 1088. DOI: 10.3390/ma17051088, PMID: 38473560; PMCID: PMC10934506.

[43] Roy R, Lee K-R. (2007). Biomedical applications of diamond-like carbon coatings: A review. *Journal of Biomedical Materials Research. Part B, Applied Biomaterials*, 83, 72–84. DOI: 10.1002/jbm.b.30768.

[44] Lin W, Klein J. (Feb 25 2022). Hydration lubrication in biomedical applications: From cartilage to hydrogels. *Accounts of Materials Research*, 3(2), 213–223. https://doi.org/10.1021/accountsmr.1c00219.

[45] Singh KK, Prabhu BR, Choudhary S, Pramanik C, John NS. (2019). Effect of graphene and MoS_2 flakes in industrial oils to enhance lubrication. *ACS Omega* 29, 4(11), 14569–14578. DOI: 10.1021/acsomega.9b01799, PMID: 31528811; PMCID: PMC6740172.

[46] Gao Q, Liu S, Hou K, Zhangpeng L, Wang J. (2022). Graphene-Based nanomaterials as lubricant additives: A review. *Lubricants*, 10(10), 273. https://doi.org/10.3390/lubricants10100273.

[47] Kumar K, Rajitha A, Rao A, Alam K, Albawi A, Gaurav. S. (2023). Smart materials for biomedical applications: advancements and challenges. *E3S Web of Conferences*, 430, DOI: 10.1051/e3sconf/202343001133.

Pranjal Pratap Singh and Vishal Chhabra

7 Soft tissue biotribology

Abstract: Soft tissue biotribology is a combined study of friction, wear, and lubrication in the context of biosystems, and the discipline is relatively new. To give the underlying mechanisms a gist, this chapter is a rigorous scientific investigation that explores the what and how of soft tissue biotribology. First, we introduce soft tissue biotribology and its relevance in the medical, biological, and engineering fields. This study investigates the rubbing processes, contact velocities, directional angulations, and flow of lubricant fluids in tissues, including skin, muscles, tendons, ligaments, and blood vessels, each of which carries out specific biomechanical functions due to differences in structural and functional performances. This chapter divides the soft tissues into several types depending on their cellular content and mechanical characteristics. For example, skin, a strong armor-like barrier, is constituted by numerous collagen and elastin fibers with varying properties of strength and flexibility. It makes muscles work as our movement engines, depending on the collagen frameworks and sliding-protein filaments. Tendons and ligaments, which are the connective bridges between the various bones of our body, help to bear tension and stabilize joints, while blood vessels, which are the roads of blood circulation, adjust pressure to give an easy way for blood to flow. Relative motion within soft tissue coverings, such as skin, depends on molecular forces, surface texture, and water content. Tribometry and atomic force microscopy are among the experimental techniques that are used in determining these forces and understanding the part played by friction in our daily movements, as presented in the chapter. Wear of soft tissues is part and parcel of life, combined with abrasion, adhesion, fatigue, and corrosion effects. From abrasive interactions, the chapter epitomizes how skin or enamel can wear out, how the material transfer is characteristic of joint cartilage, and how stress can cause micro-cracks in tendons and ligaments.

Keywords: Biotribology, friction, wear, lubrication, tensions, ligaments, tribometry

7.1 Introduction

Soft tissue biotribology is a branch of science that investigates friction, wear, and lubrication in biological tissues and their structures for required body functions, including cartilage, blood vessels, and skin. The newly developed science of soft tissue biotribology

Acknowledgments: I would like to extend my deepest gratitude to my friend Mr. Sachin Kumar for his invaluable guidance with BioRender and other tools that were essential to this chapter. A heartfelt thanks to my friend Ms. Neelam Sharma for her help with the proofreading process. Their contributions have been truly instrumental in the completion of this work.

https://doi.org/10.1515/9783111563176-007

evolved from classical tribology, which concerns mechanical systems, and received its name from Dowson only in 1970. It is an emerging field because it applies knowledge in material science, cellular biology, and biomedical engineering in addressing surgical procedures and designing medical products that will enhance efficacy. Consequently, the biotribology of soft tissues is significant since it impacts surgical procedures and the fabrication of artificial implants. Progress in medical techniques in recent years has called for high standards in the assessment of soft tissue mechanics. This research is critical for the reduction of trauma during surgeries and the improvement of the life and function of replacements and implanted materials. Furthermore, with advancements in robot-assisted surgery techniques, the precision of cutting force for enhancing the safety of patients and the performance of technical procedure remains a crucial aspect to model. However, there are some formidable obstacles in this area: variability in the research methods used and the absence of common requirements for assessing the effectiveness of an application. A case in point is biotribocorrosion, which is the simultaneous wear and corrosion of a material in a biological environment, making the selection of the proper biomaterial for surgical use very challenging. The diagram of the synovial joint and its positioning in the human body is shown in Fig. 7.1. Synovial lubrication works as the biolubrication for analyzing the biotribology in the human body. Daily research advancements point to the need for researchers and clinicians to come together more often to address these issues and help convert their research findings into useful tools that may improve the patient's care. In a nutshell, soft tissue biotribology aims to be at the vanguard of biomedical engineering evolution in the future by directing advanced mechanical property enhancement of biological tissues using nanotechnology, smart machine learning, and advanced material science technology to create gap-specific solutions for tissue repair and regeneration [1, 2].

7.2 Historical background

The exploration of soft tissue biotribology has evolved alongside our comprehension of the broader field since Dowson first coined the term in 1970 [1]. Having its roots in conventional tribology, this field deals with friction, wear, and lubrication involving biological surfaces that are crucial for various tissues, including articular cartilage, blood vessels, and skin [3]. Initially, the study focused mainly on defining the mechanical characteristics of soft tissue that form part of the connective, muscles, epithelial, and nervous tissues [4]. As surgery progressed, the need to reduce tissue injury during surgical procedures increased the focus on the cutting elements and thermal characteristics surrounding soft tissue handling. This required the authors to investigate the structure and response of soft tissues more comprehensively, in terms of age, hydration, and mechanical load [1, 4]. Originally, the focus of the field was on analyzing biopolymers and composites that resembled natural tissues, and over the years, it extended toward artificial and composite

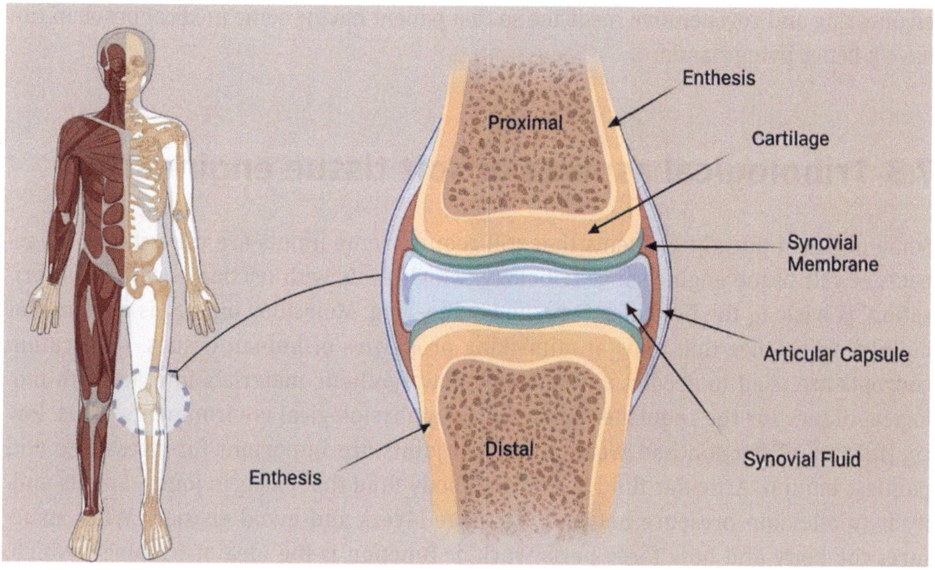

Fig. 7.1: Illustration of synovial joint and its positioning in the human body (created using BioRender).

materials with designed properties. This has, in turn, given impetus to the medical implants and prosthetics, their ability as well as their compatibility within the human body. Innovations in technology, particularly in imaging and computational modeling have enabled improved understanding of the microstructural and biomechanical nature of soft tissues and enhanced the thermoplasticity of treatments and surgeries. These considerations readily translated into the study of soft tissues, continued to progress as engineering principles were incorporated into the field. Mechanical cutting technology joined hands with other heat-assisted cutting technologies that use heat to mainly arrest blood flow and cut through tissues, affecting all surgical technologies [4]. It was objective enough to show that with the advent of robotic-assisted surgery, it was also necessary to improve the modeling of cutting forces to increase the effectiveness of surgical procedures [4, 5]. Over time, the increased application of biotribology was registered especially in the clinical area to improve on the machinery. The mechanics of joint wear and human joint lubrication strategies had emergent importance in enhancing not only the patient's quality of life but also the durability of the manufactured medical implements [2]. Presently, biotribology is a blend of material science, cellular biology, and biomedical engineering, advancing daily to focus on the difficulty of reconstructing tissue during surgery [3, 6]. Further, developments in the imaging technologies with computational models have given enhanced perspectives on the behavior of soft tissues under certain circumstances. This has led to advancements in surgical instruments and procedures, which inextricably make surgical operations precise and productive. The advanced researching in biotribology is being done incessantly which further enhance the possibility in means of tissue engineering and regenerative medicine so that patient having some medical problem can have a better living standard.

7.3 Tribological aspects of soft tissue engineering

Some of the important elements that still require some inputs are the tribological aspects of soft tissue engineering. Tribology, which deals with friction, wear, and lubrication, is basic to the field of soft tissue engineering. Molecular biology is a branch of biomedical science that aims at mimicking properties of human tissues. Lubrication concepts are used to predict and enhance how synthetic materials interact with biological tissues for the requisite functionality in physiological environments. Thus, low coefficients of friction and wear in natural joints are important for stress-free and painless motion. Articular fluid is a normal body fluid that exists in joints, and its ability is to alleviate pressure between cartilage layers and avoid erosion. What structures the body and how these parts work or function is the idea of tribology, which, when designing soft tissues, or, for instance, when developing implants, ought to be emulated. In designing tissues, the material should also have a relatively low friction value and high wear rate since the tissues will undergo cyclical motion daily [7]. Soft biological tissues, characterized by their lack of calcium deposits, include connective, muscle, epithelial, and nervous tissues. These tissues, composed of various fibers and ground substances, exhibit unique mechanical properties influenced by several factors, enhancing the study and applications of biotribology, particularly in surgical methods and biomedical implants. Some of the properties are as follows:

– **Characteristics of soft biological tissues:** Soft biological tissue, on the other hand, is defined as any body tissue that does not contain calcium deposits and therefore encompasses connective, muscle, epithelial, and nervous tissue. Besides cells, these tissues consist of a ground substance, which is formed of a diverse mixture of collagen, elastic, and reticular fibers. In hard tissues, there are properties to consider, such as age, hydration, osmolarity, as well as loading conditions, and these result in the variation of the mechanical properties of these tissues and as result, research in on the cutting of soft tissues is tasks [4]. Incorporation of engineering principles in the research of soft tissues has enhanced development of the discipline of biotribology immensely. Scientists and engineers started using sharp cutting tools, including knife, scalpel, and laser cutting, while others used the energy cutting where heat was used to coagulate and cut the tissues. Such methods have been proved to play an important role in surgery, and such positivism is explained by the fact that they enhance minimal tissue destruction and precision [4]. For instance, electrosurgery, the utilization of high-frequency electrical current to divide tissue and control hemorrhaging is now normal in many surgeries. It also helps in preventing infection allied to surgeries, and in reducing the time required for healing. Likewise, laser surgery that involves the use of light beams to both incise and coagulate tissue, is highly accurate and is normally applied in operations considered sensitive, like instance, eye and brain operations [4]. The increasing use of robotic-assisted surgery has added more value to force modeling in cutting them in order to increase the safety and effectiveness of the operation. Surgical robots, including the patented da Vinci Surgical System, enhance the ability of surgeons in delivering refined procedures and control. Such systems actually require accurate prediction of cutting forces to facilitate the manner that these components are conceived in order to engage the tissues in a safe way that avoids cutting them inadvertently [4, 5]. Over time, the focus moved more and more to the application of biotribological concepts in clinical practice. A major avenue of research that has been investigated includes friction and wear in joint replacements. For instance, due to gained knowledge of studying the wear mechanisms of artificial hip and knee joints, new and improved materials were developed for these implants as well as improved strategies of lubrication of these implants. This study has helped save patient lives and also improves the quality of life of people with anterior cruciate ligament reconstruction [2]. Today, biotribology lies in the crossroads of material science, cellular biology, and biomedical engineering, constantly developing to deal with problems of tissue repair and regeneration in surgeries. Modern technologies, including MRI and CT scans, along with mathematical models, have given a crisp view of how tissues containing the organism behave under different circumstances. These tools embody a way by which such mechanical interactions in tissues can be modeled and analyzed, thereby enhancing the design of improved instruments to be used in surgery. For instance, by developing modeling algorithms, the reaction of tissues to different instruments to facilitate better tool design that increases healing rates. Furthermore, tissue engineering has developed opportunities to design synthetic and combined materials that successfully resemble natural tissues. They are employed in the fabrication of medical implants and prosthetics, improving the materials' performance internal to the human body [8].

– **Thermal properties and cutting mechanisms:** Cutting methods for indulgent tissues generally pass into two categories: mechanical press cutting and energy cutting off. Mechanically skillful cutting calls for traditional shaft like scalpels and scissors, while energy cutting methods use heat to facilitate cutting and achieve hemostasia (the unconscious process of stopping bleeding). Energy cutting methods, such as electrosurgery and optical maser OR, operate on the rationale of elevating tissue temperature to make cutting easy going and to control bleeding. The performance of these cut method acting is influenced by various factors, including the type of material being cut, the initial malleable forces applied to the tissue, and the condition of the tissue itself. For example, increase initial tensile force-out can shorten both the cut force required and the fracture huskiness of the tissue, making it easy to cut through [4]. This is especially important in surgical contexts where precision and minimizing tissue harm are crucial.

– **Fracture toughness:** This paper shows that crack resistance or crack arrest capability is a crucial parameter of soft biological tissue. It means the measure of a material sample's resistance to breakage. In connection with the breakage of softer tissues, cutting fracture strength tends to be smaller than the material strength of the tissue. This is because the tissue material is very compliant and unruly, and hence the quantitative correspondence between cutting energies and material failure mechanisms is not easily captured [1, 6]. For example, tissue features such as muscle and connective tissue are opaque and more difficult to model due to their fibrillar structure, and their stress response is not the same as that of homogeneous materials.

– **Multifactorial wear mechanisms:** In biological milieus, wear phenomena result from mechanical, chemical, and biological interactions. Knowledge of these multifactorial interactions is important for the long-term stability of medical devices that integrate effectively with the body. These forces deal with the interaction of interfacial roughness, such as friction and abrasion forces that bring about wear out of the material. For instance, in joint replacements, the loading and movement cycles cause abrasive actions on the implant material and its surrounding surfaces; the generated wear debris can stimulate inflammatory reactions. Chemical factors relate to the behaviors of the implant materials in the biosystems of the body. Corrosion, as well as oxidation, changes the physical characteristics on the surface of implants and, therefore, their function and life cycle. For instance, implantable metals are susceptible to corrosion by body fluids, which reduces their mechanical properties and biocompatibility. Biological is the behavior of the body to the implant and includes immune responses and tissue reactions. Wear particles are known to stimulate an inflammatory reaction which results in osteolysis and ultimately implant loosening. Finally, it is possible to note that the stability and functioning of the implant strongly depend on the fusion of tissues with the implant. Further studies are needed to understand the profile of interfaces and the related tissue reactions before wear-related complications can be well addressed. For instance, the application of coatings or texturing surfaces improves the wear properties, and compatible implantation of biomaterials. Thus, by analyzing these diverse wear mechanisms, the goal is based on making medical devices more durable and thus benefit overall patient care [2, 9].

7.4 Wear mechanism in soft tissue engineering

An understanding of wear mechanisms is crucial for achieving the goal of producing tissue engineering constructs that will have a long service life. Wear mechanisms include:

- **Adhesive wear:** This is a process where adjacent surfaces are initially in mechanical contact and stick together and then separable again, thereby moving different amounts of their material onto each other. With regard to implants, adhesive wear can result in the delamination of the implant material or the surrounding tissue.
- **Abrasive wear:** Abrasive wear is one type of wear that occurs when hard particles or rough surfaces rub against a milder surface and keep tearing the latter away. In soft tissue engineering, it might be either between the manufactured material and the native tissue, which may deteriorate and lead to the shortening of the implant's durability.
- **Fatigue wear:** Fatigue wear is characterized by cyclic stress and strain, whereby material is subjected to a mechanical load and then unloaded until failure occurs due to microscopic cracks. This kind of wear is of much significance in applications like joints, which involve constantly moving and stressing parts.
- **Corrosive wear:** This is the process whereby various materials are corroded chemically, which is powerfully augmented by the presence of body fluids. In soft tissue engineering specifically, the materials used have to be resistant to corrosion because the structures have to work for a long time and maintain their performance inside the human body [7].

7.4.1 Correlation between tribological characteristics and wear mechanism

A proper understanding of tribological aspects and wear mechanisms is indeed essential for the development of some soft tissue engineering solutions. It is important to know that wear behaviors are governed by tribological characteristics, including friction and lubrication. For instance, good lubrication helps to decrease the amount of wear through adhesive and abrasive wear mechanisms, hence increasing the life of the implant or engineered tissue. On the other hand, a lack of proper lubrication or unsuitable material choice can result in high friction, high wear rates, and high tendencies toward failure. By controlling certain factors, including material selection and the application of appropriate lubricants, engineers can control the tribological properties that are inherent in wear mechanisms, thus improving soft tissue replacements. The study of typical tribological aspects and wear mechanisms offers the possibility to increase the durability and efficiency of engineered soft tissue and, therefore, enhance the quality of life for patients [10].

7.5 Experimental techniques

7.5.1 Surface modification methods

In particular, surface treatments are indispensable for improving the properties of biomaterials applied to soft tissue engineering. Different types of surface modification methods also contain different characteristics, providing different kinds of merits and demerits that would influence their extent of applicability for mass production or medical usage. The present research is primarily concerned with developing new safeguard coating approaches that overcome the existing approaches. For example, emerging techniques, including the alkalinity-activated solid-state dewetting technique, can make quasi-periodic nanoscale structures on titanium alloys that normally are hard to achieve under standard surface processes [1]. Nevertheless, the expenses associated with industrial scaling and the efficiency of these coating procedures remain significant limitations, which allow only a few new technologies to enter clinical trials and even fewer to reach commercialization, while the majority of the technologies are in the preclinical stage [1, 11–13].

7.5.2 Mechanical modification methods

The application of mechanical modification techniques in soft tissue surgeries has been improved by the integration of assisted technologies. One such technique is the high-frequency vibration technique used in conjunction with ordinary surgical tools with the aim of improving their efficiency, especially for soft tissues. It is particularly popular in orthopedic surgery at the moment, but its focus on other surgeries is also pushing more attention to the micro-cutting of soft tissues [4].

7.5.3 Biological modification methods

Biological surface modifications are also complementary to mechanisms for enhancing tissue integration and biocompatibility. Recent advances have been directed toward improving the existing investigations to exploit the phenomena to reveal how coating behavior affects cell and tissue response. These are the advanced techniques of nanolithography which again are beneficial but much of the time lack rigorous in vitro/in vivo biological assessment; therefore, their practical application is questionable [1].

7.5.4 Measurement techniques

If the tensile stress is to be calculated in soft tissues, the cross-sectional areas of the specimen have to be measured accurately. The classical contact methods are prone to inaccuracy; hence, non-contact techniques are favored. These comprise, for example, the shadow amplitude method and laser micrometer systems that allow performing accurate measurements while preserving the rest condition of the tissues [5].

7.6 Applications of biotribology

Several applications of biotribology are listed in Fig. 7.2. A detailed elaboration is outlined below.

7.6.1 Joint lubrication and wear reduction

- **Synovial joints**: Biotribology studies the lubrication mechanisms in synovial joints, such as the knee and hip, to understand how synovial fluid reduces friction and wear. This knowledge helps in designing better joint replacements and improving the longevity of prosthetic joints.
- **Cartilage health**: Research in biotribology focuses on understanding the wear mechanisms of articular cartilage and developing materials and treatments to prevent or repair cartilage damage.

7.6.2 Skin tribology

- **Friction and wear**: Biotribology examines the frictional properties of human skin to develop better skincare products, such as moisturizers and protective creams, that minimize skin damage and irritation [1].
- **Wound healing**: Understanding the tribological properties of skin helps in designing dressings and treatments that promote faster and more effective wound healing by reducing friction and shear forces on the skin [1].

7.6.3 Tendon and ligament mechanics

- **Tendon health**: Biotribology investigates the mechanical properties and wear behavior of tendons and ligaments to develop treatments for tendonitis and ligament injuries [1].
- **Rehabilitation**: Knowledge of tendon and ligament tribology is used to design rehabilitation protocols and devices that minimize stress and wear on these tissues during recovery.

7.6.4 Biomedical implants

- **Implant design**: Biotribology researches the interaction between implants and surrounding tissues to design implants with improved biocompatibility and reduced wear [1].
- **Material selection**: The study of biotribology helps in selecting materials for implants that mimic the natural properties of biological tissues, reducing the risk of rejection and improving implant performance [1].

7.6.5 Biomimetic materials

- **Inspiration from nature**: Biotribology draws inspiration from natural systems, such as the self-lubricating surfaces of certain plants and animals, to develop new materials and coatings that reduce friction and wear in medical devices.
- **Bioinspired solutions**: These biomimetic materials are used in various medical applications, such as creating self-cleaning surfaces for surgical instruments and developing drag-reducing coatings for catheters.

7.6.6 Tissue engineering

- **Tissue regeneration**: Biotribology contributes to tissue engineering by studying the tribological properties of engineered tissues to ensure they can withstand the mechanical stresses and movements of the body.
- **3D printing**: The field also explores the use of 3D printing technologies to create scaffolds and implants with optimized tribological properties for better integration with the body [16].

7.6.7 Current research trends

a. **Advancements in ARP techniques:** With these shifts in clinical practice, recent literature on augmented regeneration procedures (ARP) reveals an increasing preference for newer approaches, including platelet-rich fibrin (PRF), injectable bone-replacing materials, and newer hydrogel systems over older allografts, xenografts, or membranes. This shift, as it is described, is made possible by progress in material science and cell biology, which allow resolving infection problems, and improve time for healing and bone restoration results [17]. Nonetheless, the absence of uniform evaluation procedures persists as an important weakness in this field, and it is recommended that similar evaluation procedures be applied in subsequent research to improve the validity and reliability of the research results [17].
b. **Integration of nanotechnology:** Introducing soft tissue biotribology of implants and applying nanotechnology are opening new areas of research. The new approaches, using nanostructured surfaces with nanopillars, nanoposts, or nanocones, show promises of achieving biocompatibility of medical implants by mechanically rupturing cells and preventing bacterial adhesion. In addition, gradient nanostructures, which are being developed for controlling the direction of migration and regeneration of cells, may offer a good potential for future directions [9].
c. **Biotribological insights:** Biotribology of medical devices is becoming a new field of focus where materials science trends merge with regenerative medicine. With an increased focus on soft biological tissues, such as their viscoelastic and anisotropic properties, recent studies have begun looking at their influence on cells, for example, their morphology and differentiation. It is crucial to comprehend these interactions to bring about the refinement of implant designs and improve clinical results with respect to osseointegration and vascularization [18].
d. **Machine learning applications:** Hybrid modeling, together with machine learning, biotribology application trends are increasing with the potential to increase predictability of different tribological events. With the collection and interpretation of significant amounts of datasets during clinical trials and experimental studies, researchers are attempting to correlate material parameters and biological cross-reactions, aiming for creating better bioproducts for soft tissues [6].

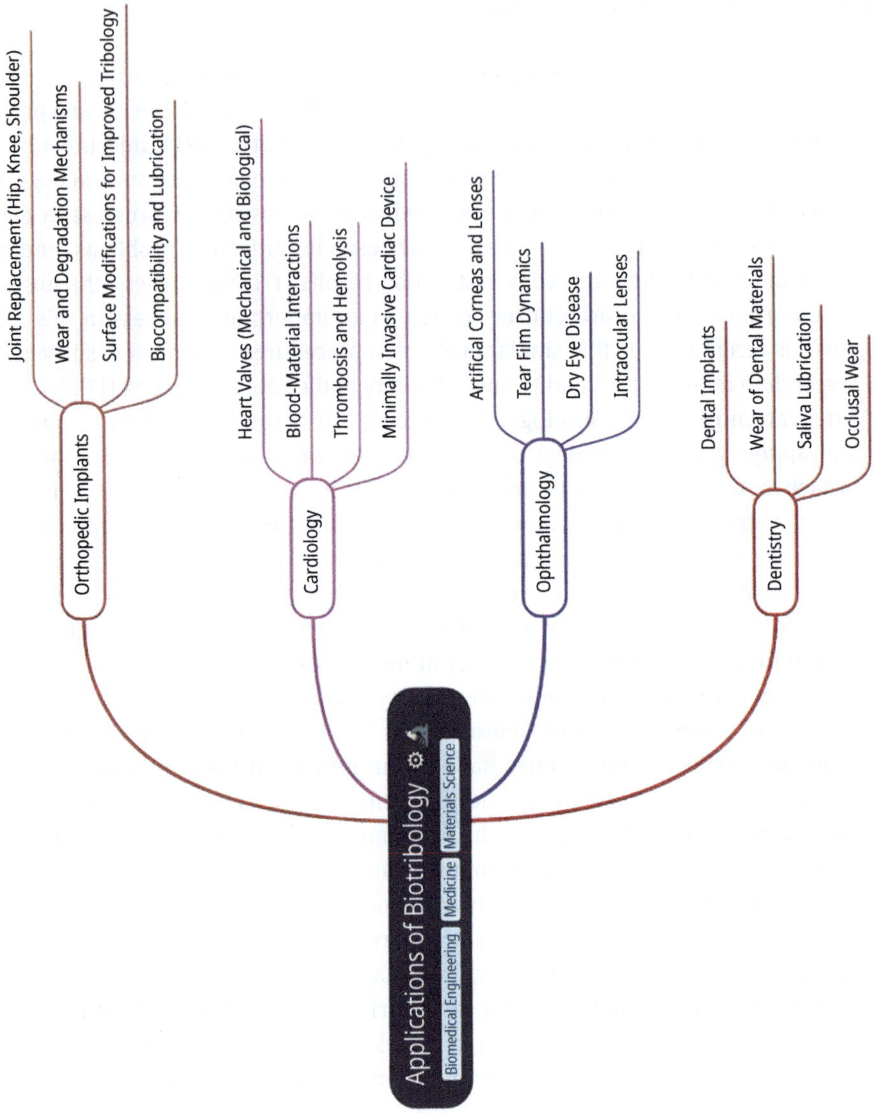

Fig. 7.2: Applications of biotribology [16].

7.7 Future directions

The biotribology science in soft tissue has notable potential for further developmental progress during the next eras, particularly as studies expand over the behavior of soft tissues and the surrounding environment in many interactions. There are expectations that a more thorough comprehension of mechanical tissue properties is attainable, and this then translates to developing interventions that can suitably alter these parameters to meet specific clinical needs [18, 19]. Simulations done in the present time did identify some gaps in the precision of simulative soft tissue cuts due to factors that include composite soft tissue structure, if present, its viscoelasticity, and anisotropy [4]. In the case of such studies, future work should consider these aspects to improve the accuracy of constructing simulation models. Furthermore, it will also be important to develop cohesive element parameters, as seen in Misra et al. and Tai et al., and so forth, which will determine how well soft tissue resists cutting and breaking [4]. An area that has also received very little attention is how to properly assess soft tissue for various applications. Future research should also consider the need to have similar evaluation procedures so that the reliability and consistency of study results can be improved. There is an enormous gap that needs to be filled for the progress of the science of soft tissue [16, 17]. Advanced surface modification techniques for soft tissue applications are probably required to allow further improvements in biocompatibility and functional integration. There may also be a trend in the field toward the use of smart materials that will respond dynamically to environmental stimuli and thereby further enhance the performance of soft tissue implants. Therefore, as this area continues to develop, it will need an ongoing combination of technological innovation and biological insight.

7.8 Challenges and limitations

One significant limitation is the simplification of tissue properties in existing simulations. Many current studies neglect the composite structure, viscoelasticity, and anisotropy of soft tissues, which can adversely affect the accuracy of simulation models used to predict cutting behaviors during surgical procedures [4, 14]. Furthermore, the parameters selected for cohesive elements, which dictate soft tissue resistance to cutting and breaking, have varied significantly across studies, complicating direct comparisons and applications [4].

a. **Tool-tissue interaction:** The way surgical instruments and soft tissue interact during cutting depends on a number of elements, including the tools' qualities, the tissues' properties, and the cutting parameters. Even though a lot of mechanical cutting techniques have been researched, there is still a lot more to shear soft tissues, and the dynamics of the full tool-tissue interaction are not fully understood. Additionally, the materials used in cutting tools may deteriorate due to wear and corrosion; these elements are crucial for assessing instrument dependability over time [9].

b. **Quality control and clinical efficacy:** Another important challenge deals with the quality assurance of medical cutting instruments. Studies show that more than 50% of surgical instruments are likely to be imperfectly made, with cracks or corrosion newly formed due to the manufacturing process or use. These quality issues raise surgical failures and medical mishaps [4]. Although promising results are obtained using new techniques for soft tissue engineering, standardized protocols and long-term clinical data are still lacking, which prevents its wide application. The clinicians are challenged by a requirement for specialized training and knowledge about the biological mechanisms involved with new materials [15, 17].
c. **Research variability:** The story-like structure of some reviews in the literature is also a limitation because, without any criterion for choosing the articles, different methods and outcome measures are again found in Chapter 16. This evaluates important outcomes, including bone regeneration and patient satisfaction with ARP of practices. Standardization of assessment protocols will further enhance the validity and comparability of research findings to incorporate clinical practice in soft tissue biotribology into perfection [17].
d. **Material considerations:** A special concern is biotribocorrosion, that is, the interaction of wear and erosion in biological environments. Knowledge about environmental factors affecting the behavior of materials and the change mechanisms in those materials is insufficient, thus making it hard to choose appropriate biomaterials for surgical applications [20, 21]. Therefore, advances in material science as well as biocompatibility are required so that soft tissue materials will be able to endure the complexities of in vivo conditions while promoting healing and minimizing adverse reactions [9].

7.9 Common materials and equipment in soft tissue biotribology

This kind of biotribology stresses rebalancing bearing designs and materials in diverse biomedical applications; thus, the selection of appropriate materials and equipment is important in engineering soft tissues and their integration. The materials and equipment, which are briefed in this section, include some of the most employed in soft tissue engineering regarding their functions and applications.

7.9.1 Common materials

When speaking of soft tissue biotribology, "common materials" mean friendly, that is, biocompatible substances and compounds being used in the engineering and regeneration of soft tissues. These materials are chosen for their biocompatibility and extracellular matrix (ECM)-like properties, their effect on the attachment and growth of cells, as well as the integration and remodeling of tissues. Natural and synthetic polymers, ceramics, and composites indeed have specific mechanical, chemical, and biological characteristics for specific tissue engineering applications [22].

a. **Hydroxyapatite (HA)**: This is a bioceramic material similar in composition to the mineral phase of bone. That is why it is commonly applied in soft tissue engineering because of its high biocompatibility and capacity to enhance cell adhesion and growth. HA has also been applied as a coating of implants to promote better osseointegration and is mostly combined with polymers to form a new composite that has better mechanical performance and retains its desired bioactivity.
b. **Collagen**: Collagen is the predominant protein of the extracellular matrix and is involved in the structural and functional properties of the tissue. Currently, it is applied in the field of soft tissue engineering because of its biocompatibility, hydrolyzing ability, and, furthermore, cell adhesive and mitogenic properties. Scaffolds derived from collagen are well-suited for skin, cartilage, and tendon tissue engineering since they afford the cells an environment conducive to migration and tissue formation.
c. **Polycaprolactone (PCL)**: PCL is a biodegradable polyester that is commonly used in soft tissue engineering due to its favorable mechanical properties and slow degradation rate. The types that can be produced include films, fibers, and scaffolds, and hence applicable to a myriad of uses. The PCL scaffolds are more useful where long-term tissue regeneration is required, say in bone and cartilage tissue engineering applications or gelatin is an inherent polymer of collagen that has extensively been used in tissue engineering and other biomedical applications due to its biocompatibility and hydrogel formation ability.
d. **Gelatin**: Gelatin is a natural polymer derived from collagen and is known for its excellent biocompatibility and ability to form hydrogels. They use it in soft tissue engineering, for example as a material to build a scaffold that will encourage cell formation and organization. Stimuli-responsive gels, such as gelatin hydrogels, are useful in pharmaceutical applications, such as drug delivery systems and wound healing, where the hydrogel creates a moist environment typical of healing skin.
e. **Alginate**: Alginate is a polysaccharide derived from brown seaweed and is widely used in soft tissue engineering due to its biocompatibility and ability to form hydrogels. Alginate has the ability to confine cells, seeds, as well as growth factors for applications, for example, in cartilage and bone tissue engineering. It also has the capability to make scaffolds that are similar to the ECM, promoting the migration of cells in tissues.

f. **Silk fibroin**: Silk fibroin is an extracellular natural protein derived from the silkworm that has impressive mechanical strength and biomimetic compatibility. It can be processed into forms such as films, fibers, and scaffolds needed for soft tissue engineering. Silk fibroin scaffolds promote cell attachment and viability, and they are highly applicable in nervous tissue engineering and skin regeneration [22, 23].

7.9.2 Common equipment

About the biotribology of soft tissue, machinery for work, investigation, techniques' designs, and common medical products fall under "common equipment." This equipment is critical for creating, stabilizing, and analyzing stem cell-derived tissues and other de novo-designed tissue constructs and scaffolds. These pieces of equipment include bioreactors, 3D printers, electrospinning setups, confocal microscopes, scanning electron microscopes, and tools required for creating micro patterns to ensure safe implantation of the biomaterials or tissue constructs. Some of the types of equipment are shown in Fig. 7.3.

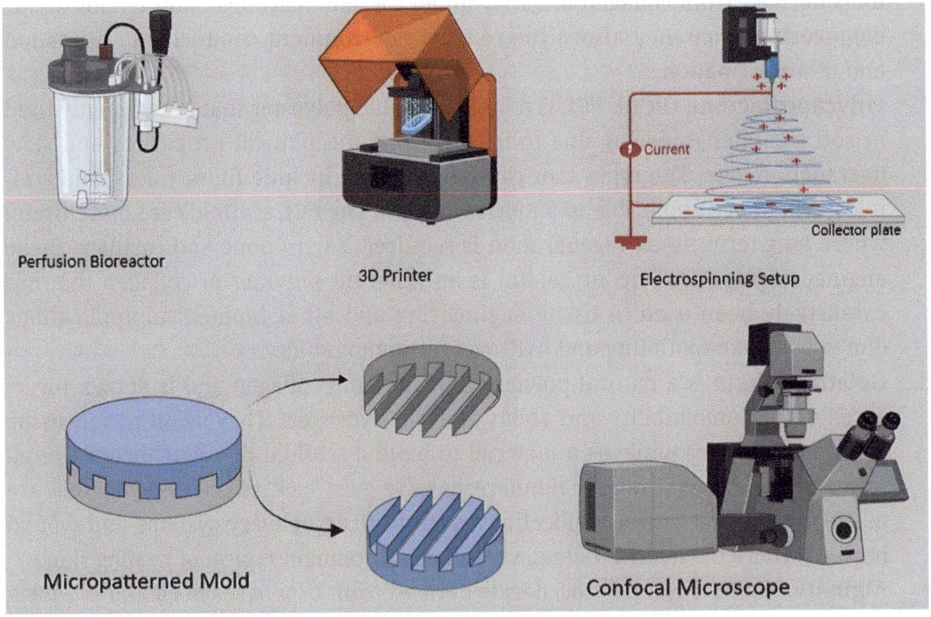

Fig. 7.3: Types of equipment used in soft tissue engineering (created using BioRender).

a. **Perfusion bioreactor:** The perfusion bioreactor, therefore, plays a very important role in soft tissue engineering. It allows the maintenance of a cell culture atmosphere free from interference but enables the constant replenishment of nutrients while eliminating waste products. This dynamic culture system directly supports the exercise of cell growth and tissue formation under physiological conditions. In the Netherlands, the perfusion bioreactor methods improve the quality and biological capacity of engineered tissue for implantation [24].
b. **3D printer**: Recently, the advancements in 3D technology of printing facilitated in changing the fabrication of the scaffold in soft tissue engineering. It enables explicit control and the fabrication of intricate shapes which can replicate those of the natural tissue. Thanks to the application of biocompatible materials, the 3D printers can develop adaptable scaffolds that relate directly to the patients and, therefore, enhance host compatibility and the functioning of implants [25].
c. **Electrospinning setup:** Electrospinning is a method applied to generate nanofibrous substrates to mimic the native ECM environment. This configuration allows the fabrication of fibers in the nanometer scale, thus promoting cell adhesion and proliferation. Electrospun scaffolds are used mostly in tissue engineering applications, for instance, in wound healing and nerve regeneration, where it is paramount to achieve proper integration of the scaffold into the tissue microenvironment.
d. **Micropattern mold:** Successful engineering of tissue requires control over cell behavior, which can be achieved by the formation of structures of a particular shape and size on the scaffold surface using micropatterning techniques. Using micropattern molds, there is the ability to create a surface that can direct cell orientation and differentiation, which is fundamental to the engineering of tissues. This technology is especially useful for tissue engineering, where controlled directional growth is required, for example, in muscles or nerve tissue.
e. **Confocal microscope:** A confocal microscope is a crucial imaging system that is used to characterize engineered tissues in terms of microstructure and mechanical characteristics at the tissue level. It gives high magnification and allows researchers to make quantitative analyses of cell distribution, scaffold structure, and tissue organization. This imaging modality is indispensable for assessing the effectiveness of tissue engineering and ascertaining whether the tissue-engineered construct fulfills the biological and mechanical requirements for implantation.
f. **Scanning electron microscope (SEM)**: Another technique employed in this study is SEM. The morphology and structure of scaffolds and tissues at high power are analyzed using SEM. These tools produce high-resolution images, which assist in the study of how the materials respond to cells, and how changes in surface characteristics may affect the behavior of cells. SEM is paramount in determining the nature of the mechanical properties of built scaffolds to ascertain their applicability to certain uses [26].

7.10 Conclusion

Thus, the intended hybridization of advanced materials and equipment in soft tissue biotribology is crucially principled to regenerative therapies. Knowledge of these materials and technologies can then help researchers make better improvements to the design and function of engineered tissues in real-life treatment benefits for patients in clinical settings. This continuous enhancement will mean that future integrated materials science and engineering advancements will help progress soft tissue biotribology and, in turn, solve the problems of tissue repair and tissue regeneration.

Abbreviation

BT	Biotribology
STB	Soft tissue biotribology
ECM	Extracellular Matrix
ACL	Anterior cruciate ligament
TJL	Tibiofemoral joint lubrication
STM	Soft tissue mechanics
FRS	Fracture resistance strength

References

[1] Zhu G, Wang G, Li JJ. (2021). Advances in implant surface modifications to improve osseointegration. *Materials Advances*, 2(21), 6901–6927. https://doi.org/10.1039/D1MA00675D.

[2] PCS Instruments. (2024, June 4). Tribology trends – Biotribology. PCS Instruments. Retrieved from https://pcs-instruments.com/articles/tribology-trends-Biotribology/

[3] Neu CP, Komvopoulos K, Reddi AH. (2008). The interface of functional biotribology and regenerative medicine in synovial joints. *Tissue Engineering Part B: Reviews*, 14(3), 235–247. https://doi.org/10.1089/ten.teb.2008.0038.

[4] Liu Z, Liao Z, Wang D, Wang C, Song C, Li H, Liu Y (2022). Recent advances in soft biological tissue manipulating technologies. *Chinese Journal of Mechanical Engineering*, 35, Article 89. https://doi.org/10.1186/s10033-022-00767-4

[5] Maiorino A, Muscolo GG (2020). Biped robots with compliant joints for walking and running performance growing. *Frontiers in Mechanical Engineering*, 6, Article 17. https://doi.org/10.3389/fmech.2020.00017

[6] Research and Markets. (2024, January 24). Soft tissue biologics product pipeline report 2023: Reveals key development stages, industry trends, regions, and companies. GlobeNewswire. Retrieved from https://www.globenewswire.com/news-release/2024/01/24/2815776/28124/en/Soft-Tissue-Biologics-Product-Pipeline-Report-2023-Reveals-Key-Development-Stages-Industry-Trends-Regions-and-Companies.html

[7] Link JM, Salinas EY, Hu JC, Athanasiou KA. (2020). The tribology of cartilage: Mechanisms, experimental techniques, and relevance to translational tissue engineering. *Clinical Biomechanics*, 79, 104880. https://doi.org/10.1016/j.clinbiomech.2019.10.016.

[8] Simson D, Subbu SK. (2024). Investigating the tribological performance of bioimplants. In Abhilash APM, Gajrani KK, Luo X. (Eds.), *Bioimplants Manufacturing* (1st ed., pp. 235–247). CRC Press, https://doi.org/10.1201/9781003509943-12https://doi.org/10.1201/9781003509943-12.

[9] Zhang Y, Chen X, Smith P. (2020). The role of mechanical forces in the regulation of articular cartilage development, aging, and degeneration. **journal of Orthopaedic Research*, 38*(6, 1235–1245. https://doi.org/10.1002/jor.24612.

[10] Davim JP. (Ed.), *(2013). Biotribology*. John Wiley & Sons, https://www.google.co.in/books/edition/Biotribology/f5W1oRZ2ZZ8C?hl=en&gbpv=0.

[11] Fiveable. (n.d.). Biomedical tribology. *Fiveable*. Retrieved from https://library.fiveable.me/friction-and-wear-in-engineering/unit-10/biomedical-tribology/study-guide/Q4xS13GU1j1bOoz4

[12] Carestream Dental. (2023, November 7). The latest trends in dental implant technologies. *Carestream Dental*. Retrieved from https://www.carestreamdental.com/en-us/training-resources/the-digital-stream/english/blog/2023/the-latest-trends-in-dental-implant-technologies/

[13] Dental Products Report. (2023, October 15). Special report: New frontiers of dental implants. *Dental Products Report*. Retrieved from https://www.dentalproductsreport.com/view/special-report-new-frontiers-of-dental-implants

[14] Yuan P, Chen M, Lu X, Yang H, Wang L, Bai T, Zhou W, Liu T, Yu S. (2024). Application of advanced surface modification techniques in titanium-based implants: Latest strategies for enhanced antibacterial properties and osseointegration. **Journal of Materials Chemistry B*, 12*(41), Article D4TB01714E. https://doi.org/10.1039/D4TB01714E.

[15] Dentist LA. (2023, January 24). The advancements of dental implants in 2023. *LA Dentist*. Retrieved from https://ladentist.com/the-advancements-of-dental-implants-in-2023/

[16] Neu CP, Komvopoulos K, Reddi AH. (2015). Elastoviscous transitions of articular cartilage reveal a mechanism for regulating tissue lubrication. **PLOS ONE*, 10*(11), Article e0143415. https://doi.org/10.1371/journal.pone.0143415.

[17] Jung H-J, Fisher MB, Woo SL-Y (2009). Role of biomechanics in the understanding of normal, injured, and healing ligaments and tendons. **BMC Sports Science, Medicine and Rehabilitation*, 1*, Article 9. https://doi.org/10.1186/1758-2555-1-9

[18] Smith J, Doe A. (2022). *Applications of Biotribology in Biomedical Systems*. Springer, Retrieved from https://link.springer.com/book/10.1007/978-3-030-51994-4.

[19] Kim S, Kim S-G (2024). Advancements in alveolar bone grafting and ridge preservation: A narrative review on materials, techniques, and clinical outcomes. **maxillofacial Plastic and Reconstructive Surgery*, 46*, Article 14. https://doi.org/10.1186/s40902-024-00425-w

[20] Levental I, Georges PC, Janmey PA. (2007). Soft biological materials and their impact on cell function. **Soft Matter*, 3*3*(3), 299–306. https://doi.org/10.1039/B610522J.

[21] Huang Y, Mucke L. (2012). Alzheimer mechanisms and therapeutic strategies. **Cell*, 148*(6), 1204–1222. https://doi.org/10.1016/j.cell.2012.02.040.

[22] Bansal G, Gautam RK, Misra JP, Mishra A. (2023). Coating methods for hydroxyapatite – A bioceramic material. In *Coating Materials: Computational Aspects, Applications and Challenges* (pp. 279–302). Singapore: Springer Nature Singapore, https://doi.org/10.1177/14644207241240623.

[23] Bansal G, Gautam RK, Misra JP, Mishra A. (2024). Tribological behavior of silver-doped eggshell-derived hydroxyapatite reinforcement in PMMA-based composite. *Proceedings of the Institution of Mechanical Engineers, Part L: Journal of Materials: Design and Applications*, 14644207241240623. https://doi.org/10.1177/14644207241240623.

[24] Plunkett N, O'Brien FJ. (2011). Bioreactors in tissue engineering. *Technology and Health Care*, 19(1), 55–69. DOI: 10.3233/THC-2011-0605.

[25] Bozkurt Y, Karayel E. (2021). 3D printing technology; methods, biomedical applications, future opportunities and trends. *Journal of Materials Research and Technology*, 14, 1430–1450. https://doi.org/10.1016/j.jmrt.2021.07.050.

[26] Nam SY, Ricles LM, Suggs LJ, Emelianov SY. (2015). Imaging strategies for tissue engineering applications. *Tissue Engineering Part B. Reviews*, 21(1), 88–102. https://doi.org/10.1089/ten.teb.2014.0180.

Priyadarsini Morampudi and Venkata Satya Prasad Somayajula

8 Biotribology of implantable medical devices

Abstract: This chapter explores the biotribological challenges associated with various implantable medical devices, including devices for skin care, artificial joints, natural joints, dental restoration, and cardiology. These devices are critical for treating or managing a wide range of medical conditions, but their long-term performance and service life are significantly influenced by tribological interactions within the body. The review provides a comprehensive insight into the understanding of biotribological behaviors, issues, and mechanisms in these devices as well as recent advancements aimed at improving their tribological properties. Additionally, it highlights the ongoing challenges and potential future directions for enhancing the biotribological performance of these essential medical implants.

Keywords: Biotribology, medical devices, friction, wear

8.1 Introduction

Tribology is often connected with mechanical systems, but it also plays a significant role in studying biological systems. Biotribology is a relatively recent area within tribology. Biotribology covers all elements of tribology related to living systems [1]. Dowson defined "biotribology" in 1970 as "the aspect of tribology concerned with biological systems." Biotribology is a fascinating and rapidly evolving field within tribology research, affecting a wide range of daily aspects such as skin blisters and artificial joints. A detailed overview of research efforts in various categories of biotribology is provided in Fig. 8.1. Biotribology encompasses not only tribology-related fields but also biochemistry, biology, pathology, clinical medicine, biomechanics, and physiology. There are several potentials for biotribology research, which increasingly benefit science, society, and healthcare. In healthcare, medical devices play a vital role and significantly improve patients' lives, but they also pose inherent safety risks, particularly at the interfaces formed during use. These interfaces, such as those in artificial joints, dental implants, and minimally invasive surgery devices, are critical to device performance. Failures like excessive wear or loosening often occur at these points, making the study of interface behavior essential for ensuring the safety and effectiveness of medical devices. Biotribology is mainly used for the above applications, and this review mainly focuses on implantable medical devices.

Biomaterials are used to create devices that can stay inside the body for a long time with minimal hassle. The materials chosen for implants need to have certain key

https://doi.org/10.1515/9783111563176-008

properties that make them suitable for specific uses. When selecting bimetallic materials, the main factors to consider are (a) being affordable, (b) having mechanical properties similar to human bones and the skull, and (c) being compatible with the body without causing harm [2, 3]. Another concern for biomedical implant materials is that they must be compatible with the human body, meaning that they should integrate smoothly without causing any harmful effects, and additionally, the materials need to be resistant to corrosion and wear. These are the properties that are essential for determining how effective the implant will be. Common implant areas include teeth, knee joints, and hips (Fig. 8.2). When selecting biomaterials, it is important to choose the ones with the mechanical properties that prevent implant failure, especially under the many cycles of stress the body undergoes. The material must also be compatible with surrounding tissues and cost-effective. In recent years, various materials have been developed to meet these criteria, but it is difficult for a single material to provide all the required properties. Materials for bioimplants are generally categorized into polymers, poly gel, metals and alloys, and ceramics.

8.1.1 Ceramics

Like bioactive glass, hydroxyapatite, ZrO_2, Al_2O_3, and other calcium and silica-based ceramics, polymers, and metals [5]. Because of their exceptional biocompatibility, these ceramics are a fantastic choice for applications involving biomedical implants. Hydroxy apatite and fluorapatite are examples for bioactive ceramics [6].

8.1.2 Polymers

A polymer is a repeating unit of molecules called a monomer, and it can be classified into two types: biodegradable polymers and nonbiodegradable polymers. Biodegradable polymers include polylactide, chitosan, alginate, and polyacetal. Materials like polypropylene, polyethylene terephthalate and poly methyl methacrylate are not biodegradable. Polymer implants are commonly used to replace artificial blood vessels, kidneys, bones, skin, contact lenses, heart valves, and also used in pacemakers [7].

8.1.3 Polymer gels

Bone is composed of natural polymers such as collagen because of their hydrophilic nature. This allows them to form hydrogels in water-based solutions, which have many desirable properties for use in bone tissue engineering [8].

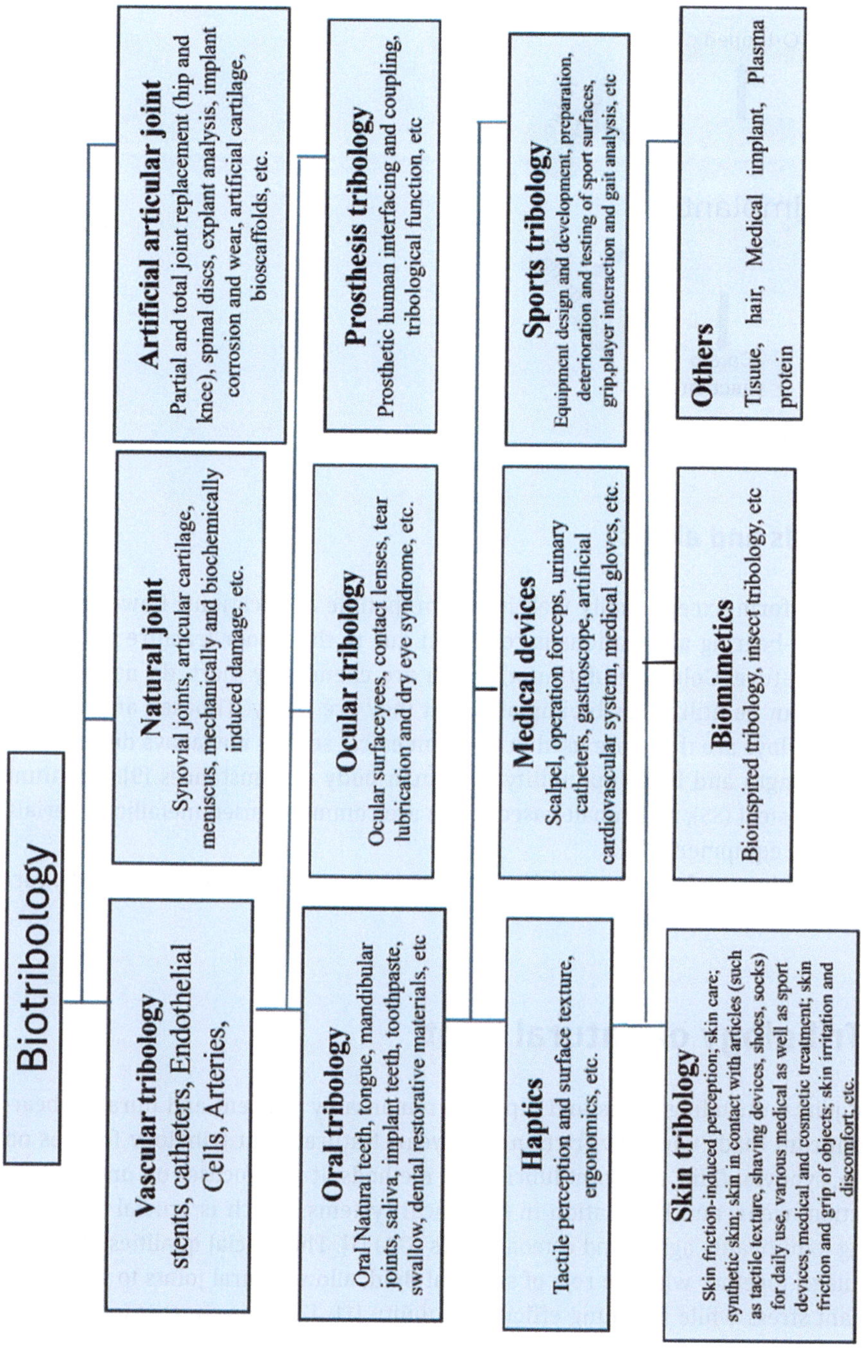

Fig. 8.1: Biotribology in major areas [1, 4].

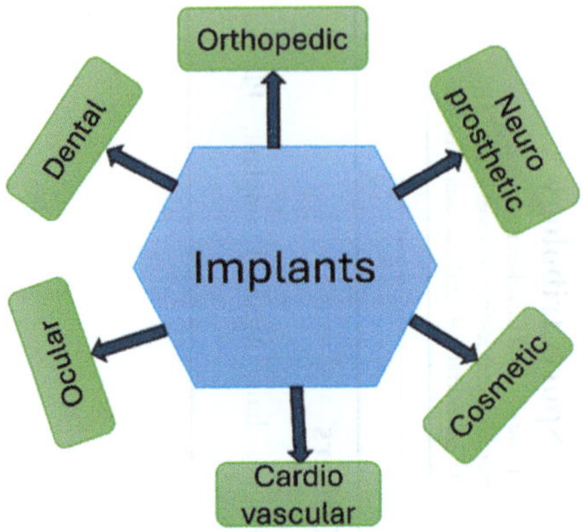

Fig. 8.2: Common implants.

8.1.4 Metals and alloys

Ceramics perform exceptionally well in biocompatible applications; however their usage in load-bearing applications is restricted due to their poor fracture toughness and brittle nature. Gold and platinum, which are chemically inert, do not corrode in situ and can be utilized as bioimplants, but they are costly. Ti-based and magnesium-based alloys are therefore used in the biomedical sectors nowadays due to their superior strength and biocompatibility in human body circumstances [9]. Titanium (Ti), stainless steel (SS), and cobalt-based alloys are commonly used metallic materials in biomedical equipment [10].

This chapter mainly focuses on the tribology in natural joints, artificial joints, cardiovascular, skin, and dental.

8.2 Tribology of natural joint

Natural joints, including knees and hips, are remarkably efficient and durable, bearing significant loads with low friction and wear. Natural joint tribology focuses on cartilage, synovial fluid, and joint lubrication methods. It also focuses on understanding friction, wear, and lubrication in biological systems, which is crucial to understanding joint health, aging, and osteoarthritis (OA) [1]. The special qualities of articular cartilage, together with the role of synovial fluid, allow natural joints to withstand significant stress while retaining efficient mobility [11, 12]. This chapter investigates the natural lubrication processes in joints and considers the implications for biomimetic materials in synthetic joint design.

8.2.1 Articular cartilage (AC)

According to Seror et al. [13], a mechanism for articular cartilage border lubrication based on their research demonstrated that hyaluronic acid (HA) molecules surface-attached to PC lipids in the gel phase provide robust boundary lubrication, as shown in Fig. 8.3. As lubricin binds HA to cartilage surfaces, PC lipids interact with them, exposing their highly hydrated phosphocholine groups, which reduce friction between opposing cartilage surfaces [14]. Several studies have shown that lipids play an important role in lubricating both natural and artificial joints. Hills [15] hypothesized that lamellar phospholipids that adsorb to articular surfaces contribute to joint lubrication. Articular cartilage is a highly specialized, avascular tissue that surrounds the ends of bones in a joint. Cartilage, which is made up of chondrocytes embedded in a matrix of collagen fibers and proteoglycans, offers a low-friction surface that can absorb compressive stresses [16]. The hierarchical arrangement of collagen in cartilage helps to reduce wear by providing mechanical resistance under different stresses. Articular cartilage composition varies based on age, joint location, and depth from the surface [17].

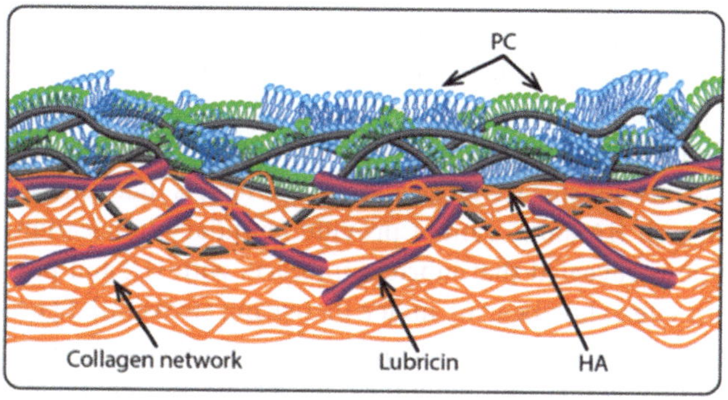

Fig. 8.3: Illustrating the proposed boundary layer on articular cartilage [14] (licensed under CC by MDPI).

8.2.2 Synovial fluid

The synovial joints of the body are excellent tribological systems. Synovial fluids contain several key elements, including HA and lubricin, which are essential for joint lubrication. Synovial fluid, a viscous liquid within the joint capsule, provides essential lubrication and nutrient transport to the avascular cartilage. It contains HA, lubricin, and phospholipids, which work in tandem to reduce friction and prevent cartilage degradation [18]. HA contributes to the fluid's viscosity, while lubricin forms a boundary layer on cartilage surfaces, preventing direct contact between articulating surfaces [14]. According to Jasmine Seror, each molecule plays a unique role, but it must be coordinated with the others. Hyaluronan, which is anchored to AC by lubricin molecules, combines with joint phosphatidylcholines for extreme joint lubrication through hydration lubrication [13].

8.3 Lubrication mechanisms in natural joints

Devices and tissues with moving surfaces, including joints, eye surfaces, and the lungs, require lubrication to function properly. In fact, lubrication failure can result in greater friction and degradation.

8.3.1 Boundary lubrication

This lubrication takes place when thin layers of lubricin adhere to cartilage surfaces, preventing direct bone contact. This mechanism is critical during low-speed or high-load joint movements such as standing or slow walking [19]. The lubrication capabilities of several molecules found in synovial fluid and cartilage have been widely explored including HA [20], lubricin [21], and phospholipids [22]. In natural joints, boundaries are lubricated by particular molecules, such as HA, lubricin, and phospholipids, which interact with one another to create an effective lubricating layer [13]. The HA-binding method was initially used to treat osteoarthritic cartilage by replacing or improving fluid-phase lubricants but surface lubrication under load is independent of synovial fluid bulk viscosity, so HA alone cannot provide boundary lubrication. Subsequent in vitro investigations revealed that lubricin interacts with HA and produce boundary lubrication of a latex-glass bearing under increasing stress [21]. Synovial joints have low friction properties due to the combination of lubrication processes including interstitial fluid pressure and boundary lubrication [11]. Boundary lubrication produces friction by sliding surfaces or molecules against each other and then dissipating energy by breaking bonds or hopping locally. Researchers like Seror et al. [13] have demonstrated that, even under high pressures (over 100 atm), a boundary layer composed of surface-bound HA combined with phosphatidylcholine vesicles can produce low friction. The mechanism, which is synergistic, involves lubricin-attaching HA to the cartilage surface, while PC lipids interact with HA, exposing highly hydrated phosphocholine head groups that facilitate efficient lubrication within the cartilage boundary. When connected to solid surfaces, phospholipids can produce reduced friction forces as well as a high-load bearing capability [22].

8.3.2 Fluid film lubrication

Reynolds initially suggested the notion of fluid film lubrication in 1886 [23]. This lubrication process is effective during dynamic activities since the synovial fluid creates a pressurized layer that protects cartilage from direct contact. The lubricating mechanism consists of two parts: hydrodynamic lubrication and electrohydrodynamic lubrication, both of which rely on the movement of the joint [24].

8.3.3 Boosted lubrication

Boosted lubrication is unique to natural joints, involving the retention of concentrated pockets of synovial fluid within the cartilage matrix. This retained fluid serves as a reservoir during high-load conditions, ensuring continuous lubrication even under substantial stress. The phrase "boosted lubrication" refers to this theory of how a weight-bearing joint functions, which relies on the creation of trapped pools of synovial fluid [25].

Age and illness, especially OA, reduce the effectiveness of natural joint lubrication. Increased friction, wear, and pain are caused by OA's effects on cartilage deterioration, synovial fluid alterations, and joint inflammation. In addition to providing information for synthetic joint materials and designs, studying the tribology of natural joints aids in determining the factors driving the course of OA [18]. Tribology of natural joints is essential for addressing joint diseases and improving prosthetic designs. By replicating the efficient lubrication systems of natural joints, synthetic replacements can achieve better wear resistance and durability, significantly benefitting patients with joint conditions [26].

8.4 Tribology in artificial joints

An artificial joint's design and longevity depend on friction, wear, and lubrication. Nowadays artificial joints have demand in hip and knee replacement due to an aging population. Resistance to wear and friction reduction are the primary tribological considerations in artificial joints, which directly affect joint longevity and patient comfort. To prevent heat generation and material degradation, artificial joints must have a low COF. Tribological challenges mainly in articulating surfaces and crucial factors to consider are the attachment to bone and the relationship between modular components.

8.4.1 Articulating surfaces

Artificial joint surfaces employ a variety of biomaterial combinations, which are generally classified into hard-on-hard and soft-on-hard (such as UHMWPE with cobalt-chromium or ZTA ceramics) pairings. Although surface treatments are required to improve resistance to wear, titanium alloys are also utilized, particularly in spinal implants [27]. The major tribological problem in hip and knee artificial joints is wear; therefore, the main focus is on improving their wear resistance. Figure 8.4 shows typical knee and hip implants.

Materials and coatings are the solutions to improve resistance to wear and minimize the friction in artificial joints. Metals like SS, tantalum, cobalt-Cr alloys, Ti and Ti alloys, Zr alloy, and uncemented implants are used. The majority of knee implants are made from plastic tibial components and oxinium and oxidized zirconium. Ceramic on plastic (UHMWPE), metal on metal, ceramic on ceramic, and metal on plastic (polyethylene or UHMWPE) materials are used for hip implants [29]. Ti alloys are used as hard tissue replacement materials, which calls for a high fatigue strength and modulus comparable to human bone [30]. Coating is also one of the solutions to reduce the friction and enhance wear resistance because of their better tribological qualities and bone-like mechanical characteristics. Joint implants employ coatings made of diamond-like carbon (DLC), tantalum TiN, and graphite-like carbon. The design parameters like size, shape, and clearances in artificial hip or knee implants are one of the approaches to reduce the wear and friction of implant materials [31]. Metal-on-metal bearings have demonstrably lower linear and volumetric wear than metal-on-UHMWPE couples, according to hip simulator testing [32]. Nevertheless, some research has demonstrated that the produced metallic or UHMWPE particles are very hazardous [33–35]. Consequently, a suitable coating enhances implanted joints' acceptability in addition to reducing wear and friction.

In biotribological systems, such as joints, the boundary lubrication is provided by synovial fluid, but its limited availability necessitates the use of other lubricants like saline, serum, and synovial fluid components. Since serum is clinically relevant, it is particularly useful for metal-on-polyethylene tests. Researchers are still focused on developing biolubricants that imitate synovial fluid and intelligent coatings that adapt to joint movements. Moreover, 3D printing might be used to create personalized implants with ideal tribological characteristics [36].

8.5 Tribology in skin

Skin provides protection, regulates body temperature, and retains water. Its biotribological properties play a crucial role in these functions. The structure of the skin affects its mechanical and frictional properties. It consists of three layers: epidermis,

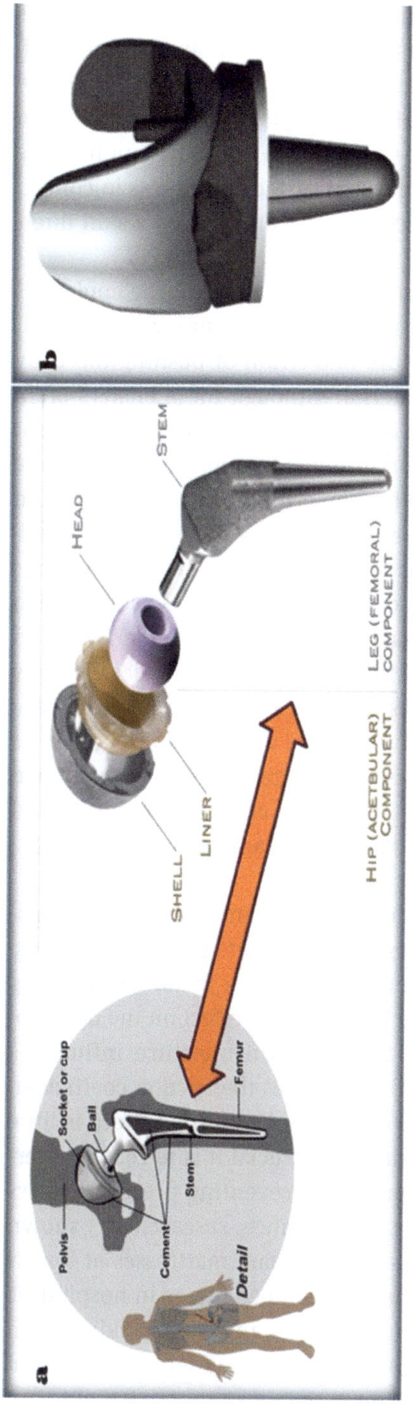

Fig. 8.4: Typical (A) hip implant and (B) knee implant [28] (licensed under CC by OP Conference Series Materials Science and Engineering).

dermis, and subcutaneous layers [37]. Human skin is exposed to various substances like textures, fabrics, lubricants, and environmental factors on a regular basis. Various factors affect the frictional behavior of human skin including, gender, stress levels, weather conditions, age, and counter materials/textiles [38, 39]. Friction between the skin is caused by two primary mechanisms: Components of deformation: these are caused by the elasticity and viscoelasticity of the skin surface when it is in contact with a foreign object. Adhesive components: an adhesion occurs when the skin interacts with a surface [40].

Another problem that can arise from skin-fabric contact is blisters. The epidermis contains tiny pockets of biological fluids called blisters. Walking, running, hiking, etc., can result in blisters on the feet due to the pressure and strain of pushing the body forward as well as friction from the footwear material. Although these blisters often produce only mild pain and discomfort, in certain cases, they might cause impairment [41]. It is advised to either use a large area fixation dressing or apply adhesive to the blister [42]. Generally speaking, the blister should be kept clean and its burden should be kept to a minimum. Despite the fact that blisters heal on their own, Brennan advises preventing friction blisters as the best course of action [43]. The more serious problem of pressure ulcers, often known as bedsores, is brought on by skin friction on garments and other materials. "Localized damage to the skin and underlying soft tissue, usually over a bony prominence or related to medical or other devices" is how the National Pressure Injury Advisor Panel defines pressure injuries. Long-term pressure or shear forces can cause localized skin and soft tissue damage, often known as bedsores or pressure ulcers, over bony regions. Tissue compression preventing blood flow and nutrition delivery causes these injuries, which are common in elderly and bedridden people. Usually, they appear on the shoulders, back of the head, heels, hips, and spine [44].

Shear loading, normal pressure loading, the microclimate, and the surface roughness of the device interface are some of the factors that contribute to pressure ulcers. By definition, a pressure ulcer indicates that normal pressure is a major contributing element [45]. Surface roughness and microclimate have a major effect on the development of pressure ulcers by affecting interface friction. Skin moisture influences changes in skin-device friction, and higher humidity increases the friction coefficient. Research on silicone materials and other medical devices indicates that there is little friction at a surface roughness of 4 μm, emphasizing how crucial it is to reduce interface shear stress in order to avoid pressure ulcers [46]. Preventing pressure ulcers may be accomplished by using various materials. For instance, research has shown that foam substitutes are more successful than traditional foam mattresses at lowering the prevalence of pressure ulcers [47]. Using low-friction bed linens in hospitals is perhaps a crucial way to reduce the friction between textiles and human skin. When the time comes to change the bed linens or flip a bedridden patient, low-friction sheets like EZ move slide sheets make it easier for them to move around. Patients can move more safely and comfortably because of the ultra-low-friction material. Patients who are bedridden frequently have pressure ulcers or bed sores as a result of the skin's shearing forces on the bedding [48].

The textiles used to make clothing can be modified or improved to lower the COF and draw moisture away from the skin. Because it eliminates the need for consumers to apply items or modify their current clothing or equipment, this solution will be the most straightforward alternative for users looking to minimize friction at the skin textile material contact. Chalk is used in some sports and activities, including lifting, to help the hands absorb moisture and perform at their best [49]. Researchers are paying more attention to biopolymers for textile production in addition to synthetic polymeric materials. Textiles have made use of a variety of biopolymers including alginate, chitosan, cyclodextrin, and sericin [50]. These biopolymers, like the polysaccharide found in polyester, woven fabric, have been investigated as textile materials for drug administration while assessing tribological comfort [51]. In addition to offering the ability to create complex designs, 3D printing technology permits the use of recycled polymers. Like biopolymer fabrics, there are not many tribological research on 3D-printed clothing or accessories when they come into contact with the skin. Kasar et al. [52] compared the dry and wet properties of 3D-printed thermoplastic polyamide and thermoplastic polyurethane materials to a skin model. The skin model's simulation of sweat absorption caused the COF to rise under damp circumstances. In order to avoid skin model softening and minimize COF, 3D-printed layers with an 80% fill density were created to catch sodium polyacrylate crystals that absorbed water. This illustrates how 3D printing may be used to create fabrics that are tribologically pleasant. Lubrication, like the use of oil in mechanical systems, can be used to reduce friction at the skin-textile contact and reduce friction and avoid blisters, lubricants such as lotions and gels are applied to the skin or material in places that are prone to rubbing. BodyGlide, Hydropel, and Nok are widely accessible products that are intended to prevent skin damage caused by friction [53].

8.6 Cardiovascular tribology

8.6.1 Ventricular assist devices

Cardiovascular devices confront tribological issues due to three primary sources of friction: fluid friction from blood flow on device surfaces, mechanical wear in moving parts such as heart valves and VADs, and friction between devices and soft tissue during insertion or operation. The most common vascular medical devices are stents and catheters; they must contact with the blood vessels in order to function properly. Tribological problems mostly affect prosthetic heart valves, intravascular stents, and capillary blood flow.

Patients with vascular heart disease are implanted with artificial heart valves, with two main types available: mechanical heart valves (MHVs) and bioprosthetic heart valves (BHVs). BHVs exhibit lower wear and friction than MHVs. Excessive mechanical wear can cause MHVs to fail including fractured struts and leaflet escape. Figure 8.5 represents the heart valves anatomy.

The tribological characteristics of MHVs, which are subject to wear and friction, can be significantly impacted by plasma alteration. Additionally, it can influence texture and morphology, influencing characteristics including surface area, patterning, roughness, and smoothness. Advanced surface modifications such as plasma treatment, ion implantation, and UV irradiation [54] and coating technologies like polysiloxanes and silicon carbide coatings, carbon coatings, titanium oxide and titanium nitride coatings, heparin coating, silver coatings, superhydrophobic surfaces, and nanocrystalline diamond coating. Considerations have emerged as viable solutions to the problems with MHVs [55].

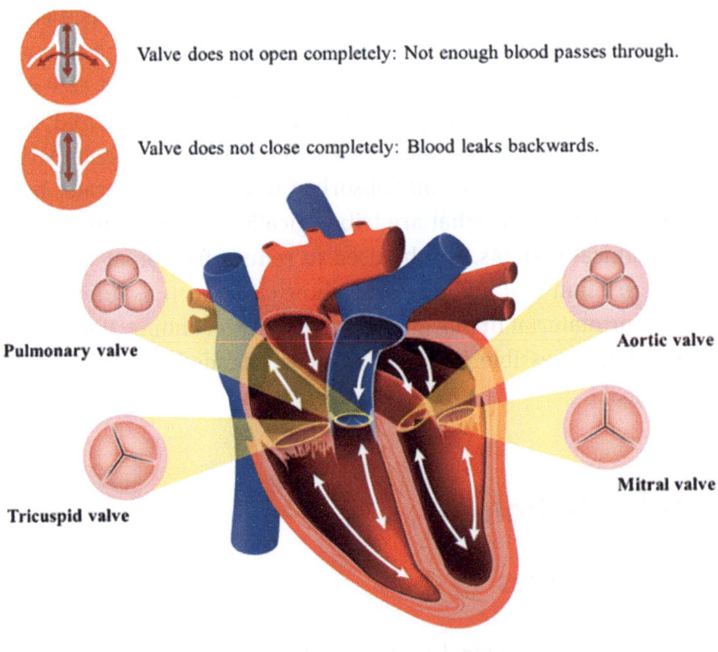

Fig. 8.5: Heart valves anatomy [55] (licensed under CC by *European Polymer Journal*).

The purpose of ventricular assist devices is to support or replace one or more heart chambers that are failing. A pulsatile pump and an advanced continuous-flow rotary pump are two examples of these pumps [56], which use bearing systems like rolling element, journal bearing, and pivot bearings to achieve high performance [57]. There are several moving parts in pulse pumps, including actuator bearings, valves, and pushers, which cause mechanical wear and friction. Blood and tissue are frequently in contact with these pumps. An impeller in a rotary pump is constrained by mechanical contact bearings, minimizing the number of components required. Depending on the direction the blood enters and exits the impeller, it can be classified as axial, radial, or mixed flow. Due to their fewer moving parts, rotary pumps are more reliable and long-lasting than other pumps [58]. Rolling element bearings use shaft seals to prevent blood from entering gaps between rolling elements, avoiding coagulation and bearing seizure. While sealed ball bearings are isolated from blood contact, friction-generated heat can still cause hemolysis. In pivot bearings without seals, blood serves as a lubricant, displaying boundary lubrication behavior [57].

8.6.2 Vascular stents and catheters

In order to revascularize (reestablish flow) a stenosed (constricted) channel, stents are often folded around a balloon catheter and inflated inside the vessel. The deployment of a catheter or stent induces friction and radial forces, potentially damaging blood vessels in two ways: by stretching the arterial wall and injuring it or by damaging and removing the endothelium [59]. Tribology is important throughout both the deployment and deployment phases. During the deployment phase, it is critical not to induce severe tissue injury since this may demand further intervention. How stenting process will be done explained by Rasmus M.F. Wagner as shown in Fig. 8.6 [4].

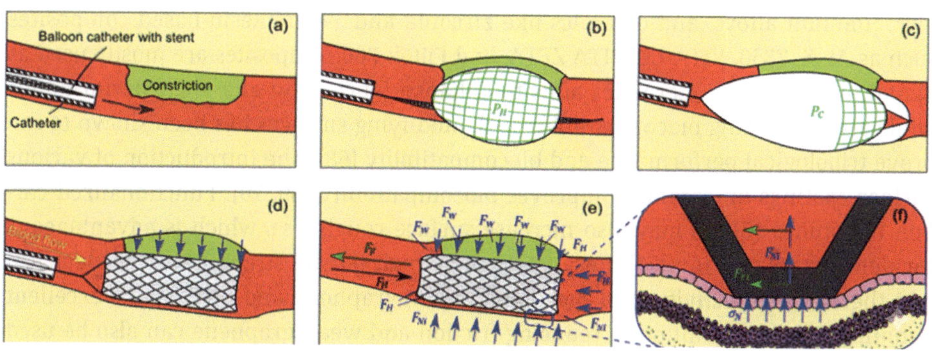

Fig. 8.6: Stenting process [45] (licensed under CC by friction).

The insertion of a catheter can damage the endothelium, resulting in complications such as restenosis. Low-friction catheter surfaces can reduce injury and improve procedural safety [60]. Polymer coatings, hydrogel coatings, and polymerization are the solutions to minimize the catheter friction [61].

8.7 Tribology in dental

Because of dentistry's rapid development and expansion, tribological examinations are becoming increasingly common. The wear of teeth is primarily caused by mastication. Mastication involves chewing food, which involves an open phase as well as a closed phase. Open occlusal pressure results in little tooth wear, while closed occlusal pressure occurs when hard particles in foods are dragged across opposing surfaces, resulting in occlusal surface wear. Tooth wear can also be caused by thegosis and bruxism [62]. Other factor for teeth wear is chemical effect due to consumption of acidic drinks and pipe smoking and chewing as well as improper brushing [63, 64]. Chewing causes severe wear on the teeth. Thus, wear resistance is an important consideration when selecting a dental replacement material like composite-based materials, metals and their alloys, polymers, and ceramics [65].

As soft drink, fruit juice, and sports drink consumption rises, teeth are increasingly exposed to acidic environments. Dental erosion, the chemical loss of hard tissue without microbial action, can result from extrinsic factors (like acidic foods and drinks) or intrinsic factors (like gastric regurgitation or disorders such as bulimia) [65]. Tooth wear is shown in Fig. 8.7 [66].

Dental implants replace teeth and restore their functionality, mechanical integrity, and dependability. Dental materials are chosen based on location, tooth size, biocompatibility, and physical/mechanical/tribological qualities. The materials used are gold, dental amalgam, vanadium surgical steel, cobalt-based and nickel-based alloys, titanium alloys, and ceramics like zirconia and Y-TZP. Resin-based composites such as AP-X, Z350, VITA LC, VITA ZETA, and Filtek P60 composites are mostly used as dental materials [66]. Ceramics aims to enhance fabrication and wear performance while also increasing biocompatibility [67]. Modifying surfaces has been shown to improve tribological performance and biocompatibility [68]. The introduction of various surface coatings on ceramics improves biocompatibility [69, 70]. Functionalized carbon nanotubes (CNTs) may also increase surface roughness, which is advantageous for dental implantation [71]. Because of their high tensile strength, better electrical and thermal conductivity, and biocompatibility, graphene and CNTs make excellent nanofillers [72]. In addition to reducing friction and wear, graphene can also be used as a coating over zirconia [73], as well as by adding filler materials to the composite matrix [74], such as bisphenol A-glycidyl methacrylate, 2-hydroxyethyl methacrylate, triethylene glycol, dimethacrylate, and urethane dimethacrylate [75].

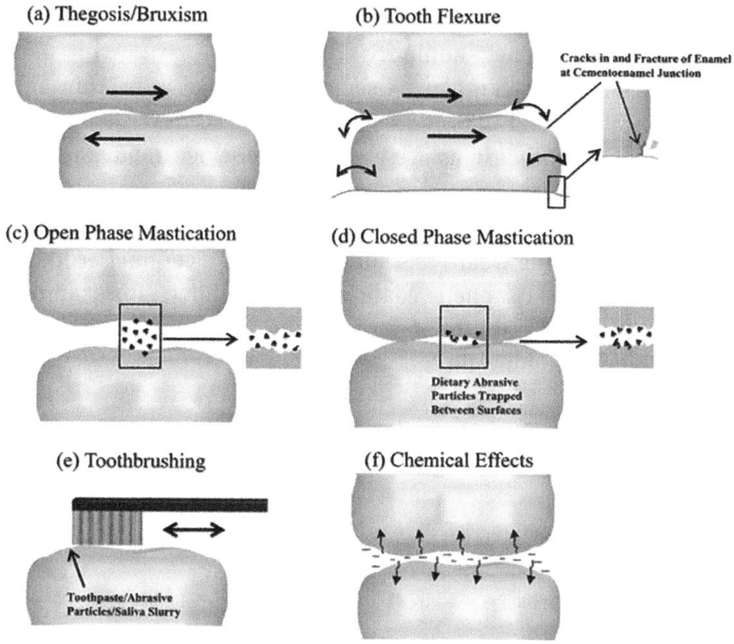

Fig. 8.7: Tooth wear is influenced by a variety of interactions [66] (licensed under CC by *Tribology International*).

8.8 Conclusions

Tribology plays a vital role in medical devices as they are used in natural joints, artificial joints, skin, cardiology, and dental. The following conclusions are drawn from this review:

- The synovial fluid and cartilage in natural joints provide wear-resistant, low-friction interfaces. Various mechanisms, including boundary lubrication (utilizing lubricin, HA, and phospholipids), fluid film lubrication, and boosted lubrication, ensure efficient movement and load distribution. Biomimetic artificial joints for improved longevity and function are designed based on these insights.
- New material science and coatings are essential for reducing friction and enhance wear in artificial joints including UHMWPE, cobalt-chrome alloys, ceramics, and DLC. By designing and using tailored materials and designs, implants become more durable and biocompatible, while biolubricants and intelligent coatings mimic natural joint lubrication efficiency.
- A person's skin friction is affected by a variety of factors including surface roughness, hydration, and materials in contact. High humidity increases friction coefficient, which can lead to pressure ulcers and blisters. In addition to reducing skin

damage and improving comfort, low-friction fabrics, biopolymers, and 3D-printed fabrics, as well as lubrication, can help minimize skin damage.

- Devices like catheters and stents can cause tissue damage due to friction and wear. To reduce complications and improve safety during procedures, enhanced surface coatings and low-friction designs are essential. Tribological advances in these devices can have a significant impact on patient outcomes.
- The field of tribology in dentistry examines tooth wear resulting from mechanical and chemical factors, emphasizing wear-resistant materials like ceramics, composites, and surface-coated metals. Graphene coatings and nanofillers, for example, improve biocompatibility and durability of dental restorations and implants by enhancing biocompatibility and durability.

References

[1] Zhou ZR, Jin ZM. (2015 Mar 1). Biotribology: Recent progresses and future perspectives. *Biosurface and Biotribology*, 1(1), 3–24.

[2] Santos G. (2017). The importance of metallic materials as biomaterials. *Adv Tissue Eng Regen Med Open Access*, 3(1), 300–302.

[3] Raghavendra GM, Varaprasad K, Jayaramudu T. (2015 Jan 1). *Biomaterials: Design, Development and Biomedical Applications. In Nanotechnology Applications for Tissue Engineering* (pp. 21–44). William Andrew Publishing.

[4] Wagner RM, Maiti R, Carré MJ, Perrault CM, Evans PC, Lewis R. (2021 Mar 1). Bio-tribology of vascular devices: A review of tissue/device friction research. *Biotribology*, 25, 100169.

[5] Shekhawat D, Singh A, Banerjee MK, Singh T, Patnaik A. (2021 Feb 1). Bioceramic composites for orthopaedic applications: A comprehensive review of mechanical, biological, and microstructural properties. *Ceramics International*, 47(3), 3013–3030.

[6] Shen JZ, Fäldt J. (2014 Jan 1). Requirements of bioactive ceramics for dental implants and scaffolds. In *Advanced Ceramics for Dentistry* (pp. 279–300). Butterworth-Heinemann.

[7] Lam MT, Wu JC. (2012 Aug 1). Biomaterial applications in cardiovascular tissue repair and regeneration. *Expert Review of Cardiovascular Therapy*, 10(8), 1039–1049.

[8] GL K, Diba M, Mikos AG. (Aug 2020). Materials design for bone-tissue engineering. *Nature Reviews Materials*, 5(8), 584–603.

[9] Hazwani MR, Lim LX, Lockman Z, Zuhailawati H. (Jan 1; 2022). Fabrication of titanium-based alloys with bioactive surface oxide layer as biomedical implants: Opportunity and challenges. *Transactions of Nonferrous Metals Society of China*, 32(1), 1–44.

[10] Hanawa T. (Dec 13 2012). Research and development of metals for medical devices based on clinical needs. *Science and Technology of Advanced Materials*, 13(6), 064102.

[11] Schmidt TA, Gastelum NS, Nguyen QT, Schumacher BL, Sah RL. (Mar 2007). Boundary lubrication of articular cartilage: Role of synovial fluid constituents. *Arthritis & Rheumatism*, 56(3), 882–891.

[12] Sakai N, Hagihara Y, Furusawa T, Hosoda N, Sawae Y, Murakami T. (Feb 1 2012). Analysis of biphasic lubrication of articular cartilage loaded by cylindrical indenter. *Tribology International*, 46(1), 225–236.

[13] Seror J, Zhu L, Goldberg R, Day AJ, Klein J. (Mar 10 2015). Supramolecular synergy in the boundary lubrication of synovial joints. *Nature Communications*, 6(1), 6497.

[14] Lin W, Liu Z, Kampf N, Klein J. (Jul 2 2020). The role of hyaluronic acid in cartilage boundary lubrication. *Cells*, 9(7), 1606.

[15] Hills BA. (Jan 1 2000). Boundary lubrication in vivo. *Proceedings of the Institution of Mechanical Engineers, Part H: Journal of Engineering in Medicine*, 214(1), 83–94.

[16] Mow VC, Kuei SC, Lai WM, Armstrong CG. Biphasic creep and stress relaxation of articular cartilage in compression: theory and experiments.

[17] Lipshitz H, Etheredge R III, Glimcher MJ. (Jun 1 1975). In vitro wear of articular cartilage. *JBJS*, 57(4), 527–534.

[18] Fragassi A, Greco A, Palomba R. (Sep 19 2024). Lubricant strategies in osteoarthritis treatment: Transitioning from natural lubricants to drug delivery particles with lubricant properties. *Journal of Xenobiotics*, 14(3), 1268–1292.

[19] Zhu L, Seror J, Day AJ, Kampf N, Klein J. (Sep 1 2017). Ultra-low friction between boundary layers of hyaluronan-phosphatidylcholine complexes. *Acta Biomaterialia*, 59, 283–292.

[20] Singh A, Corvelli M, Unterman SA, Wepasnick KA, McDonnell P, Elisseeff JH. (Oct 2014). Enhanced lubrication on tissue and biomaterial surfaces through peptide-mediated binding of hyaluronic acid. *Nature Materials*, 13(10), 988–995.

[21] Jay GD, Torres JR, Warman ML, Laderer MC, Breuer K. (Apr 10 2007). The role of lubricin in the mechanical behavior of synovial fluid. *Proceedings of the National Academy of Sciences*, 104(15), 6194–6199.

[22] Wang M, Liu C, Thormann E, Dedinaite A. (Dec 9 2013). Hyaluronan and phospholipid association in biolubrication. *Biomacromolecules*, 14(12), 4198–4206.

[23] Reynolds O IV. (Dec 1886). On the theory of lubrication and its application to Mr. Beauchamp tower's experiments, including an experimental determination of the viscosity of olive oil. *Philosophical Transactions of the Royal Society of London*, 31(177), 157–234.

[24] Ateshian GA. (Jun 19 2009). The role of interstitial fluid pressurization in articular cartilage lubrication. *Journal of Biomechanics*, 42(9), 1163–1176.

[25] Walker PS, Dowson D, Longfield MD, Wright VE. (Nov 1968). " Boosted lubrication" in synovial joints by fluid entrapment and enrichment. *Annals of the Rheumatic Diseases*, 27(6), 512.

[26] Murakami T. (2023). *Biotribology of Natural Joints* (pp. 81–149). DOI: 10.1016/B978-0-12-823669-7.00002-8.

[27] Fisher J, Jin Z, Tipper J, Stone M, Ingham E. (Dec 2006). Tribology of alternative bearings. *Clinical Orthopaedics and Related Research*, 453, 25–34. DOI: 10.1097/01.blo.0000238871.07604.49.

[28] Solanke S, Gaval V. (2021). Most commonly used metallic biomaterials for plasma sprayed hydroxyapatite coatings. *IOP Conference Series: Materials Science and Engineering* , 1168(1), 012013. DOI: 10.1088/1757-899X/1168/1/012013.

[29] Sahoo P, Das SK, Paulo Davim J. (2019). *Tribology of Materials for Biomedical Applications*. DOI: 10.1016/B978-0-08-102174-3.00001-2.

[30] Fini M, Giavaresi G, Torricelli P, Borsari V, Nicoli Aldini N, Giardino R, et al. (Jan 2001). Biocompatibility and osseointegration in osteoporotic bone. *The Journal of Bone & Joint Surgery*, 83-B (1), 139–143. DOI: 10.1302/0301-620X.83B1.0830139.

[31] Ching HA, Choudhury D, Nine MJ, Abu Osman NA. (2014). Effects of surface coating on reducing friction and wear of orthopaedic implants. *Science and Technology of Advanced Materials*, 15(1), DOI: 10.1088/1468-6996/15/1/014402.

[32] Rieker C, Kottig P. (2002). In vivo tribological performance of 231 metal-on-metal hip articulations. *HIP International*, 12(2), 73–76. DOI: 10.1177/112070000201200202.

[33] Green TR, Fisher J, Matthews JB, Stone MH, Ingham E. (Sep 2000). Effect of size and dose on bone resorption activity of macrophages by in vitro clinically relevant ultra high molecular weight polyethylene particles. *Journal of Biomedical Materials Research*, 53(5), 490–497. DOI: 10.1002/1097-4636(200009)53:5. <490::AID-JBM7>3.0.CO;2-7.

[34] Smith RA, Hallab NJ. (Apr 2010). In vitro macrophage response to polyethylene and polycarbonate-urethane particles. *Journal of Biomedical Materials Research A*, 93(1), 347–355. DOI: 10.1002/jbm.a.32529.

[35] Matthews JB, Green TR, Stone MH, Wroblewski BM, Fisher J, Ingham E. (Oct 2000). Comparison of the response of primary human peripheral blood mononuclear phagocytes from different donors to challenge with model polyethylene particles of known size and dose. *Biomaterials*, 21(20), 2033–2044. DOI: 10.1016/S0142-9612(00)00089-2.

[36] Okolie O, Stachurek I, Kandasubramanian B, Njuguna J. (2021). Material challenges and opportunities in 3D printing for hip implant applications. *Recent Progress in Materials*, 4(1), 1. DOI: 10.21926/rpm.2201004.

[37] Yousef H, Alhajj M, Fakoya AO, Sharma S. (2024). *Anatomy, Skin (Integument), Epidermis* Treasure Island (FL): StatPearls Publishing.

[38] Gerhardt LC, Strässle V, Lenz A, Spencer ND, Derler S. (2008). Influence of epidermal hydration on the friction of human skin against textiles. *Journal of the Royal Society Interface*, 5(28), 1317–1328. DOI: 10.1098/rsif.2008.0034.

[39] Cua AB, Wilhelm KP, Maibach HI. (1990). Elastic properties of human skin: Relation to age, sex, and anatomical region. *Archives of Dermatological Research* , 282(5), 283–288. DOI: 10.1007/BF00375720.

[40] Bailey J, Oliveri A, Levin E. (2013). 基因的改变NIH Public Access. *Bone*, 23(1), 1–7. DOI: 10.1177/1099800411434151.

[41] Sulzberger MB, Cortese TA, Fishman L, Wiley HS. (1966). Studies on blisters produced by friction. I. Results of linear rubbing and twisting technics. *Journal of Investigative Dermatology*, 47(5), 456–465. DOI: 10.1038/jid.1966.169.

[42] Janssen L, Allard NAE, Ten Haaf DSM, Van Romburgh CPP, Eijsvogels TMH, Hopman MTE. (2018). First-aid treatment for friction Blisters: 'Walking into the right direction?'. *Clinical Journal of Sport Medicine*, 28(1), 37–42. DOI: 10.1097/JSM.0000000000000424.

[43] Brennan FHJ, Jackson CR, Olsen C, Wilson C. (2012). Blisters on the battlefield: The prevalence of and factors associated with foot friction blisters during Operation Iraqi Freedom I. *Military Medicine*, 177(2), 157–162. DOI: 10.7205/milmed-d-11-00325.

[44] Mervis JS, Phillips TJ. (2019). Pressure ulcers: Pathophysiology, epidemiology, risk factors, and presentation. *Journal of the American Academy of Dermatology*, 81(4), 881–890. DOI: 10.1016/j.jaad.2018.12.069.

[45] Zhang X, Zhang Y, Jin Z. (2022). A review of the bio-tribology of medical devices. *Friction*, 10(1), 4–30. DOI: 10.1007/s40544-021-0512-6.

[46] Klaassen M, De Vries EG, Masen MA. (2019). Friction in the contact between skin and a soft counter material: Effects of hardness and surface finish. *Journal of the Mechanical Behavior of Biomedical Materials*, 92(September 2018), 137–143. DOI: 10.1016/j.jmbbm.2019.01.006.

[47] Stannard D. (2012). Support surfaces for pressure ulcer prevention. *Journal of PeriAnesthesia Nursing*, 27(5), 341–342. DOI: 10.1016/j.jopan.2012.07.007.

[48] Baernholdt M, Yan G, Hinton ID, Cramer E, Dunton N. (2020). Effect of preventive care interventions on pressure ulcer rates in a national sample of rural and urban nursing units: Longitudinal associations over 4 years. *International Journal of Nursing Studies* , 105, 103455. DOI: 10.1016/j.ijnurstu.2019.103455.

[49] MacFarlane MJ, Theobald P. (2021). Skin tribology in sport. *Biosurface Biotribology*, 7(3), 113–118. DOI: 10.1049/bsb2.12015.

[50] Shahid-ul-Islam, Shahid M, Mohammad F. (2013). Green chemistry approaches to develop antimicrobial textiles based on sustainable biopolymers – A review. *Industrial & Engineering Chemistry Research*, 52(15), 5245–5260. DOI: 10.1021/ie303627x.

[51] Gerhardt L-C, Lottenbach R, Rossi RM, Derler S. (2013). Tribological investigation of a functional medical textile with lubricating drug-delivery finishing. *Colloids Surf B Biointerfaces*, 108, 103–109. DOI: 10.1016/j.colsurfb.2013.01.055.

[52] Kasar AK, Chan A, Shamanaev V, Menezes PL. (2022). Tribological interactions of 3D printed polyurethane and polyamide with water-responsive skin model. *Friction*, 10(1), 159–166. DOI: 10.1007/s40544-020-0472-2.

[53] Hashmi F, Kirkham S, Nester C, Lam S. (2016). The effect of topical anti-blister products on the risk of friction blister formation on the foot. *Journal of Tissue Viability*, 25(3), 167–174. DOI: 10.1016/j.jtv.2016.04.002.

[54] Ul Ahad I, et al. (2014). Surface modification of polymers for biocompatibility via exposure to extreme ultraviolet radiation. *Journal of Biomedical Materials Research Part A*, 102(9), 3298–3310. DOI: 10.1002/jbm.a.34958.

[55] Yousefi S, Borna H, Rohani Shirvan A, Wen C, Nouri A. (January 2024). Surface modification of mechanical heart valves: A review. *European Polymer Journal*, 205, 112726. DOI: 10.1016/j.eurpolymj.2023.112726.

[56] Timms D. (2011). A review of clinical ventricular assist devices. *Medical Engineering & Physics*, 33(9), 1041–1047. DOI: 10.1016/j.medengphy.2011.04.010.

[57] Sundareswaran KS, Reichenbach SH, Masterson KB, Butler KC, Farrar DJ. (2013). Low bearing wear in explanted HeartMate II left ventricular assist devices after chronic clinical support. *ASAIO J*, 59(1), 41–45. DOI: 10.1097/MAT.0b013e3182768cfb.

[58] Travis AR, et al. (2007). Vascular pulsatility in patients with a pulsatile- or continuous-flow ventricular assist device. *Journal of Thoracic and Cardiovascular Surgery*, 133(2), 517–524. DOI: 10.1016/j.jtcvs.2006.09.057.

[59] Laroche D, Delorme S, Anderson T, DiRaddo R. (2006). Computer prediction of friction in balloon angioplasty and stent implantation. *Lect Notes Comput Sci (Including Subser Lect Notes Artif Intell Lect Notes Bioinformatics)*, 4072, LNCS:1–8. DOI: 10.1007/11790273_1.

[60] Uyama Y, Tadokoro H, Ikada Y. (1989). Catheter materials by graft polymerization. *Biomaterials*, 12, 71–75.

[61] Weiss D, Gefen A, Einav S. (2016). Modelling catheter–vein biomechanical interactions during an intravenous procedure. *Computer Methods in Biomechanics and Biomedical Engineering*, 19(3), 330–339. DOI: 10.1080/10255842.2015.1024667.

[62] Mair LH, Stolarski TA, Vowles RW, Lloyd CH. (1996). Wear: Mechanisms, manifestations and measurement. Report of a workshop. *Journal of Dentistry*, 24(1–2), 141–148. DOI: 10.1016/0300-5712(95)00043-7.

[63] Lewis R, Dwyer-Joyce RS. (2005). Wear of human teeth: A tribological perspective. *Proceedings of the Institution of Mechanical Engineers, Part J: Journal of Engineering Tribology*, 219(1), 2–19. DOI: 10.1243/135065005x9655.

[64] Amaechi BT, Higham SM. (2005). Dental erosion: Possible approaches to prevention and control. *Journal of Dentistry*, 33(3 SPEC. ISS), 243–252. DOI: 10.1016/j.jdent.2004.10.014.

[65] Carvalho A, Pinto P, Madeira S, Silva FS, Carvalho O, Gomes JR. (March 2020). Tribological characterization of dental restorative materials. *Biotribology*, 23, 100140. DOI: 10.1016/j.biotri.2020.100140.

[66] Zheng Y, Bashandeh K, Shakil A, Jha S, Polycarpou AA. (November 2021 2022). Review of dental tribology: Current status and challenges. *Tribology International*, 166, 107354. DOI: 10.1016/j.triboint.2021.107354.

[67] Kelly JR. (1997). Ceramics in restorative and prosthetic dentistry. *Annual Review of Materials Science*, 27(1), 443–468. DOI: 10.1146/annurev.matsci.27.1.443.

[68] Ranc H, Servais C, Chauvy PF, Debaud S, Mischler S. (2006). Effect of surface structure on frictional behaviour of a tongue/palate tribological system. *Tribology International*, 39(12), 1518–1526. DOI: 10.1016/j.triboint.2006.01.017.

[69] Liu D, et al. (2013). A new modified laser pretreatment for porcelain zirconia bonding. *Dental Materials*, 29(5), 559–565. DOI: 10.1016/j.dental.2013.03.002.

[70] Henningsen A, et al. (2018). Changes in surface characteristics of titanium and zirconia after surface treatment with ultraviolet light or non-thermal plasma. *European Journal of Oral Sciences*, 126(2), 126–134. DOI: 10.1111/eos.12400.

[71] Kou W, Akasaka T, Watari F, Sjögren G. (2013). An In vitro evaluation of the biological effects of carbon nanotube-coated dental Zirconia. *ISRN Dentistry*, 2013, 1–6. DOI: 10.1155/2013/296727.

[72] Chen H, Müller MB, Gilmore KJ, Wallace GG, Li D. (2008). Mechanically strong, electrically conductive, and biocompatible graphene paper. *Advanced Materials*, 20(18), 3557–3561. DOI: 10.1002/adma.200800757.

[73] Li H, et al. (2014). Microstructure and wear behavior of graphene nanosheets-reinforced zirconia coating. *Ceramics International*, 40(8 PART B), 12821–12829. DOI: 10.1016/j.ceramint.2014.04.136.

[74] Ramalho A, Braga De Carvalho MD, Antunes PV. (2013). Effects of temperature on mechanical and tribological properties of dental restorative composite materials. *Tribology International*, 63, 186–195. DOI: 10.1016/j.triboint.2012.04.025.

[75] Okulus Z, Voelkel A. (2017). Mechanical properties of experimental composites with different calcium phosphates fillers. *Materials Science and Engineering C*, 78, 1101–1108. DOI: 10.1016/j.msec.2017.04.158.

P. Rajasekar, Swetha P., Suwathi S., and S. Rishikesavan

9 Biomaterials and surface modifications

Abstract: Biomaterials are biocompatible materials that are used for implantation and incorporation with the living system to replace or support the living tissues or organs. Biomaterials are broadly classified into ceramic, metallic, polymeric, composite, and semiconductors. The unique features of biomaterials include biocompatibility, inertness, biodegradability, and bioabsorbable. These ideal properties can be modified or tuned through various surface modification methods for distinct applications including healthcare. As surface modification influences various biomaterial behaviors, this chapter aims to discuss the available surface-modifying techniques with technological advancements. Furthermore, the applications of biomaterials are also discussed extensively.

Keywords: Biomaterials, surface modification, techniques, applications

9.1 Introduction

The term "biomaterial" was originally developed in 1967 [1]. Initially, biomaterial means that materials of synthetic or natural origin contact with blood, tissue, and biological fluids and thereby perform intended functions (diagnostic, therapeutic, storage, etc.) without affecting living organisms [2]. Alternatively, biomaterials also mean that any substance (other than drugs) or combination of substances (synthetic or natural) can be used to treat a whole or part of a system for any time for augmenting or replacing any tissue, organ, or function [3]. Further, biomaterials mean that materials are employed for substitutions [4].

Biomaterials are predominantly considered in the healthcare sector for restoring physiological processes that are affected by infections/diseases, aging, traumas, or accidents. For example, biomaterials are widely used in heart and dental valves, drug delivery systems, wound closure devices, and other medical implants. Moreover, biomaterials can also be considered for the replacement of important organs like kidneys, hearts, and skeletal components (knees, hips, and fingers). Additionally, the role of biomaterials is also increasing in the fields of tissue engineering and regenerative medicine, specifically for the development of new wound dressings, nerve stimulation devices, and artificial joints.

Biomaterials can be either inorganic (metals) or organic (ceramics and biodegradable substances) in nature. Primarily, materials used in biomaterials are metals, polymers, ceramics, and composites. For example, titanium, stainless-steel-based metals, nylon, and carbon [4]. Generally, biomaterials are either in one of two categories, namely nonresorbable materials (not degraded or absorbed in the tissue environ-

https://doi.org/10.1515/9783111563176-009

ment) or resorbable materials (absorbed in the tissue environment). On use, these materials can interact with biological systems and exhibit desirable physical, chemical, and biological properties [5]. The key properties include elasticity, yield strength, Young's modulus, stress and strain resistance, durability, permeability, and refractive index. Furthermore, the following features are essential for considering biomaterials in modern medicine as food and drugs or an implant. Biomaterials must be nontoxic and biocompatible, that is, it should not induce undesirable inflammatory, allergic, or pyogenic reactions, should support targeted drug delivery, and ensure non-carcinogenicity in the host system, should involve repair and reconstruction, should depict sufficient mechanical strength, osseointegration, and good blood compatibility. Apart from these, degradability is also frequently preferred [6–8]. Particularly for orthopedic applications, the implants must offer a surface for encouraging cell proliferation and differentiation and must prevent degradation.

Besides these, the corrosion of metal-based biomaterials can release micro- and nanoscale metal ions, which can cause inflammation and tissue damage in humans [9]. Moreover, the oral/injection of biomaterials may absorb selectively and can be reduced to the micro- and nanoscale level before reaching targets through the circulation [10]. Further, the reaching of biomaterials to specific targets may interact with cells, proteins, and organelles, leading to the production of interfaces due to biophysiochemical interactions, which may cause positive or negative effects [11]. A higher dose of biomaterials may be required for an effective systemic distribution, which may cause major issues like degradation and poor renal clearance. Additionally, the use of macroscale drug delivery devices only enables the controlled release of bioactive substances [6]. Though the biomaterials are considered for substitution or replacement of impaired tissue/organs/skeletal components, their negative impacts on interaction with biological fluids or biomolecules and working performance must be fine-tuned. It has been stated that the surface modification of materials can boost their functionality and biological performance. Also, fight against thrombogenicity, improve corrosion resistance, regulate cell adhesion and proliferation, etc. [12]. Therefore, this chapter aims to explain various biomaterials and their surface modification techniques and applications in detail.

9.2 Criteria for the selection of biomaterial scaffolds

9.2.1 Biotolerance

Biotolerance or material biocompatibility refers to a material's ability to interact with the biological environment without eliciting toxic or adverse immune responses. An ideal scaffold material should support the integration of new bone tissue with the surrounding host tissue and stimulate osteogenesis/new bone formation. The scaffold materials must encourage cell adhesion, growth, differentiation, and inflammation. Commonly used biocompatible materials include hydroxyapatites, tricalcium phosphate, and related calcium phosphate-based ceramics. These materials mimic bone mineral components and thereby support the attachment and growth of osteoblasts [13].

9.2.2 Biodegradability

Scaffolds should be biodegradable in a controlled way in accordance with the rate of new tissue development. During degradation, scaffolds should not produce any toxic by-products. Materials must provide adequate structural support during tissue regeneration. For example, materials like poly(lactic-*co*-glycolic acid) (PLGA) and polycaprolactone (PCL) are commonly used because of their stable and safe degradation profile. The degradation rate of these biomaterials can be engineered for specific tissue applications by altering crystallinity, polymer blends, molecular weights, and copolymer ratios [14].

9.2.3 Mechanical properties

The tensile strength, stiffness, and elasticity of the scaffold should be comparable to the tissue of the target (i.e., materials should allow physiological loads and functional integration within dynamic mechanical environments like bone and cardiac tissue). The ideal range of bending strength and elastic modulus for scaffolding material designed for human cortical bone is 100–150 MPa and 7–25 GPa, respectively [15].

9.2.4 Microbes inhibitory property

Scaffolds should possess antimicrobial properties. For example, biomaterial incorporated with antimicrobial agents like silver nanoparticles can inhibit the growth of bacteria on the material surface after implantation. Modifying the surface properties of materials can also inhibit the adhesion and growth of microbes on the scaffold surface. This feature is crucial for scaffolds used in engineering tissues like skin, dental, and bone tissue [16].

9.2.5 Stability

Under physiological conditions, the chemical and physical stability of the scaffold is a major determinant of scaffold integrity until tissue formation is complete. The stability factors related to chemical aspects include chemical composition, pH, enzymatic activity, and biodegradability. Composite materials with cross-linking agents promote stability while maintaining biocompatibility [17].

9.2.6 Microstructure

In tissue engineering, the microstructure of biomaterial scaffolds is crucial because it has a direct impact on cell behavior and tissue regeneration. Porosity, shape, and pore size are important structural characteristics that affect cell adhesion, proliferation, and differentiation. The microstructure of scaffolds should have interconnected pores larger than 100 µm for efficient vascularization and nutrient transport [18]. Apart from this, nanopatterned surfaces also offer enhanced features and functions. Therefore, the nano- and microscale topographies of scaffolds can closely mimic the structural complexity of the natural extracellular matrix (ECM) and thereby improve protein adsorption and cell attachment. Cellular interactions and physiological signaling are maintained over time [19]. Therefore, the relationship between a scaffold's microarchitecture and biodegradation kinetics is ultimately essential for successful tissue regeneration [20].

9.3 Biomaterials

Biomaterials are categorized into four types: metals, polymers, ceramics, and composites (Fig. 9.1).

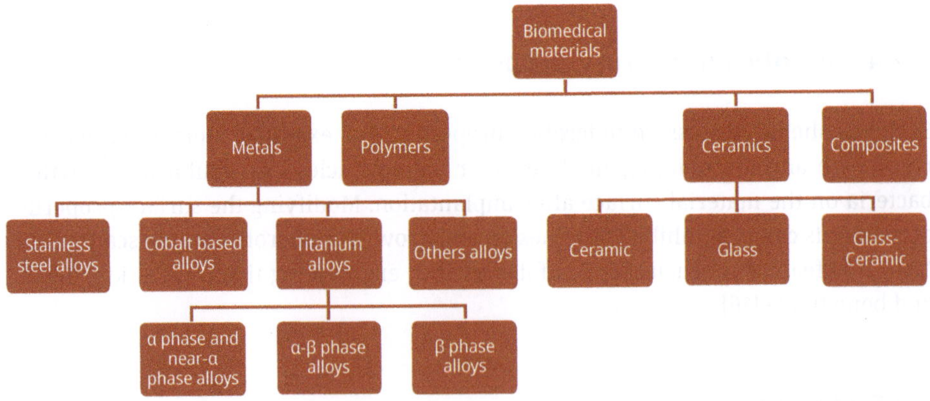

Fig. 9.1: Classification of biomaterials [21].

9.3.1 Metals

Metal-based biomaterials possess excellent thermal and electrical conductivity along with mechanical properties. These features are mainly due to the free mobile electrons of metals, which favor a fast transfer of electricity, heat, and strong atomic bonding. Furthermore, metal ions can move without breaking their structure; thus, this property enables metals to bend and deform without alteration. Metal-based biomaterials also show corrosion resistance. Therefore, these materials are commonly used in orthopedics, such as replacement of hard tissues like hips and knees, dental implants, and spinal fixation devices [22]. Apart from these, the use of metal alloys has also proved their corrosion resistance. For example, stainless steel with 18% Cr and 8% Ni and 316 stainless steel with molybdenum were developed, both showing improved corrosion resistance [23]. Then, the carbon content of 316 stainless steel was also reduced from 0.08% to 0.03% to obtain increased corrosion resistance, which introduced a new grade of stainless steel [24].

9.3.2 Inorganic biomaterials

Inorganic biomaterials include gold and silica nanoparticles, which are studied extensively in various disease treatments. For example, gold nanorods are used in photothermal therapy along with chemotherapy, which modulates the tumor microenvironment and thereby promotes cancer treatment [25]. Gold nanoparticle-coated cancer-specific T-cell receptors are used in imaging to determine T-cell distribution, migration, and kinetics [26].

Moreover, silica nanoparticles are considered another inorganic biomaterial that has been studied for anticancer properties when doped with zinc, calcium, and magnesium. These mesoporous silica nanospheres have been shown to stimulate anticancer properties by increasing immune cells (CD_4^+ and CD_8^+ T-cells) in the spleen [27]. Furthermore, silica nanoparticles are also used as carriers for cellular immunotherapy. For example, specific amino acid-coated silica nanoparticles can induce IL1 and IFN production in dendritic cells [28]. Additionally, silica nanoparticles are used for the development of vaccines against viruses and bacteria [29].

9.3.3 Ceramics

Bioceramic biomaterials are made from metallic and nonmetallic elements. These elements are held together by covalent bonds/ionic bonds or both [30]. Therefore, ceramics are stiff, hard, and brittle [31]. Furthermore, atoms or elements in ceramics cannot slip past one another. To be designated as bioceramics, they must fulfil various characteristic features, including being biocompatible, biofunctional, nontoxic, nonallergic, carcinogen-free, and noninflammatory [32]. Examples include oxides (magnesium oxide, silica or silicon dioxide, and aluminum oxides), ionic salts (NaCl, CsCl, and ZnS, which can form polycrystalline aggregates), and carbon-based materials (diamond). Ceramics and their composites are widely used in medical devices and implants because they are resistant to corrosion and have high compression strength. However, they have low tensile strength and low fracture toughness. Bioceramics are also used in dental applications because they are inert to saliva and offer remarkable strength and appearance [33]. Furthermore, the role of bioceramics in gene therapies, cancer therapies, and drug delivery has also been documented [34]. Ceramic implants are classified into three types: (i) biodegradable or resorbable ceramics, which are inert and consist of calcium phosphate and calcium aluminate; (ii) surface reactive or bioactive ceramics, which are semi-inert and consist of glass ceramics and hydroxyapatites; and (iii) nonabsorbable ceramics (relatively inert) which include alumina, zirconia, and carbons [35].

9.3.4 Composites

Biomaterials that contain two or more materials are called composite biomaterials. Composite biomaterials offer changes in characteristics such as elastic modulus, flexibility, and biocompatibility compared to the individual materials [36]. These are strong, stiff, and lightweight materials; therefore, they are widely used in dental and orthopedic applications [37].

9.3.5 Polymeric biomaterials

Naturally or artificially obtained polymers are either nonbiodegradable or biodegradable. Naturally obtained biodegradable polymers include collagen, starch, and chitin, which are widely used in biomaterials. Most synthetic polymers are nonbiodegradable and are used in dental materials, medical implants, and disposable materials. Furthermore, the physical and mechanical qualities of synthetic polymers are identical to those of human soft tissues; therefore, they are used as biomedical materials in humans. For example, the synthetic polymeric materials such as polypropylene, polymethylmethacrylate, polyethylene, polyurethane, and polyethyleneterephthalate are currently used in various medical applications [38]. Polymers can be manufactured as sheets, films, fibers, and synthetic latex. Beyond this, polymers can be processed easily. They are available at a reasonable cost and also possess desirable mechanical and physical properties [2]. The disadvantages of polymers include water and protein absorption during human use, susceptibility to surface contamination, difficulty in sterilization, release of residual compounds, susceptibility to biodegradation, structural breakdown, and wear. The use of synthetic polymers can cause environmental pollution and pose challenges for waste management [39].

9.3.6 Lipid-based biomaterials

The best example of a lipid-based biomaterial is the liposome, which is widely used and commonly employed to deliver drugs. Liposomes can carry both hydrophilic and hydrophobic drugs to specific targets. Hydrophilic drugs are loaded in the central cavity of liposomes, whereas hydrophobic drugs can be localized within the bilayer [40]. Various drugs such as chemotherapeutic drugs, antibiotics, painkillers, and hormones have been administered using liposomes [41]. Furthermore, liposomes are also used in immunotherapy and serve as nanocarriers specifically for clinical diagnosis and treatment [42].

9.4 Strategies or methods of surface modification

Surface modifications involve physicochemical alterations through the addition of external substances over the material's surface, which can alter surface atoms, molecules, topography, or introduce overcoating modifications. The surface property of solid materials governs their functional properties. The important surface property is morphology. Most surface modification techniques can modify the morphology of materials, which eventually modulate the interactions of materials with tissues, biocompatibility, immune responses, and specific functional properties [43].

Generally, the surface modifications of materials, that is, physicochemical alterations of materials are performed through chemical methods (self-assembled monolayers (SAMs) and nanostructure coatings), which control hydrophobicity and specific biomolecule interactions, and physical methods (ion implantation and laser treatment), which alters the energy of the surface and chemical composition. Furthermore, overcoating modifications of materials are performed through various methods like covalent and noncovalent coatings, thin-film deposition, and grafting [44]. The choice of surface modification techniques depends upon the nature of the materials' applications. A thick coating may negatively impact the mechanical and functional qualities of materials. Therefore, the coating of layer thickness ranging between 10 and 100 nm can preserve the mechanical and chemical performance of materials along with uniformity, durability, and stability [45]. Furthermore, the dynamic character of materials should be examined through topography, chemistry, and surface energy analyses since the exposure of materials to biological fluids [46]. Recently, biomedical studies have focused on biomaterials with 3D bioprinting bioink for the reconstruction of human tissues (Fig. 9.2). Moreover, the recent technologies like deep learning, artificial intelligence, and machine learning have made progress in the development of biomaterials [47]. For example, alginate patches (repair lungs), zinc stents and smart wound dressings (to monitor recovery condition), supramolecular biomaterials (respond to physiological alterations), and injectable biomaterials (targeted drug delivery) are considered as most advanced materials of research.

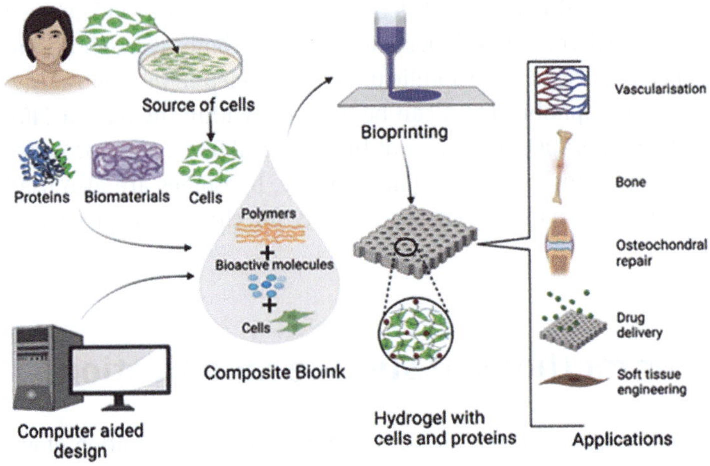

Fig. 9.2: Roles of bioprinted material generated from composite bioink [48] (reprint under CC by 4.0 permission).

9.4.1 Physical methods

9.4.1.1 Ion implantation

This method is a sophisticated physical modification technique widely used to increase resilience and wear. It is carried out at a low temperature. Since this technique combines beam-line and plasma absorption ion implantation techniques, it is considered as a hybrid technique [49]. It involves attaching a marker [silver isotope (109 Ag), identify and track the material's deterioration] on the stainless steel alloy, SS316L, using high intensity. Because of high intensity employed during the implantation, the silver ions deeply penetrated or implanted into the surface of SS316L [50]. The link between isotopic markers and material deterioration has been demonstrated through several similar experiments.

In another experiment, both Ag and Ti ions were implanted on textiles using metal vapor vacuum arc for studying their antibacterial properties and durability. The ions implanted in medical fabrics were washed 30 times and tested for antibacterial activity after each wash to determine the extent of ion adherence to the fibers. Additionally, the rust resistance ability of Ag and Ti ions implanted in polyester fabrics was also studied and revealed that TiO_2 ion-implanted polyester cloth maintained over 85% antimicrobial property [51]. This implies that it has a strong adhesion of antibacterial ions to the fabrics and their long-lasting antimicrobial action. Apart from these, the fabrics like polyester, polyolefin, synthetic polyamide, natural cellulose fiber, and polymer elastomer are also subjected to ion implantation and showed excellent properties including enhanced antimicrobial properties, hydrophobicity, biocompatibility, mechanical strength, wound healing, and durability with reduced friction [51].

9.4.1.2 Laser surface modification

This method preserves the original properties of the material by modifying only the surface of the metal. Laser-exposed surfaces showed higher hardness compared to the bulk material and showed lower friction coefficients. The laser treatment-associated risk, like surface cracking, can be eliminated by preheating the material using a supplementary heat source or by using a defocused laser beam. This process is widely used in industries because it reduces the laser beam effect on the materials. It involves applying of laser with high power density on the material's surface. The surface temperature of the materials rises, resulting in absorption of high-energy beams and local melting of material [52]. The heat from the surface is transferred into the bulk of the material. The quick solidification of the melted area can develop finer grains in the changed zone. Using this method can create a hard ultrafine structure on the material surface without changing its chemical composition. This procedure also offers properties like reduced erosion and wear resistance [53]; for example, the process of laser surface texturing of zirconia-resin cement bonding using a surface transition machine (Fig. 9.3). Moreover, pure Ti [54], alloys such as Ti-6Al-4V [55], AISI 316L stainless steel [56], and AISI 304L stainless steel [57] were also subjected to LSM treatment, which showed various biomedical applications including improved osseointegration, increased wettability with cell response, hardness, and a shift in corrosion potential, respectively.

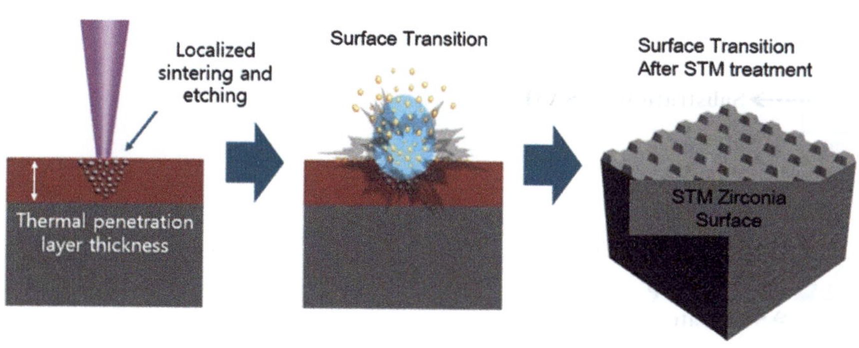

Fig. 9.3: The process of laser surface texturing of zirconia-resin cement bonding using a surface transition machine [58] (reprint under CC by 4.0 permission).

9.4.2 Chemical methods

9.4.2.1 Self-assembled monolayer (SAM)

This method was first introduced in 1945. SAMs are molecular assemblies used in biomaterials for surface modifications and functional enhancement [59]. This process involves the gathering of unidirectional, ultra-thin molecular layers on a material surface through various interactions including covalent and noncovalent bonding, electrostatic, hydrophobic, and van der Waals interactions. Generally, SAMs are prepared on gold or hydroxy-terminated substrates. Initially, the gold surface is cleaned using inorganic acids (HCl and H_2SO_4). Subsequently, the surface-cleaned gold is placed in a solution-containing SAM precursor, which facilitates the self-assembly process through the above-listed interactions. SAMs formed using this method are chemi-

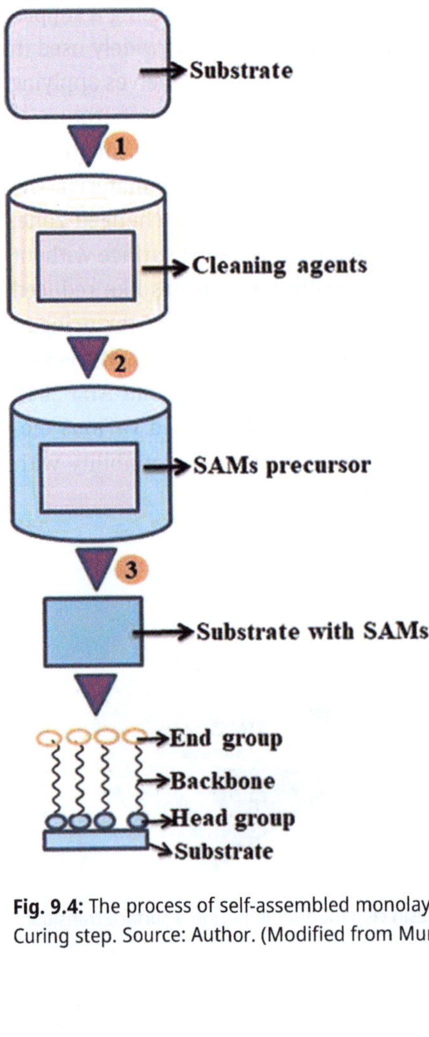

Fig. 9.4: The process of self-assembled monolayer synthesis. 1- Cleaning step, 2- Self-assembly step, 3- Curing step. Source: Author. (Modified from Murugan et al., [61].

cally and thermally stable [60]. Additionally, this procedure can also generate a closely packed and stable monolayer structure (Fig. 9.4). A wide variety of SAMs with different functional groups have been synthesized for different applications. The formation of SAMs can be confirmed using various analytical techniques such as scanning tunnel microscope, atomic force microscope, and Auger electron spectroscopy.

The biocompatibility of polymer and metal surfaces can be improved through the process of surface coatings and films. It involves the introduction of secondary chemical functionalities or the addition of layers on the surface of materials. The chemical methods such as Langmuir-Blodgett (LB) films, spin and dip coating, layer-by-layer (LbL) assembly, chemical vapor deposition (CVD), and physical vapor deposition are widely used for surface coating of materials and fabricating thin films in recent days. Here, the thickness of the resultant layer is regulated through factors such as gravity, surface tension, and viscosity [62].

9.4.2.2 Spin coating

Spin coating involves applying a thin film onto the surface of the rotating substrate. Primarily, it is used for the fabrication of semiconductors.

9.4.2.3 Dip coating

It is a simple method. It involves simple immersion and pulling out/withdrawal of material from a coating solution for coating a thin film on the material surface. However, it is difficult to generate uniformly coated films through the process of dip coating or dip casting. Nevertheless, these methods are considered for the effective modification of biomaterial surfaces. For example, the integration of synthetic peptide onto the surface of orthopedic implants has been successfully demonstrated through these methods [63]. Furthermore, these techniques can also be carried out after the surface activation of materials through methods like plasma treatment.

9.4.2.4 Hybrid multifunctional coating

Hybrid multifunctional coating involves coating or applying multifunctional materials onto the surface of biomaterials like stents or implants in the form of a single layer. This material coating can fulfill multiple needs; for example, it can perform various biological and mechanical functions simultaneously. Stents are commonly used as a drug delivery system to restore blood vessels and reduce the need for oral anticoagulants. Drug delivery through stents can be achieved either by directly attaching the drug to the stent surface or by incorporating the drug into the porous structure of the stent. Pores on the stent surface can be created using techniques such as soft lithography, photolithography, and prolonged plasma etching. For example, drug-infused porous stents serve as advanced and effective drug delivery systems. Additionally, acid-treated stainless steel stents develop a porous surface that enables drug release at the site of implantation [64].

9.4.2.5 Langmuir-Blodgett (LB)

It involves the transfer of a thin LB film, generated from amphiphiles, onto a solid surface. It is a unique technique used to functionalize polymeric biomaterial surfaces. LB film is a densely packed and monolayer arrangement of both hydrophilic heads and hydrophobic heads able to deposit and cross-link onto the polymeric material's surface. This technique helps to create a deposited thin protein film on substrates via nonspecific and noncovalent adhesion. Generally, this method involves dipping and removing of substrate/material from the intended coating solution. As a result, the desired molecules are deposited as a thin film on the surface materials at the interface between the liquid and air. For LB film deposition, the chemical treatment of polymeric materials is essential to attain effective deposition. This process is regulated through various intricate forces such as electrostatic, intermolecular, and capillary forces [65]. Several studies have shown that LB films enhance cell attachment and improve hemocompatibility while reducing platelet adhesion [66].

9.4.2.6 Layer-by-layer (LbL)

It is a simple and versatile method that can be used to modify a wide range of surfaces, shapes, and biomaterials, including metals and polymers. This method helps in the building of self-assembled nanostructured films ranging from simple 2D multilayers to complex 3D porous structures. This process is mainly governed by electrostatic interactions between the components and the material surface. However, other forces such as hydrogen bonding and hydrophobic interactions may also be involved. For example, polyanions can be adsorbed onto a positively charged material surface, forming a layer with a thickness ranging from a few nanometers to several micrometers. By repeating this process with alternating positive and negative layers, the multilayered film with the desired dimensions can be attained [65]. For example, chitosan is a positively charged polymer used in this method to kill bacteria through its bacterial membrane binding and damaging actions [67]. Moreover, the stable and effective multilayer surface coatings have been developed using chitosan in combination with negatively charged polymers such as hyaluronic acid or heparin [68]. Additionally, the compounds like 3-aminopropyltriethoxysilane or dopamine are commonly used as a linker to anchor chitosan onto the surface of materials [69].

9.4.2.7 Chemical vapor deposition (CVD)

It involves basic principles of chemical reactions between the gaseous phase reagents and the surface of the material. As a result, a nonvolatile coating is deposited onto the surface of the material or substrate. Currently, various modified CVD techniques are employed including plasma-enhanced CVD (PECVD), plasma-assisted CVD (PACVD), low-pressure CVD (LPCVD), atmospheric pressure CVD (APCVD), and laser-enhanced CVD (LECVD). Among these, the PECVD has not been developed further due to the plasma-induced activation of precursors and their compatibility with gas-phase transport [71]. These modified methods of CVD offer numerous advantages, including uniform coating, boosted levels of biocompatibility, chemical stability, wear, hardness, and corrosion resistance [72]. The chemical and physical properties of the coating are determined by precursor chemistry. Furthermore, other factors such as gas type, processing time, and plasma power also affect the features of the coating.

9.4.2.8 Plasma techniques

These are a group of surface modification methods that involve plasma treatment (i.e., partially ionized gas consisting of ions, neutral particles, and electrons) to modify the surface of a material without altering its bulk properties. Plasma treatment involves several steps, starting with keeping or placing the material in a vacuum chamber filled with a specific gas, either oxygen or nitrogen. Then, a high-frequency voltage is applied to generate plasma or reactive oxygen species (ions and electrons). Further interaction of these particles on the material's surface, causing modifications like etching, cleaning, or functional groups addition. Plasma methods are highly efficient in altering the physicochemical and biological properties of materials. Particularly, the plasma treatment of polymeric materials has shown desirable outcomes including augmented cell adhesion and proliferation. These effects are mainly attributed to the increased hydrophilicity and wettability of the surface-modified materials. Furthermore, plasma techniques can generate novel surfaces on metallic or polymeric materials that can fulfill various expected properties including biocompatibility, corrosion resistance, and tailored drug delivery [73]. Moreover, the antibacterial properties of metal and polymer implants can be improved through other methods such as grafting, ion implantation, and sol-gel coating. However, the plasma-based methods are useful because they are simple, efficient, and cost-effective. All these methods can improve the function and biocompatibility of materials without affecting their stability and internal strength.

Recently, various methods have been developed to prevent biocorrosion (i.e., metal damage caused by microorganisms). These involve attaching or applying protective or antimicrobial polymer coatings onto metal surfaces using traditional techniques such as LbL sol-gel methods, covalent bonding, or free-radical graft polymerization [74]. These methods are also used to introduce linkers on renewable composite materials for further attachment of functional biomolecules. A bioinspired coating has been developed, where biomimetic linkers are grafted onto a biocompatible polymer, enabling the attachment of natural bioagents. For a specific biomedical application, metal surfaces have been functionalized using catecholic anchors for the simultaneous integration of antifouling hydrophilic brushes and antibacterial agents (enzymes, peptides, or lysozyme) [75].

The attachment of bioactive compounds to polymeric surfaces primarily occurs through three major mechanisms: covalent bonding, electrostatic adsorption, and ligand-receptor interactions. Among these, the principle of electrostatic (noncovalent) adsorption is widely used in applications such as the avidin-biotin system, antimicrobial textiles, and drug delivery. On the other hand, covalent bonding can establish a stronger and stable attachment or bond between the polymer surface and the bioactive compound. This method also helps preserve the activity, stability, and functional lifespan of biomolecules. Therefore, it is commonly used in medical devices such as vascular implants and catheters [76]. Additionally, it also helps prevent accidental contamination in active food packaging [77]. Moreover, the surface-immobilized antimicrobial peptides (indolicidin, β-defensins, magainins, and cecropin A) obtained through the process of either adsorption or LbL coating or covalent bonding can destroy microbes and serve as potent antimicrobial coatings for medical devices [78].

9.4.2.9 Other chemical methods for biomaterial surface modification

Several approaches have been followed for the surface modification of biomaterials, which include grafting techniques, sol-gel treatment, and additional techniques [70].

9.5 Applications

The application of biomaterials is constantly increasing in regenerative medicine.

9.5.1 Surface-modified biomaterials as drug delivery systems

Biomaterials are widely used to design drug delivery systems. Here, therapeutic agents are encapsulated either through the process of encapsulation, attachment, or entrapment with polymers/biomaterials/matrices to achieve controlled release. This offers a higher efficiency of drugs with minimal side effects, and the matrix protects the drugs from degradation, enhances bioavailability, and facilitates prolonged release in chronic conditions. For example, biodegradable polymer nanoparticles can be loaded with the drugs and administered intravenously for sustained release at the site of action. Generally, the biomaterials like liposomes, polymers, hydrogels, microparticles, and nanoparticles are used to encapsulate and deliver drugs [79].

The 3D hybrid scaffold of collagen-grafted-chitosan-glucan fiber was prepared using a freeze-drying method. The physicochemical properties (percentage of swelling, hydrolytic stability, and elasticity) and antibacterial properties of the 3D hybrid bioscaffold were enhanced through the chemical modification of collagen and chitosan-glucan fiber [80]. A "grafting-to" approach was performed for the surface polymer chain modification of liposomes [81]. Surface modification of biomaterials enables tailoring of physicochemical properties, structural characteristics, interaction attributes, etc. (Tab. 9.1) [82].

Tab. 9.1: The role of surface-modified materials in drug delivery.

Biomaterial	Surface modification	Applications	References
Mesoporous silica nanoparticles	PEGylation	Suitable carriers for drug delivery; improve the solubility of hydrophobic drugs	[83]
Biopolymer-liposome hybrid system	Surface-modified liposome	Used for controlled delivery of bioactive compounds	[84]
Hydroxyapatite nanoparticles and chitosan/hydroxyapatite composites	Combining nano-HAp with chitosan matrix	Orthopedic, osteoconductive, dental, and drug delivery applications	[85]
Chitosan and its derivatives	PEGylation	Improves the stability and solubility; used for drug and gene delivery	[86]
PLGA nanoparticles	Ligand conjugation	Used as a drug delivery system to target a specific cell or tissue, or organ	[87]
Gold nanoparticles	Surface functionalization using thiol-terminated polyethylene glycol	Enhances the stability in various media and serves drug delivery system	[88]
Ligand bases the dendritic system	Surface functionalization using multiple copies of ligands	Site-specific delivery of anticancer drugs	[89]
Carbon nanotubes	Functionalized by bioactive peptides, proteins, nucleic acids, and drugs	Used to deliver the cargo to cells and organs	[90]

9.5.2 Surface-modified biomaterials in tissue engineering

Biocompatible materials like ceramics, hydrogels, and polymers are used to make scaffolds to support tissue regeneration and cell growth. Scaffold acts as a structural framework for cells to attach, proliferate, and differentiate, leading to the formation of fully functional, replaced tissues and bones [91]. The roles of surface-modified biomaterials in tissue engineering are described in Tab. 9.2.

Tab. 9.2: Functions of surface-modified biomaterials in tissue engineering.

Biomaterial	Surface modification	Applications	References
Polymeric biomaterials	Surface functionalization of hydrophilic functional groups on polymeric biomaterials through the process of plasma treatment.	Promotes cell proliferation, anchorage, and spreading. Therefore, these kinds of biomaterials are useful in the development of tissue engineering scaffolds, vascular prostheses, and implantable devices.	[92]
3D-printed gelatin methacryloyl scaffolds	UV photo-cross-linking-mediated surface functionalization of scaffolds using strontium-doped nanosized hydroxyapatite.	Shows a strong potential in bone tissue regeneration or bone repair.	[93]
3D-printed porous polyetherketoneketone implants	Single-step immobilization of hydroxyapatite (HAp) particles on the surface of implants.	Shows increased calcium and phosphorus contents, better stem cell adhesion, and promotes osteogenic differentiation.	[94]
Mesoporous bioactive glasses	Zwitterionic zero charge surface functionalization.	Serve as advanced biomaterials in bone tissue regeneration.	[95]
Ovine forestomach matrix	Ionic silver functionalization of ovine forestomach matrix.	Prevents biofilm formation, no cytotoxicity over mammalian fibroblasts, and antimicrobial effects. Retains the structure and function of the matrix, and hence useful in tissue repair.	[96]

Tab. 9.3: Functions of surface-modified biomaterials in biomedical implants.

Biomaterial	Surface modification	Applications	References
Polymeric biomaterials	Plasma treatment-mediated surface modification introduced hydrophilic functionalities on the biomaterial surface.	Useful in the development of implantable devices, scaffolds, and vascular prostheses.	[92]
Titanium	Silver nanocoating on the titanium surface is performed via direct current (DC) sputtering.	Serves as a prophylactic surface against periodontal pathogens, depicted excellent cytocompatibility and osseointegration.	[98]
Titanium-based implants	Surface functionalization of titanium implants is performed through various methods such as acid etching, chemical vapor deposition, sol-gel method, layer-by-layer self-assembly, electrochemical techniques, and chemical grafting.	Display enhanced antibacterial activity and osteointegration.	[99]
Ti-6Al-4V bioalloy	Surface topography, chemical composition, and wettability of alloys can be modified using a picosecond laser. Specifically, two hybrid structures containing groove patterns with periodic and nonperiodic substructures were generated on the bio alloy using a picosecond laser.	Picosecond-induced two-hybrid groove structures showed increased hydroxyapatite growth, protein adsorption, and cell adhesion. Together, contribute to a higher cell density, implant longevity, and integration in the body.	[100]
316 LVM stainless steel	Strategy plasma spraying is performed for hydroxyapatite coating and hydroxyapatite-magnesium double-layer coatings.	Sequences of studies revealed that Mg degradation generates surface porosities that enable bone growth, toxic metal release (nickel), and increased biocompatibility and bioactivity of stainless steel substrate.	[101]
Ti6Al4V alloy	Chemical composition, surface topography, and wettability of alloys can be modified. Alloy surface is modified by using a femtosecond laser.	Showed superior antibacterial activity. Therefore, it is considered as an infection-free dental and orthopedic implant.	[102]
Titanium discs	Human bone morphogenetic protein 7 adsorbed onto the surface of poly (ethyl acrylate) coated titanium discs.	Increases mineralization, proliferation, cell adhesion, and osteogenic markers expression, and thereby promotes osseointegration.	[103]
45S5 Bioglass	Surface modification using a femtosecond laser.	Enhances biointegration potential by increasing surface area and leading to faster hydroxyapatite growth.	[104]

9.5.3 Surface-modified biomaterials in biomedical implants

Implants such as artificial joints, pacemakers, and stents should be biocompatible to ensure proper interaction with the tissues and do not trigger an immune response [97]. Titanium alloy and polymers are major substrates/matrices for developing biocompatible implants. The functions of various surface-modified biomaterials in implants are discussed in Tab. 9.3.

9.5.4 Surface-modified biomaterials in imaging

Biomedical imaging is an important tool for the identification and treatment of various diseases including cancer. Recently, imaging techniques have also been used to characterize biomaterials and their interactions with biological systems. Biomaterials are used as contrast agents in different imaging techniques like ultrasound and magnetic resonance imaging. These materials can enhance contrast and thereby facilitate the visualization of specific structures or tissues. Consequently, significant advancements have been made in engineering or modulating material properties for use in various imaging applications. For example, lignin synthesized with carbon dots or metal ions exhibited fluorescence ranging from 300 to 800 nm. Therefore, it is considered for various near-infrared applications including biomedical imaging [105]. Furthermore, the nano biomaterials obtained from biomass have shown protein binding, antibiofilm activity, and potential as biological probes in bioimaging studies [106]. Several surface-modified biomaterials (liposomes, hydroxyapatite-based scaffold, polymeric micelles, silica nanoparticles, gold nanoparticles, quantum dots, and iron oxide nanoparticles) are used in various imaging techniques as contrast agents, carriers of radiotracers, and fluorescent probes to enhance molecular visualization [107–110].

9.5.5 Role of surface-modified biomaterials in vaccine development

Viral infections pose a major threat to human health. Biomaterials play a crucial role in the development of vaccines, particularly mRNA vaccines. Although effective vaccines exist for diseases like measles, there are still many diseases that require targeted and more efficient vaccines. Biomaterials have proven to be excellent vaccine carriers and adjuvants due to their immunomodulatory properties, which enhance the immunogenicity of vaccines [111]. Biomaterials can be engineered into different structures such as nanogels, micelles, microneedles, and hydrogels for use in vaccination. Among these, lipid nanoparticles are effective as carriers, protect vaccine components, and facilitate delivery, thereby triggering a robust immune response [112]. Furthermore, biomaterials facilitate the controlled release of immunogens, stabilize antigens, and serve as adjuvants to enhance immune responses. For example, the biomaterials such as lipid nanoparticles, polymeric nanoparticles, microneedles, and hydrogels have been utilized in vaccine development for various diseases like malaria, hepatitis B, cancer, influenza, acquired immunodeficiency syndrome, tuberculosis, and coronavirus disease 2019 (COVID-19) [113].

Numerous polymeric particles (PEG, PLGA, dextran, PCL, self-assembled peptides, and chitosan) have been developed as carriers for vaccines [114]. Moreover, liposomal vaccines composed of the MUC1 peptide, adjuvant (monophosphoryl lipid A), and lipids [dimyristoyl phosphatidylglycerol and dipalmitoylphosphatidylcholine] were shown to increase survival in patients with advanced unresectable non-small cell lung cancer [115]. Similarly, a lipid-based mRNA vaccine induced the production of interferon-α and elicited robust effector and memory T-cell responses by targeting dendritic cells [116]. Liposome-based mRNA vaccines have also been developed for treating COVID-19 infection. Liposomes preserve mRNA from degradation by nucleases and facilitate its entry into cells via endocytosis [117]. Furthermore, protein antigens conjugated with gold nanoparticles can induce human T-cell proliferation. Gold nanoparticles also serve as immunological vaccine adjuvants. It has been observed that the surface-engineered gold nanorods effectively induce both cellular and humoral immunity during HIV treatment [118]. An injectable polymer-based receptor binding domain (RBD) vaccine formulated with an adjuvant has been developed against SARS-CoV-2 infection. This system enables the sustained release of both the vaccine and adjuvant and thereby enhances the recruitment of antigen-presenting cells, stimulating adaptive immunity and boosting the production of anti-RBD IgG antibodies [119].

The major house dust mite allergen (Der f2)-loaded hollow mesoporous silica nanoparticles prevented allergic inflammation and served as a vehicle for allergen-specific immunotherapy [120]. Similarly, dendritic mesoporous silica nanoparticles (DMSNs) were developed for delivering peptide-based subunit vaccines (B2T vaccine against foot and mouth disease virus), and it was observed that B2T-DMSNs elicited a robust immune response without the need for any adjuvants [121].

9.5.6 Advances in biomaterial applications

i) *Molecular dynamics in biomaterials characterization*
The mechanical properties of hexagonal hydroxyapatite have been determined by molecular dynamics simulations. Specifically, these simulations highlight the role of pore flaws in maintaining structural integrity [122].
ii) *Hybrid materials in cancer treatment*

Novel hybrid materials like fluorescent mesoporous silica-based compounds are synthesized and characterized. These fabricated hybrid materials show exceptional anticancer properties and antioxidant activities, suggesting new possibilities in cancer treatment [123].

iii) *Light-controlled prodrug activation*
This field integrates material sciences and therapeutics. Light-controlled prodrug activation methods have been developed to achieve targeted drug release and minimize off-target side effects. Example: External stimuli like light mediated activation and targeted release of platinum IV prodrugs [124].

iv) *Nano platforms in photodynamic therapy*
Mesoporous silica acts as a shell to boost reactive oxygen species production. This property enables effective photodynamic therapy and improved therapeutic performance. Example: $LiYF_4$ nanoplatforms [125].

v) *3D printing*
The link between biomaterials and regenerative medicine has been explored through 3D printing. This holds significant promise for personalized medicine [126].

vi) *Advances in tissue engineering and drug delivery*
Hydrogels are useful in tissue engineering and regenerative medicine due to their resemblance to the properties of natural ECM, like adjustable mechanical strength, water retention capacity, biocompatibility, and encapsulating ability [127]. ECM-derived and decellularized hydrogels obtained from pig urinary bladder and small intestine submucosa provide a biomimetic environment that supports cell attachment, growth, and differentiation. Therefore, these hydrogels are tailored for particular regenerative purposes, including cartilage regeneration using synthetic melanin nanoparticles and kartogenin encapsulated in hyaluronic acid [128].

Furthermore, biomaterials are studied alone or in combination with ECM components, cells, growth factors, neurotrophic factors, or angiogenic factors for potential therapeutic applications in various neurological conditions like stroke, Alzheimer's disease, Parkinson's disease, spinal cord injury, and traumatic brain injury [129].

In statin therapy, polymer nanocarriers are used to deliver drugs directly to the target site to improve drug efficiency without side effects. This reflects an expanded direction toward material-based and personalized therapies [130].

Recently, nanotechnology has played a major role in drug delivery. These systems use targeting molecules like antibodies and peptides to direct drugs to specific cells and thereby reduce side effects and improve drug efficiency. Furthermore, pH and temperature-sensitive carriers are used to release drugs at specific target sites with increased precision and safety. Biodegradable polymers are also employed to address patient-specific issues with minimal side effects. Moreover, poly(2-hydroxyethyl methacrylate) hydrogels have been synthesized and used in intraocular lenses and controlled drug release systems [131, 132].

9.6 Conclusion

Biomaterials have received considerable attention in modern science. Four types of biomaterials, such as metals, polymers, ceramics, and composites, are used for various purposes. The integration of biomaterials into biological systems can serve as a tool for therapeutic intervention, prosthetic replacement, regenerative therapies (i.e., biomaterials can repair, replace, and regenerate damaged tissues and organs), and drug delivery. Specifically, biomaterials must fulfill various desirable attributes such as biotolerance, biodegradability, mechanical properties, antimicrobial properties, microstructure, and stability for biological applications. Additionally, biomaterials must interact with the biological environment and should not cause any undesirable immune responses or toxicity. To minimize side effects associated with biomaterials and to achieve desired biological responses, various surface modification procedures have been developed.

The surface modification of biomaterials is performed through physical and chemical methods. Procedures, like ion implantation and laser treatment, are considered physical methods. Whereas strategies such as SAMs, grafting, Langmuir-Blodgett films, LbL assembly, CVD, and plasma techniques fall under chemical methods. These methods introduce unique functionalities or fine structure onto the surface of biomaterials, thereby improving properties such as biocompatibility, corrosion resistance, mechanical integrity, and antimicrobial activity.

The surface-modified biomaterials have been successfully applied in numerous biomedical fields including targeted drug delivery, tissue engineering, biomedical implants, imaging, vaccine development, and other applications. In drug delivery systems, surface-modified biomaterials facilitate the controlled and sustained release of therapeutic agents, thereby improving treatment efficiency. Similarly, the different biomaterials subjected to surface modifications have played a significant role in the successful regeneration of tissues, bones, cartilage, and skin through enhanced cell adhesion, proliferation, differentiation, and vascularization. Furthermore, surface-modified biomaterials are used in imaging and diagnostics for disease detection and real-time monitoring. In vaccine development, surface-modified materials serve as adjuvants, carriers of immunogens, stabilizers, and immunogenic boosters.

Moreover, recent technologies, including nanotechnology, 3D bioprinting, and molecular simulations, are enabling the development of personalized, smart, and multifunctional biomaterials. The convergence of biomaterial science and surface engineering is driving a new era of advanced medical solutions and personalized healthcare.

References

[1] Marin E, Boschetto F, Pezzotti G. (2020). Biomaterials and biocompatibility: An historical overview. *Journal of Biomedical Materials Research Part A*, 108(8), 1617–1633.

[2] Parida P, Mishra SC. (2012). UGC sponsored national workshop on innovative experiments in physics. Biomater Med, 9(10), 110–134.

[3] Helmus MN, Gibbons DF, Cebon D. (2008). Biocompatibility: Meeting a key functional requirement of next-generation medical devices. *Toxicologic Pathology*, 36(1), 70–80.

[4] Raghavendra GM, Varaprasad K, Jayaramudu T. (2015). Biomaterials: Design, development and biomedical applications. In *Nanotechnology Applications for Tissue Engineering* (pp. 21–44). William Andrew Publishing.

[5] Hudecki A, Kiryczyński G, Łos MJ. (2019). Biomaterials, definition, overview. In *Stem Cells and Biomaterials for Regenerative Medicine* (pp. 85–98). Academic Press.

[6] Wu G, Li P, Feng H, Zhang X, Chu PK. (2015). Engineering and functionalization of biomaterials via surface modification. *Journal of Materials Chemistry B*, 3(10), 2024–2042.

[7] Patel NR, Gohil PP. (2012). A review on biomaterials: Scope, applications & human anatomy significance. *International Journal of Emerging Technology and Advanced Engineering*, 2(4), 91–101.

[8] Huebsch N, Mooney DJ. (2009). Inspiration and application in the evolution of biomaterials. *Nature*, 462(7272), 426–432.

[9] Liao Y, Pourzal R, Wimmer MA, Jacobs JJ, Fischer A, Marks LD. (2011). Graphitic tribological layers in metal-on-metal hip replacements. *Science*, 334(6063), 1687–1690.

[10] Parveen S, Misra R, Sahoo SK. (2017). Nanoparticles: A boon to drug delivery, therapeutics, diagnostics and imaging. In *Nanomedicine in Cancer* (pp. 47–98).

[11] Nel, A. E., Mädler, L., Velegol, D., Xia, T., Hoek, E. M., Somasundaran, P., ... & Thompson, M. (2009). Understanding biophysicochemical interactions at the nano -bio interface. *Nature materials*, 8(7), 543–557.

[12] Ratner BD, Bryant SJ. (2004). Biomaterials: Where we have been and where we are going. *Annual Review of Biomedical Engineering*, 6(1), 41–75.

[13] Paltanea G, Manescu V, Antoniac I, Antoniac A, Nemoianu IV, Robu A, Dura H. (2023). A review of biomimetic and biodegradable magnetic scaffolds for bone tissue engineering and oncology. *International Journal of Molecular Sciences*, 24(5), 4312.

[14] Du B, Yin H, Chen Y, Lin W, Wang Y, Zhao D, Wang G, He X, Li J, Li Z, Luo F. (2020). A waterborne polyurethane 3D scaffold containing PLGA with a controllable degradation rate and an anti-inflammatory effect for potential applications in neural tissue repair. *Journal of Materials Chemistry B*, 8(20), 4434–4446.

[15] Heidari F, Razavi M, Bahrololoom ME, Bazargan-Lari R, Vashaee D, Kotturi H, Tayebi L, Helmus MN, Gibbons DF, Cebon D. (2008). Biocompatibility: Meeting a key functional requirement of next-generation medical devices. *Toxicologic Pathology*, 36(1), 70–80.

[16] Patil S, Singh N. (2019). Antibacterial silk fibroin scaffolds with green synthesized silver nanoparticles for osteoblast proliferation and human mesenchymal stem cell differentiation. *Colloids and Surfaces B: Bio Interfaces*, 176, 150–155.

[17] Rezwan K, Chen QZ, Blaker JJ, Boccaccini AR. (2006). Biodegradable and bioactive porous polymer/inorganic composite scaffolds for bone tissue engineering. *Biomaterials*, 27(18), 3413–3431.

[18] Chen X, Fan H, Deng X, Wu L, Yi T, Gu L, Zhou C, Fan Y, Zhang X. (2018). Scaffold structural microenvironmental cues to guide tissue regeneration in bone tissue applications. *Nanomaterials*, 8(11), 960.

[19] Resende RR, Fonseca EA, Tonelli FM, Sousa BR, Santos AK, Gomes KN, Guatimosim S, Kihara AH, Ladeira LO. (2014). Scale/topography of substrates surface resembling extracellular matrix for tissue engineering. *Journal of Biomedical Nanotechnology*, 10(7), 1157–1193.

[20] Chen Y, Zhou S, Li Q. (2011). Microstructure design of biodegradable scaffold and its effect on tissue regeneration. *Biomaterials*, 32(22), 5003–5014.

[21] Omarov S, Nauryz N, Talamona D, Perveen A. (Dec 28 2022). Surface modification techniques for metallic biomedical alloys: A concise review. *Metals*, 13(1), 82.

[22] Park JB, Bronzino JD. (2002). *Biomaterials: Principles and Applications*. crc press.

[23] Ibrahim MZ, Sarhan AA, Yusuf F, Hamdi M. (2017). Biomedical materials and techniques to improve the tribological, mechanical and biomedical properties of orthopedic implants– A review article. *Journal of Alloys and Compounds*, 714, 636–667.

[24] Eliaz N. (2019). Corrosion of metallic biomaterials: A review. *Materials*, 12(3), 407.

[25] Li D, Zhang M, Xu F, Chen Y, Chen B, Chang Y, Zhong H, Jin H, Huang Y. (2018). Biomimetic albumin-modified gold nanorods for photothermo-chemotherapy and macrophage polarization modulation. *Acta Pharmaceutica Sinica B*, 8(1), 74–84.

[26] Meir R, Shamalov K, Betzer O, Motiei M, Horovitz-Fried M, Yehuda R, Popovtzer A, Popovtzer R, Cohen CJ. (2015). Nanomedicine for cancer immunotherapy: Tracking cancer-specific T-cells in vivo with gold nanoparticles and CT imaging. *ACS Nano*, 9(6), 6363–6372.

[27] Wang X, Li X, Ito A, Sogo Y, Watanabe Y, Tsuji NM, Ohno T. (2017). Biodegradable metal ion-doped mesoporous silica nanospheres stimulate anticancer Th1 immune response in vivo. *ACS Applied Materials & Interfaces*, 9(50), 43538–43544.

[28] Kakizawa Y, Lee JS, Bell B, Fahmy TM. (2017). Precise manipulation of biophysical particle parameters enables control of proinflammatory cytokine production in presence of TLR 3 and 4 ligands. *Acta Biomaterialia*, 57, 136–145.

[29] Mechler-Dreibi ML, Almeida HM, Sonalio K, Martines MA, Petri FA, Zambotti BB, Ferreira MM, Storino GY, Martins TS, Montassier HJ, Sant'Anna OA. (2021). Oral vaccination of piglets against *Mycoplasma hyopneumoniae* using silica SBA-15 as an adjuvant effectively reduced consolidation lung lesions at slaughter. *Scientific Reports*, 11(1), 22377.

[30] Subedi MM. (2013). Ceramics and its importance. *Himalayan Physics*, 4, 80–82.

[31] Jodati H, Yılmaz B, Evis Z. (2020). A review of bioceramic porous scaffolds for hard tissue applications: Effects of structural features. *Ceramics International*, 46(10), 15725–15739.

[32] Shekhawat D, Singh A, Banerjee MK, Singh T, Patnaik A. (2021). Bioceramic composites for orthopaedic applications: A comprehensive review of mechanical, biological, and microstructural properties. *Ceramics International*, 47(3), 3013–3030.

[33] Sharanraj V, Ramesha CM, Kavya K, Kumar V, Sadashiva M, Chandan BR, Naveen Kumar MZ. (2021). As a biocompatible biomaterial used in dental implants. *Advances in Applied Ceramics*, 120(2), 63–68.

[34] Ghiasi B, Sefidbakht Y, Mozaffari-Jovin S, Gharehcheloo B, Mehrarya M, Khodadadi A, Rezaei M, Ranaei Siadat SO, Uskoković V. (2020). Hydroxyapatite as a biomaterial–a gift that keeps on giving. *Drug Development and Industrial Pharmacy*, 46(7), 1035–1062.

[35] Kumar P, Dehiya BS, Sindhu A. (2018). Bioceramics for hard tissue engineering applications: A review. *International Journal of Applied Engineering Research*, 13(5), 2744–2752.

[36] Kaur G, Kumar V, Baino F, Mauro JC, Pickrell G, Evans I, Bretcanu O. (2019). Mechanical properties of bioactive glasses, ceramics, glass-ceramics and composites: State-of-the- art review and future challenges. *Materials Science and Engineering: C*, 104, 109895.

[37] Park GE, Webster TJ. (2005). A review of nanotechnology for the development of better orthopedic implants. *Journal of Biomedical Nanotechnology*, 1(1), 18–29.

[38] Shirvan AR, Nouri A, Wen C. (2021). Structural polymer biomaterials. In *Structural Biomaterials* (pp. 395–439). Woodhead Publishing.

[39] Zhang S, Wang J, Yan P, Hao X, Xu B, Wang W, Aurangzeib M. (2021). Non-biodegradable microplastics in soils: A brief review and challenge. *Journal of Hazardous Materials*, 409, 124525.

[40] Xing H, Hwang K, Lu Y. (2016). Recent developments of liposomes as nanocarriers for theranostic applications. *Theranostics*, 6(9), 1336.

[41] Allen TM, Cullis PR. (2013). Liposomal drug delivery systems: From concept to clinical applications. *Advanced Drug Delivery Reviews*, 65(1), 36–48.
[42] Watanabe S, Yuba E, Akazawa T, Wijewardana V, Kakihara Y, Azuma A, Hagimori K, Kanegi R, Hatoya S, Inoue N, Inaba T. (2022). Potent adjuvant effect elicited for tumor immunotherapy by a liposome conjugated pH-sensitive polymer and dendritic cell-targeting Toll-like-receptor ligand. *Vaccine*, 40(10), 1448–1457.
[43] Asadollahi S, Profili J, Farzaneh M, Stafford L. (2019). Development of organosilicon-based superhydrophobic coatings through atmospheric pressure plasma polymerization of HMDSO in nitrogen plasma. *Materials*, 12(2), 219.
[44] Anderson JM. (2001). Biological responses to materials. *Annual Review of Materials Research*, 31(1), 81–110.
[45] Anderson JM, Rodriguez A, Chang DT. (2008). Foreign body reaction to biomaterials. In *Seminars in Immunology* (Vol. 20, No. 2, pp. 86–100). Academic Press.
[46] Agarwal R, García AJ. (2019). Surface modification of biomaterials. In *Principles of Regenerative Medicine* (pp. 651–660). Academic Press.
[47] Gokcekuyu Y, Ekinci F, Guzel MS, Acici K, Aydin S, Asuroglu T. (2024). Artificial intelligence in biomaterials: A comprehensive review. *Applied Sciences*, 14(15), 6590.
[48] Troy E, Tilbury MA, Power AM, Wall JG. (2021). Nature-based biomaterials and their application in biomedicine. *Polymers*, 13(19), 3321.
[49] Huang R, Liu L, Li B, Qin L, Huang L, Yeung KW, Han Y. (2021). Nanograins on Ti-25Nb- 3Mo-2Sn-3Zr alloy facilitate fabricating biological surface through dual-ion implantation to concurrently modulate the osteogenic functions of mesenchymal stem cells and kill bacteria. *Journal of Materials Science & Technology*, 73(31), 44.
[50] Malvi B, Chaudhari R, Balasubramanian C, Kumar A, Asokan K, Das S, Paliwal M, Misra SK. (2022). Ion implantation of 109Ag stable isotope as a tracer in SS316L biomedical implant for failure detection. *Materials Today Communications*, 33, 104563.
[51] Akpek A. (2021). Analysis of surface properties of Ag and Ti ion-treated medical textiles by metal vapor vacuum arc ion implantation. *Coatings*, 11(1), 102.
[52] Anthony TR, Cline HE. (1977). Surface rippling induced by surface-tension gradients during laser surface melting and alloying. *Journal of Applied Physics*, 48(9), 3888–3894.
[53] Jeyaprakash N, Yang CH, Kumar DR. (2020). Laser surface modification of materials. *Practical Applications of Laser Ablation*, 61–81.
[54] Braga ND, Ferreira NG, Cairo CA. (2007). Obtenção de titânio metálico com porosidade controlada por metalurgia do pó. *Química Nova*, 30, 450–457.
[55] Hao L, Lawrence J, Li L. (2005). Manipulation of the osteoblast response to a Ti–6Al–4V titanium alloy using a high-power diode laser. *Applied Surface Science*, 247(1–4), 602–606.
[56] Singh R, Martin M, Dahotre NB. (2005). Influence of laser surface modification on corrosion behavior of stainless steel 316L and Ti-6Al-4V in simulated biofluid. *Surface Engineering*, 21(4), 297–306.
[57] Akgun OV, Inal OT. (1995). Laser surface melting and alloying of type 304 L stainless steel: Part I microstructural characterization. *Journal of Materials Science*, 30, 6097–6104.
[58] Yoon JY. (2025). Improving Zirconia–Resin cement bonding through laser surface texturing: A comparative study. *Prosthesis*, 7(1), 19.
[59] Georger JH Jr, Stenger DA, Rudolph AS, Hickman JJ, Dulcey CS, Fare TL. (1992). Coplanar patterns of self-assembled monolayers for selective cell adhesion and outgrowth. *Thin Solid Films*, 210, 716–719.
[60] Schlenoff JB, Li M, Ly H. (1995). Stability and self-exchange in alkanethiol monolayers. *Journal of the American Chemical Society*, 117(50), 12528–12536.

[61] Murugan R, Molnar P, Rao KP, Hickman JJ. (2009). Biomaterial surface patterning of self-assembled monolayers for controlling neuronal cell behaviour. International journal of biomedical engineering and technology, 2(2), 104–134.

[62] Rashidi H, Yang J, Shakesheff KM. (2014). Surface engineering of synthetic polymeric materials for tissue engineering and regenerative medicine applications. *Biomaterials Science*, 2, 1318–1331.

[63] Reyes CD, Petrie TA, Burns KL, Schwartz Z, García AJ. (2007). Biomolecular surface coating to enhance orthopaedic tissue healing and integration. *Biomaterials*, 28(21), 3228–3235.

[64] Govindarajan T, Shandas R. (2014). A survey of surface modification techniques for next-generation shape memory polymer stent devices. *Polymers*, 6(9), 2309–2331.

[65] Scholz M. (2014). Biofunctionalized wound dressings for advanced wound care. *Biofunctional Surface Engineering*, 1, 251.

[66] Tirrell M, Kokkoli E, Biesalski M. (2002). The role of surface science in bioengineered materials. *Surface Science*, 500(1–3), 61–83.

[67] Krajewska B, Wydro P, Jańczyk A. (2011). Probing the modes of antibacterial activity of chitosan. Effects of pH and molecular weight on chitosan interactions with membrane lipids in Langmuir films. *Biomacromolecules*, 12(11), 4144–4152.

[68] Shu Y, Ou G, Wang L, Zou J, Li Q. (2011). Surface modification of titanium with heparin-chitosan multilayers via Layer-by-Layer Self-Assembly technique. *Journal of Nanomaterials*, 2011(1), 423686.

[69] Dumitriu S, & Popa V. (Eds.). (2013). *Polymeric Biomaterials: Medicinal and Pharmaceutical Applications* (Vol. 2). CRC Press.

[70] Rashidi H, Yang J, Shakesheff KM. (2014). Surface engineering of synthetic polymer materials for tissue engineering and regenerative medicine applications. *Biomaterials Science*, 2(10), 1318–1331.

[71] Merche D, Vandencasteele N, Reniers F. (2012). Atmospheric plasmas for thin film deposition: A critical review. *Thin Solid Films*, 520(13), 4219–4236.

[72] Desmet T, Morent R, De Geyter N, Leys C, Schacht E, Dubruel P. (2009). Nonthermal plasma technology as a versatile strategy for polymeric biomaterials surface modification: A review. *Biomacromolecules*, 10(9), 2351–2378.

[73] Wu S, Liu X, Yeung A, Kw Y, Kao RY, Wu G, Hu T, Xu Z, Chu PK. (2011). Plasma-modified biomaterials for self-antimicrobial applications. *ACS Applied Materials & Interfaces*, 3(8), 2851–2860.

[74] Siedenbiedel, F., & Tiller, J. C. (2012). Antimicrobial polymers in solution and on surfaces: overview and functional principles. *Polymers*, 4(1), 46-71.

[75] Yuan S, Wan D, Liang B, Pehkonen SO, Ting YP, Neoh KG, Kang ET. (2011). Lysozyme-coupled poly (poly (ethylene glycol) methacrylate)– stainless steel hybrids and their antifouling and antibacterial surfaces. *Langmuir*, 27(6), 2761–2774.

[76] Goddard JM, Hotchkiss JH. (2007). Polymer surface modification for the attachment of bioactive compounds. *Progress in Polymer Science*, 32(7), 698–725.

[77] Banerjee I, Pangule RC, Kane RS. (2011). Antifouling coatings: Recent developments in the design of surfaces that prevent fouling by proteins, bacteria, and marine organisms. *Advanced Materials*, 23(6), 690–718.

[78] Onaizi SA, Leong SS. (2011). Tethering antimicrobial peptides: Current status and potential challenges. *Biotechnology Advances*, 29(1), 67–74.

[79] Geszke-Moritz M, Moritz M. (2024) Biodegradable polymeric nanoparticle-based drug delivery systems: comprehensive overview, perspectives and challenges. *Polymers*, 16(17), 2536.

[80] Hu X, Wang T, Li F, Mao X. (2023). Surface modifications of biomaterials in different applied fields. *RSC Advances*, 13(30), 20495–20511. DOI: 10.1039/d3ra02248j.

[81] Takahashi D, Koda Y, Sasaki Y, Akiyoshi K. (2018 Aug). Design and synthesis of PEGylated amphiphilic block oligomers as membrane anchors for stable binding to lipid bilayer membranes. *Polymer Journal*, 50(8), 787–797.

[82] Raval N, Kalyane D, Maheshwari R, Tekade RK. (2019). Surface modifications of biomaterials and their implication on biocompatibility. In *Biomaterials and Bionanotechnology* (pp. 639–674). Academic Press.

[83] Koohi Moftakhari Esfahani M, Alavi SE, Cabot PJ, Islam N, Izake EL. (2022). Application of mesoporous silica nanoparticles in cancer therapy and delivery of repurposed anthelmintics for cancer therapy. *Pharmaceutics*, 14(8), 1579.

[84] Tan C, Wang J, Sun B. (2021). Biopolymer-liposome hybrid systems for controlled delivery of bioactive compounds: Recent advances. *Biotechnology Advances*, 48, 107727.

[85] Boudemagh D, Venturini P, Fleutot S, Cleymand F. (2019). Elaboration of hydroxyapatite nanoparticles and chitosan/hydroxyapatite composites: A present status. *Polymer Bulletin*, 76(5), 2621–2653.

[86] Casettari L, Vllasaliu D, Castagnino E, Stolnik S, Howdle S, Illum L. (2012). PEGylated chitosan derivatives: Synthesis, characterization and pharmaceutical applications. *Progress in Polymer Science*, 37(5), 659–685.

[87] Danhier F, Ansorena E, Silva JM, Coco R, Le Breton A, Préat V. (2012). PLGA-based nanoparticles: An overview of biomedical applications. *Journal of Controlled Release*, 161(2), 505–522.

[88] Manson J, Kumar D, Meenan BJ, Dixon D. (2011). Polyethylene glycol functionalized gold nanoparticles: The influence of capping density on stability in various media. *Gold Bulletin*, 44, 99–105.

[89] Agarwal A, Saraf S, Asthana A, Gupta U, Gajbhiye V, Jain NK. (2008). Ligand based dendritic systems for tumor targeting. *International Journal of Pharmaceutics*, 350(1–2), 3–13.

[90] Bianco A, Kostarelos K, Prato M. (2005). Applications of carbon nanotubes in drug delivery. *Current Opinion in Chemical Biology*, 9(6), 674–679.

[91] Alipour M, Baneshi M, Hosseinkhani S, Mahmoudi R, Jabari Arabzadeh A, Akrami M, Mehrzad J, Bardania H. (2020). Recent progress in biomedical applications of RGD-based ligand: From precise cancer theranostics to biomaterial engineering: A systematic review. *Journal of Biomedical Materials Research Part A*, 108(4), 839–850.

[92] Rondón J, Gonzalez-Lizardo A. (2025). Plasma-treated polymeric biomaterials for improved surface and cell adhesion. *Interactions*, 1(11), 12.

[93] Codrea CI, Baykara D, Mitran RA, Koyuncu AC, Gunduz O, Ficai A. (2024). 3D-bioprinted gelatin methacryloyl-strontium-doped hydroxyapatite composite hydrogels scaffolds for bone tissue regeneration. *Polymers*, 16(13), 1932.

[94] Goreninskii S, Akimchenko I, Vorobyev A, Konoplyannikov M, Efremov Y, Sudarev E, Zvyagin A, Bolbasov E, Tverdokhlebov S. (2024). Single-step method for the immobilization of hydroxyapatite on 3D-printed porous polyetherketoneketone implants. *arXiv Preprint*, www.arXiv:2501.08133.

[95] Pontremoli C, Izquierdo-Barba I, Montalbano G, Vallet-Regí M, Vitale-Brovarone C, Fiorilli S. (2020). Strontium-releasing mesoporous bioactive glasses with anti-adhesive zwitterionic surface as advanced biomaterials for bone tissue regeneration. *Journal of Colloid and Interface Science*, 563, 92–103.

[96] Karnik T, Dempsey SG, Jerram MJ, Nagarajan A, Rajam R, May BC, Miller CH. (2019). Ionic silver functionalized ovine forestomach matrix–a non-cytotoxic antimicrobial biomaterial for tissue regeneration applications. *Biomaterials Research*, 23(1), 6.

[97] Oate JI, Comin M, Braceras I, Garcia A, Viviente JL, Brizuela M, Garagorri N, Peris JL, Alava JI. (2001). Wear reduction effect on ultra-high-molecular-weight polyethylene by application of hard coatings and ion implantation on cobalt chromium alloy, as measured in a knee wear simulation machine. Surface and Coatings Technology, 142–144, 1056–1062.

[98] Madiwal V, Rajwade J. (2024). Silver-deposited titanium as a prophylactic 'nano coat' for peri-implantitis. *Nanoscale Advances*, 6(8), 2113–2128.

[99] Yuan, P., Chen, M., Lu, X., Yang, H., Wang, L., Bai, T., ... & Yu, S. (2024). Application of advanced surface modification techniques in titanium-based implants: Latest strategies for enhanced antibacterial properties and osseointegration. *Journal of Materials Chemistry B*, 12(41), 10516–10549.

[100] Kedia S, Checker R, Sandur SK, Nilaya JP. (2023 Oct). Picosecond laser-induced hybrid groove structures on Ti-6Al-4V bio-alloy to accelerate osseointegration. *Journal of Biomedical Materials Research Part B: Applied Biomaterials*, 111(10), 1775–1784.

[101] Rezaei A, Golenji RB, Alipour F, Hadavi MM, Mobasherpour I. (2020). Hydroxyapatite/hydroxyapatite-magnesium double-layer coatings as potential candidates for surface modification of 316 LVM stainless steel implants. *Ceramics International*, 46(16), 25374–25381.

[102] Shaikh, S., Kedia, S., Singh, D., Subramanian, M., & Sinha, S. (2019). Surface texturing of Ti6Al4V alloy using femtosecond laser for superior antibacterial performance. *Journal of Laser Applications*, 31(2). 245–278.

[103] Al-Jarsha M, Moulisová V, Leal-Egaña A, Connell A, Naudi KB, Ayoub AF, Dalby MJ, Salmerón-Sánchez M. (2018). Engineered coatings for titanium implants to present ultralow doses of BMP-7. *ACS Biomaterials Science & Engineering*, 4(5), 1812–1819.

[104] Shaikh, S., Kedia, S., Singh, A. K., Sharma, K., & Sinha, S. (2017). Surface treatment of 45S5 Bioglass using femtosecond laser to achieve superior growth of hydroxyapatite. *Journal of laser applications*, 29(2). 1231–1267.

[105] Mishra M. (2023). ch. 10. In Joseph K, Wilson R, George G, Appukuttan S. (Eds.), *Lignin-based Materials Health Care and Medical Applications* (pp. 223–246). The Royal Society of Chemistry.

[106] Kosuru RY, Srinivasan V, Jhonsi MA. (2024). Biomass-derived potential nano-biomaterials: Protein binding, anti-biofilm activity and bio-imaging. *Journal of Molecular Structure*, 1300, 137155.

[107] Elendu C, Amaechi DC, Elendu TC, Amaechi EC, Elendu ID, Omeludike JC, Omeludike EK, Onubogu NC, Ogelle EC, Meduoye OOM, Oloyede PO, Ezeh CP, Esangbedo IJ, Adigwe AC, Akuma NM, Okafor SU. (2025). Essential information about nanotechnology in cardiology. *Annals of Medicine and Surgery* (Lond), 87(2), 748–779.

[108] Menagadevi, M., Nirmala, M., Thiyagarajan, D., & Somasundram, D. (2026). Biomaterials and their applications. Biomedical Materials & Devices, 4, 46–6.

[109] Ackun-Farmmer MA, Overby CT, Haws BE, Choe R, Benoit DS. (2021). Biomaterials for orthopedic diagnostics and theranostics. *Current Opinion in Biomedical Engineering*, 19, 100308.

[110] Xing LI, Yang LU. (2014). Surface modifications technology of quantum dots based biosensors and their medical applications. *Chinese Journal of Analytical Chemistry*, 42(7), 1061–1069.

[111] Wei Q, Liu Y, Zhang G. (2021). Research progress and challenges in vaccine development against classical swine fever virus. *Viruses*, 13(3), 445.

[112] Lundstrom K. (2021). Viral vectors for COVID-19 vaccine development. *Viruses*, 13(2), 317.

[113] Zhuo Y, Zeng H, Su C, Lv Q, Cheng T, Lei L. (2024). Tailoring biomaterials for vaccine delivery. *Journal of Nanobiotechnology*, 22(1), 480.

[114] Wibowo D, Jorritsma SH, Gonzaga ZJ, Evert B, Chen S, Rehm BH. (2021). Polymeric nanoparticle vaccines to combat emerging and pandemic threats. *Biomaterials*, 268, 120597.

[115] Butts C, Socinski MA, Mitchell PL, Thatcher N, Havel L, Krzakowski M, Nawrocki S, Ciuleanu TE, Bosquée L, Trigo JM, Spira A. (2014). Tecemotide (L-BLP25) versus placebo after chemoradiotherapy for stage III non-small-cell lung cancer (START): A randomised, double-blind, phase 3 trial. *The Lancet Oncology*, 15(1), 59–68.

[116] Kranz LM, Diken M, Haas H, Kreiter S, Loquai C, Reuter KC, Meng M, Fritz D, Vascotto F, Hefesha H, Grunwitz C. (2016). Systemic RNA delivery to dendritic cells exploits antiviral defence for cancer immunotherapy. *Nature*, 534(7607), 396–401.

[117] Gregoriadis G. (2021). Liposomes and mRNA: Two technologies together create a COVID-19 vaccine. *Medicine in Drug Discovery*, 12, 100104.

[118] Xu L, Liu Y, Chen Z, Li W, Liu Y, Wang L, Liu Y, Wu X, Ji Y, Zhao Y, Ma L. (2012 Apr 11). Surface-engineered gold nanorods: Promising DNA vaccine adjuvant for HIV-1 treatment. *Nano Letters*, 12(4), 2003–2012.

[119] Gale EC, Powell AE, Roth GA, Meany EL, Yan J, Ou BS, Grosskopf AK, Adamska J, Picece VC, d'Aquino AI, Pulendran B. (2021). Hydrogel-based slow release of a receptor-binding domain subunit vaccine elicits neutralizing antibody responses against SARS-CoV-2. *Advanced Materials*, 33(51), 2104362.

[120] Peng X, Liang Y, Yin Y, Liao H, Li L. (2018). Development of hollow mesoporous silica nanoparticles vaccine to protect against house dust mite induced allergic inflammation. *International Journal of Pharmaceutics*, 549(1–2), 115–123.

[121] An W, Defaus S, Andreu D, Rivera-Gil P. (2021). In vivo sustained release of peptide vaccine mediated by dendritic mesoporous silica Nanocarriers. *Frontiers in Immunology*, 12, 684612.

[122] Snyder AD, Salehinia I. (2023). Effect of pore defects on uniaxial mechanical properties of bulk hexagonal hydroxyapatite: A molecular dynamics study. *International Journal of Molecular Sciences.*, 24(2), 1535.

[123] Tudose M, Culita DC, Baratoiu-Carpen RD, Mitran RA, Kuncser A, Romanitan C, Popescu RC, Savu DI. (2022). Novel antitumor agents based on fluorescent benzofurazan derivatives and mesoporous silica. *International Journal of Molecular Sciences*, 23(24), 15663.

[124] Spector D, Pavlov K, Beloglazkina E, Krasnovskaya O. (2022) Recent Advances in Light-Controlled Activation of Pt(IV) Prodrugs. Int J Mol Sci. 2022 Nov 22;23(23):14511. doi: 10.3390/ijms232314511. PMID: 36498837; PMCID: PMC9739791.

[125] Ho TH, Yang CH, Jiang ZE, Lin HY, Chen YF, Wang TL. (2002). NIR-triggered generation of 28 reactive oxygen species and photodynamic therapy based on mesoporous. Silica Hoffman AS. Hydrogels for biomedical applications. *Advanced Drug Delivery Reviews*. 43, 1–2.

[126] Lam EH, Yu F, Zhu S, Wang Z. (2023). 3D bioprinting for next-generation personalized medicine. *International Journal of Molecular Sciences*, 24(7), 6357.

[127] Möller L, Krause A, Dahlmann J, Gruh I, Kirschning A, Dräger G. (2011). Preparation and evaluation of hydrogel-composites from methacrylated hyaluronic acid, alginate, and gelatin for tissue engineering. *The International Journal of Artificial Organs*, 34(2), 93–102.

[128] Chen C, Huang S, Chen Z, Liu Q, Cai Y, Mei Y, Xu Y, Guo R, Yan C. (2023). Kartogenin (KGN)/synthetic melanin nanoparticles (SMNP) loaded theranostic hydrogel scaffold system for multiparametric magnetic resonance imaging guided cartilage regeneration. *Bioengineering & Translational Medicine*, 8(1), e10364.

[129] Khan J, Rudrapal M, Bhat EA, Ali A, Alaidarous M, Alshehri B, Banwas S, Ismail R, Egbuna C. (2021). Perspective insights to bio-nanomaterials for the treatment of neurological disorders. *Frontiers in Bioengineering and Biotechnology*, 9, 724158.

[130] Montelione N, Loreni F, Nenna A, Catanese V, Scurto L, Ferrisi C, Jawabra M, Gabellini T, Codispoti FA, Spinelli F, Chello M. (2023). Tissue engineering and targeted drug delivery in cardiovascular disease: The role of polymer nanocarrier for statin therapy. *Biomedicines*, 11(3), 798.

[131] Kobryń J, Raszewski B, Zięba T, Musiał W. (2023). Modified potato starch as a potential retardant for prolonged release of lidocaine hydrochloride from methylcellulose hydrophilic gel. *Pharmaceutics*, 15(2), 387.

[132] Simeonov M, Kostova B, Vassileva E. (2023). Interpenetrating polymer networks of poly (2-hydroxyethyl methacrylate) and poly (*N,N*-dimethylacrylamide) as potential systems for dermal delivery of dexamethasone phosphate. *Pharmaceutics*, 15(9), 2328.

Rekha Goswami, Sargam Bahl, Naveen Negi, and Kamal Kant Joshi

10 Nanotribology at biological interfaces

Abstract: Nanotribology at biological interfaces is a field of science researching friction, wear, and lubrication that happens at the nanoscale between biological materials or surfaces. The knowledge of nanotribology can aid in the development of advanced biomedical devices, as well as various drug delivery systems and tissue engineering scaffolds. Biological systems often employ means to minimize friction and wear, such as the hydration lubrication mechanism and some biomolecules, like lubricin or hyaluronic acid, typically found in synovial fluid. The study of these and other related mechanisms can lead to the design of novel nanostructured materials that emulate natural biological lubrication to render medical implants more biocompatible and durable. Moreover, it can assist in biotechnology to optimize biosensors, microfluidic devices, and targeted drug delivery systems for improved operation in biological environments. With nanotechnology continuously progressing, knowledge of nanotribology at biological interfaces will prove to be even more crucial in the future.

Keywords: Biological interfaces, lubrication, nanotribology, nanotechnology

10.1 Introduction

Nanotribology refers to the study of friction, wear, and lubrication phenomena at the nanometer scale. When applied to biological interfaces, this field focuses on understanding how these processes occur at the interfaces of biological systems, such as cell membranes, tissues, and proteins. Biological interfaces are often soft, dynamic, and complex, making them unique in the context of nanotribology. The study is crucial for applications like medical implants, drug delivery systems, and biomimetic devices. There are various key components of nanotribology at biological interfaces, such as nanoscale friction, which plays a crucial role in biological interfaces in case of the biological interface's frictional tendency. Biological interfaces exhibit frictional behavior strongly influenced by molecular forces, including van der Waals interactions, hydrogen bonding, and hydration effects. Friction at the molecular level is impacted by the presence of hydration layers, which can mediate sliding motions between surfaces [1, 2]. In biological systems, wear mechanisms can result in structural component changes or molecular degradation, including the disintegration of protein assemblies or harm to cellular membranes. This is especially important when it comes to medical implants, as reducing wear extends the life of components like knee or hip replacements. Natural lubricants that provide low-friction surfaces, such as mucus in different organs or synovial fluid in joints, are frequently used in biological systems. These

https://doi.org/10.1515/9783111563176-010

fluids can dynamically adjust to shifting mechanical strains because of their complicated rheological characteristics, which include shear-thinning behavior. Biological interaction surfaces are often rough, squishy, and viscoelastic at the nanoscale. In contrast, non-biological tribology usually involves hard surfaces. The inherent elasticity of biological tissues, combined with the roughness of the nanoscale, creates distinct wear and frictional behaviors that change depending on applied loads. Nanoscale studies of tribology show that materials synthesized to minimize friction and wear in joint replacements or other biomedical implants can be inadvertently corrosive. Nanoparticles are used for drug delivery, and they travel through very intricate biological settings. The study also has the potential to be directly applicable in developing drug delivery platforms that offer better performance with fewer side effects. For the development of robust, low-wear systems, such as those needed for prosthetics and robotics, materials that can mimic cartilage or other low-friction reducing surfaces found in nature are now under investigation. Biological interfaces in nanotribology drag nature to machines in medicine and bioengineering. Understanding the basic physicochemical rules controlling nanoscale transactions in biological systems allows researchers to develop materials and devices that are more reliable, highly biocompatible, and energy-efficient [3].

10.2 Nanoscale characterization of biological surfaces and interfaces

Knowledge of the mechanical, chemical, and physical properties of biological systems at the nanometer and atomic scale is an obligatory prerequisite for the nanoscale characterization of biological surfaces and interfaces. Since it is crucial to explain the interactions on the outer surface of the materials, more focus is given to performance, biocompatibility, and longevity of the materials in such areas as drug delivery and biomedical implant development. Plasma membranes of cells, proteins, tissues, and other biological surfaces and interfaces are neither simple nor static but are highly ordered and complex, and exhibit dramatically different properties compared with inanimate materials. Due to their amorphous nature and irregular geometry and due to soft, viscoelastic nature-speaking nature together with more complex surface characteristics, they demand improved characterization methods. In general, the following characterization techniques are used:

10.2.1 Atomic force microscopy (AFM)

AFM is among the most common techniques for nanoscale biological surface characterization. Besides quantifying mechanical properties such as rigidity, flexibility as well as coating strength, it provides topographical analysis. The topographical properties of tissues, protein-protein, and cell membranes have been enhanced through the use of AFM. At physiological conditions, living cell elasticity and stiffness were determined with AFM that revealed some information on the mechanical properties and cellular behavior [4].

10.2.2 Scanning electron microscopy (SEM)

The use of a dense beam of electrons aims at scanning biological surfaces to provide detailed imaging. SEM is occasionally combined with cryogenic techniques (cryo-SEM) to study complex biological samples in more native states. It is used to display the nanoscale surface morphology and microstructural features. Cryo-SEM is applied to study artifacts of tissues or cells under their native state without need of GR, maintenance the hydrated condition of biological material. A previous study demonstrated the remarkable mechanical adaptability of arapaima (*Arapaima gigas*) scales into tribological properties, especially the protectability and flexibility of arapaimas. Data obtained show that arapaima scales possess distinct surfaces, depending on the area being considered. In particular, while the exposed parts of the scales contain grooves and a circulus, the covered parts are characterized by the presence of bumps. The surface features of the arapaima scales give rise to different coefficients of friction depending on the tested area or the relative motion direction. The tangential forces experienced on the arapaima scales vary with the tribological properties, which in turn directly affect the armor strength of the arapaima [5].

10.2.3 X-ray photoelectron spectroscopy (XPS)

XPS is a surface-sensitive technique, and it provides one way to analyze the elemental composition of the top 10–100 Å into biological surfaces with information about the chemical state aspects of these atoms. It is a powerful technique as it provides information on the presence of some functional groups, proteins, lipids, and other biomolecules that are located in or confined to interfaces. XPS has been employed to understand the surface chemistry of biomaterials existing as implants and is used for different purposes related with their biocompatibility. XPS can provide an alternative approach to film thickness determination, involving a large-area analysis with few limitations regarding sample chemistry [6, 7].

10.2.4 Surface plasmon resonance

This investigation of biological molecules at surfaces makes use of surface plasmon resonance (SPR). Such a gadget, which is surface plasmons, can be used for a physical sensor in terms of determining binding events such as protein-protein or adsorption on the cell wall in real time because the wave is very much sensitive to variations in density that occur next to the interface. In addition, SPR has also been employed in the study of the kinetics of binding between receptors and their ligands, such as membrane proteins of several cells. SPR imaging is a relatively recent development after the development of the basic mode of SPR with the ability of imaging facilitating high-throughput analysis to study the binding properties of up to thousands of biological molecules simultaneously. This implies that SPR imaging can be beneficial for several types of biological work, such as drug discovery, proteome work, antibody production, and pathway characterization, particularly when coupled with protein chips [7].

10.3 Molecular mechanisms of friction and wear in biological systems

In biological systems, intricate molecular interactions between proteins, lipids, cells, and tissues cause friction and wear. Because biological surfaces are usually hydrated, soft, and dynamic, these mechanisms are fundamentally different from those in non-biological systems. Friction in biological systems is controlled by a variety of molecular mechanisms that allow organisms to reduce wear and tear on tissues, facilitate movement, and protect delicate surfaces. Unlike synthetic materials, biological surfaces are typically soft, hydrated, and covered by a range of complex macromolecules such as proteins, polysaccharides, and lipids. The following mechanisms are primarily responsible for controlling friction in biological interfaces:

10.3.1 Hydration lubrication

Fluid-film lubrication is one of the primary anti-friction mechanisms in a number of biological applications. It is characterized by the presence of a water layer at interfaces, thus forming a slippery surface between two surfaces. Hydration lubrication depends on water molecules with its ordered structure associated with hydrophilic groups on biological surfaces, such as charged or polar-like groups. Water molecules are structured to avoid contact between the surfaces, and in the middle, there is an actual water layer which has a dramatically lower friction. This kind of lubrication is prevalent in biological systems where there is a need to avoid much surface wear while, at the same time, keeping such systems quite elastic. In synovial joints (e.g., knees and hips), water molecules are tightly bound to the surface of cartilage, forming a hydration layer that facilitates low-friction sliding [8]. Hyaluronic acid (HA), present in the synovial fluid, interacts with water molecules to create these protective hydration layers, keeping the surfaces of the cartilage from direct contact during joint movement. HA was one of the first molecules whose nanotribological performance was investigated. It was soon realized [9] that, under physiological conditions, HA neither adsorbs nor lubricates the most common surfaces used in Surface Force Apparatus experiments. Therefore, in order to study friction between HA layers, strategies to increase physisorption were employed. This included the ionic binding of the negatively charged HA to (i) positively charged surfactant bilayers supported on mica substrates, and (ii) negatively charged mica surfaces via Ca^{2+} bridges [10]. Normal force measurements showed that these strategies could provide HA layers on the mica surfaces. However, it was also shown that the layers failed in lubricating the surfaces except, perhaps, for very low applied loads. Moreover, it appeared that HA molecules could become forced out from the gap between the surfaces at relatively low pressures (>5 atm). Benz and co-workers [11] used two different strategies: (i) Biotinylated HA was bound to a supported lipid bilayer covered with streptavidin, and (ii) HA was directly covalently bound to a lipid bilayer-coated mica surface. None of these systems, with friction coefficients of 0.27 and 0.19, respectively, demonstrated outstanding lubricity.

10.3.2 Boundary lubrication

Boundary lubrication occurs when a thin molecular film of lubricating agents coats the surface, reducing friction between two sliding surfaces under high-pressure or low-speed conditions. Boundary lubrication is facilitated by the adsorption of molecules such as lubricin, glycoproteins, and phospholipids on surfaces. These molecules form a thin, highly viscous fluid film that isolates surfaces from each other, even under large loads. The explicit, occurring on the surface of the film, decreases the frictional forces by providing an easy way for the surfaces of the components to slide over one another.

Boundary lubricants are particularly important when hydration lubrication is insufficient due to pressure or reduced water content. Lubricin, a glycoprotein found in synovial joints, plays a crucial role in boundary lubrication. It adheres to the cartilage surface and provides protection under conditions of boundary lubrication, such as when the joint is under significant mechanical load. A load-bearing function has also been proposed for lubricin. Jay et al. [12] employed AFM to show that lubricin adsorbed on opposing model hydrophobic surfaces, thereby reducing their adhesion. They proposed that the adhesion reduction mechanism was responsible for the role played by lubricin in the reduction in wear observed in macroscale ex vivo measurements between cartilage surfaces. Zappone et al. [13] measured normal and friction forces, by means of SFA, between lubricin layers adsorbed on various model planar hydrophilic and hydrophobic surfaces. They also used the interferometric setup of the SFA to characterize wear on the surfaces. They found a net repulsive normal force in all cases, supporting the anti-adhesive properties of lubricin layers. Effective boundary lubricants can be achieved by fusing polymer brushes with the highly hydrated nature of certain monomers. A schematic representation of brush lubrication is shown in Fig. 10.1, with permission taken from [14]. Moreover, the synovial fluid's lubricin and HA are now recognized as the main ingredients in efficient joint lubrication and wear resistance. According to the research using both artificial and animal models, lubricin and HA molecules may cooperate to create a very beneficial lubricating effect. Recent findings, however, indicate that lubricin may be much more effective than HA at protecting articular joints [15].

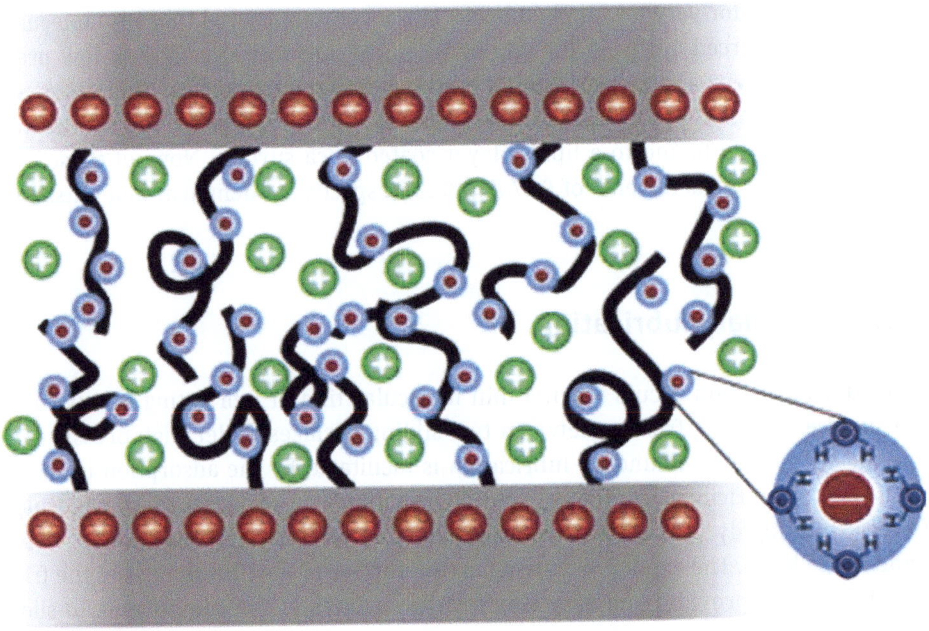

Fig. 10.1: A schematic representation of brush lubrication, reproduced with permission from ref. [14].

10.3.3 Protein and polysaccharide sliding

In tissues such as epithelial layers or joint linings, proteins and polysaccharides can reduce friction by providing a soft, flexible, and hydrated interface. The sliding of these macromolecules on surfaces helps regulate friction. Many biological surfaces are coated with glycoproteins, such as mucins or proteoglycans, which have flexible, hydrated polymer chains. These chains provide a slippery, gel-like layer that facilitates low-friction sliding. The flexibility of these molecules allows them to move across surfaces without inducing significant frictional resistance. Additionally, their water-retaining properties enhance hydration lubrication, further reducing friction. In mucosal tissues, mucins play an important role in reducing friction. They form a gel-like coating over epithelial cells, such as in the gastrointestinal tract, which allows smooth movement of tissues and protection from wear. Mucins also retain water, which contributes to hydration lubrication [16]. Systems with low friction do not always have low wear, or vice versa, even though wear and friction are often caused by the same reasons. For many biological surfaces that include mucinous glycoproteins, mechanical wear is a practical concern. For instance, wear of articular cartilage and the surfaces of the mouth and eyes are especially important. Few researches have examined how mucins and mucinous glycoproteins reduce wear, despite the common belief that they do so [17, 18].

Hydrophilic polymer coatings have been demonstrated by researchers to enhance the surface properties of artificial implants. Their interfacial tribological characteristics are impacted by the biomolecules they invariably encounter due to their in vivo use. A previous study using albumin, globulin, aggrecan, and HA as lubricants examined the friction behaviors of poly(vinylphosphonic acid) (PVPA)-modified Ti6Al4V and polytetrafluoroethylene balls. To determine the cause of the disparity in friction, a quartz crystal microbalance was used to examine the interaction qualities and dynamic adsorption features of the biomolecules and PVPA molecules. Albumin and globulin were found to increase the friction coefficient of PVPA-modified Ti6Al4V. The proteins disturbed the superlubricity (friction coefficient < 0.01) of the PVPA-PBS (phosphate-buffered saline) system. The influence of albumin was much more distinct than that of globulin. Polysaccharides, with their excellent hydration properties and polymer structure, had an unstable dynamic interaction or zero adsorption with PVPA molecules and hardly changed the superlubricity of the PVPA and PBS system. The friction coefficient improved with the addition of polysaccharides as compared to the PVPA-PBS system, suggesting that the PVPA coating's surface properties were also impacted. When aggrecan was added, the PVPA-PBS system's superlubricity vanished, but HA did not render it ineffective [19].

10.3.4 Molecular adhesion and friction

Molecular adhesion forces, such as van der Waals interactions, hydrogen bonding, and electrostatic interactions, influence friction by creating or breaking bonds between surfaces at the molecular level. Adhesion can either increase or decrease friction depending on the strength of the molecular interactions. Weak adhesion (such as that mediated by van der Waals forces) can allow sliding with minimal friction, while strong adhesive forces (such as those involving hydrogen bonds) can lead to higher friction by increasing the resistance to movement [20]. Biological systems use these adhesion forces to control friction under different conditions, particularly in tissues where cellular adhesion is necessary, like in epithelial and endothelial layers. In the movement of cells, forces are regulated by proteins such as integrins that attach the cells to the ECM. These adhesion molecules allow the cell to "cling" to the substrate, but they can switch on and off to regulate the friction necessary to allow movement of a cell when required. In biological systems, the molecular interactions between proteins, lipids, cells, and tissues rub against each other and wear out. These mechanisms are distinct from those in non-biological systems since most biological surfaces are hydrated, soft, and dynamic [21].

10.3.5 Viscoelastic damping

Some biological tissues exhibit viscoelastic behavior, which means that they have both elastic and viscous properties. This combination helps them dissipate mechanical energy, reducing friction under dynamic loading conditions. Viscoelastic damping occurs when biological materials, such as soft tissues, absorb and dissipate energy as they deform under stress. This reduces the overall frictional forces by converting mechanical energy into heat or by spreading the load across a larger surface area. The viscosity component of the tissue's response helps in regulating friction by allowing slow, fluid-like movement under stress, while elasticity helps recover the original shape after deformation. Cartilage tissue in joints exhibits viscoelastic properties, allowing it to absorb impact and reduce friction during movement. The collagen fibers in cartilage provide elasticity, while the proteoglycan matrix retains water and contributes to viscosity, combining to reduce wear. These molecular mechanisms of friction in biological systems are crucial for maintaining the integrity of tissues and ensuring the smooth functioning of processes such as joint movement, cell motility, and mucosal protection. Each mechanism operates at a different scale and under different conditions, allowing biological systems to achieve efficient, low-friction interactions [22].

Wear in biological systems is a gradual breakdown or deterioration of a material and/or a component of a biological system, consequent of the application of mechanical force on the material or component over an extended period of time, leading to molecular changes. As we know, it is different from synthetic systems because it is living, dynamic, and capable of healing itself. However, generally, when the forces are uninterrupted or pathological, no matter how minimal they are, wear occurs, thereby inflicting harm on the tissues and compromising functions.

10.3.6 Protein denaturation and unfolding

Proteins are used in building up tissues of almost all animal body parts, such as the skin, cartilage, and tendons. Mechanical stress is also associated with the process of protein denaturation, which blocks native conformation and exposes the hydrogenated core of the protein. This denaturation can wear, alter the functional properties of tissues, and propagate protein aggregation. The complexity of connective tissues can be described as fibrous proteins like collagen or elastin become flattened under mechanical stress. Precisely, this affects other molecules or aggregation, thereby exposing hydrophobic regions. These changes hamper the stiffness of cells and mechanical wearing by diminishing the solidity of the protein-containing solution. In pathological situations where unfolded proteins accumulate, this will promote tissue degradation. As already mentioned, the collagen fibers in cartilage can become denatured because of the constant mechanical pressure, for example, during physical exercise or in osteoarthritis. The laying down of collagen damages the ECM, and the brunt is borne when the wearing and tearing of the joint increase [23].

10.3.7 Fatigue wear

Fatigue wear on biological material happens as a result of periodic or regular mechanical stressing in which micro-damage gradually builds up. This process is particularly evident in load-bearing tissues, namely bones, tendons, and ligaments, given the fact that they experience cyclical loading and unloading associated with movement. Fatigue wear is characterized by microcrack and micro-tear accumulations in tissues. These are typically invisibly small and then grow and connect up, with the tissue gradually receiving gross harm and the final failure. Specific microstructures, such as proteins in tendons like collagen and mineralized structures such as hydroxyapatite in bone, are reported to be vulnerable to fatigue wear. In tendons, repetitive mechanical loading can cause micro-damage to occur in the tissue. Over time, this leads to a situation in tendinopathy at which the tissue grows weaker, assumes a more rigid structure, and ceases to function effectively [24, 25].

10.3.8 Mechanical disruption of lipid bilayers

The fluid mosaic model that describes cell membranes as composed of lipid/protein layers is vulnerable to mechanical breakdown. The mechano-tension damage in cells and tissues, in terms of mechanical pressure, would damage their cellular membranes and, in the process, cause cell death or functional compromise. Mechanical stress may alter the lipid bilayers by applying a force that stretches or compresses the membrane until it tears. Pressure after pressure and cycle after cycle of shear forces can cause the bilayers to create transient pores or tears. This permits the movement of ions, which can interfere with the normal functioning of cells, together with other molecules. In the extreme instance, mechanical disruption can result in the rupture of the membrane and death of the cell. In blood vessels, the endothelial barrier of the vascular walls is under shear forces by movements of the blood. In the course of a given time, it becomes possible that such cell membranes get worn out due to mechanical stress and the consequent complications that result in atherosclerosis that breaks the vascular integrity [26, 27].

10.3.9 Abrasive wear

Abrasive wear is a process whereby, during relative motion between two surfaces, a softer biological surface is worn by a harder material. This type of wear becomes particularly important in tissues such as teeth, bones, or cartilage, considering the fact that such tissues are constantly subjected to mechanical forces from contact with foreign objects as well as from other tissues and organs. In abrasive wear, the hard particles or surfaces abrade, cut, or plow through the softer biological tissue. This may involve processes such as the removal of some molecules from the surface and the gradual deterioration of tissue. This is particularly detrimental in dental systems, where enamel, which is considered the hardest biological material, erodes due to contact with food particles and brushing. There are abrasive wears of dental enamel that occur as a result of contact with food and help cause abrasive wear if hard elements or acidic particles are part of the food. That, in turn, can result in wear and exposure of dentin, and corresponding heightened sensitivity or caries [28].

10.3.10 Corrosive wear

Corrosive wear is a process whereby material is removed from biological surfaces through chemical interaction with the surroundings. This has special implication in tissues that experience acidic, enzymatic, or oxidative stress. In corrosive wear, other active chemicals like enzymes, acids, or free radicals alter the molecular structure of the biological tissues. For example, in the digestive tract, the low pH of stomach acid destroys the mucosal lining in case of the breakdown of protective features. Similarly, oxidative stress can degrade lipid membranes, proteins, and DNA in cells, leading to wear and eventual cell death. In the oral cavity, corrosive wear can occur when acidic foods or beverages erode dental enamel. The chemical interaction between the enamel and the acidic components results in the demineralization of enamel, which weakens the tooth and makes it more susceptible to mechanical wear [29, 30].

10.3.11 Adhesive wear

Adhesive wear occurs when two surfaces adhere to each other and are then pulled apart, causing material from one surface to transfer to the other. In biological systems, this can result from strong molecular interactions between surfaces, such as hydrogen bonding or covalent bonding between proteins and other biomolecules. When surfaces adhere to each other through molecular interactions, separating them can cause surface material to be torn away and transferred to the opposing surface. In biological systems, adhesive wear can be modulated by the presence of lubricating molecules like proteoglycans, which help reduce direct molecular adhesion between surfaces [31]. In the absence of adequate lubrication, adhesive wear can lead to tissue damage and increased friction. In synovial joints, adhesive wear can occur when the lubricating boundary layer of lubricin is compromised, allowing direct contact between cartilage surfaces. This leads to the transfer of cartilage material between the opposing surfaces and accelerates joint degeneration, a hallmark of osteoarthritis [32].

Biological systems have evolved unique strategies to reduce friction, primarily through hydration lubrication, boundary lubrication, and the sliding of hydrated biopolymers. These mechanisms depend on the presence of water and specialized molecules like lubricin and HA, which form protective layers over tissue surfaces, minimizing direct contact and friction. Despite these protective strategies, biological tissues are still subject to wear, particularly when mechanical stress becomes excessive or sustained. Wear mechanisms, such as protein denaturation, fatigue, abrasion, corrosion, and adhesion, play critical roles in tissue degradation. Under mechanical stress, proteins like collagen can unfold, leading to the weakening of the tissue matrix, while repetitive loading can cause fatigue wear, where micro-damage accumulates and coalesces into larger defects. Skin has high levels of self-repair when faced with wear; however, any wear beyond this level of repair will make it irreparable for tissue-level systems. Knowledge of these mechanisms is not only useful in studying the biomechanical behavior of biological tissues but also useful in designing artificial composites that follow friction and wear behavior similar to natural tissues.

10.4 Applications of nanotribology in biomedicine and biotechnology

Nanotribology, which concerns the friction, wear, and lubrication at the nanoscale, has important applications in biomedical and biotechnology. On this scale, molecular forces dominate, and knowledge of how surfaces at the atomic/molecular level operate has opened up areas of activity as diverse as drug delivery systems, medical implants, and tissue engineering. In biomedical applications, nanotribology plays its role in the development of materials and coatings that have minimal biological interaction with tissues, thereby lowering the level of friction and wear with biomedical devices, inclusive of joint implants, stents, and catheters. The ability to control surface interactions at the nanoscale guarantees that materials utilized in these applications mimic the characteristics of biological tissues, reducing patient suffering and enhancing medical device integration. Nanotribological theory is combined with engineering to create micro instruments and systems for controlled environments, such as biosensors, nanorobots, and microfluidics. These technologies take advantage of the control of adhesion or friction at atomic levels so that more effective biological detection, cell or biomolecule transfer, or drug delivery is possible.

Another area where nanotribological research has had one of the most significant influences in these fields is in regenerative medicine, where nanotribology helps build scaffolds that resemble the extracellular matrix and, at the same time, helps in the regeneration of tissues without harming the cell structure because of mechanical loads. With the progress in nanotechnology, as seen in previous years, the application of nanotribology in biomedicine and biotechnology is anticipated to find solutions in the existing problems or present opportunities for new innovations in healthcare and life sciences.

10.4.1 Enhancing the performance and longevity of biomedical implants

Dental implants, hip and knee replacements, cardiovascular stents, and other biomedical implants undergo mechanical stresses at all times. Nanotribology offers an understanding of decreasing the wear and friction between such devices and tissues, thereby reducing the devices wearing out and enhancing patient experiences. These signal actions of friction and wear at the interface of implants and biological tissues, including bone or cartilage, cause the failure of the implant in the long run. Using nanotechnology, surface engineering techniques make it possible to produce coatings and new material systems that provide low friction and high wear protection. For example, nanostructured coatings, namely diamond-like carbon (DLC) and titanium nitride, have recently been used in joint implants to decrease friction and wear, as well as to replicate synovial joint lubrication. DLC coatings in artificial hip joints have been reported to result in lower wear rates, high biocompatibility, and high durability, resulting to fewer revision surgeries [32, 33].

10.4.2 Drug delivery systems

Nanotribology is particularly useful in developing drug delivery methods that function with less motion resistance within micro and macro-biological conditions. The knowledge on surface interactions at the nanoscale has thus advanced the formulation of smart and controlled drug carriers, including liposomes, micelles, and nanoparticles. As a consequence, the general applicability of drug carriers can be attributed to the features making it possible to control adhesion and friction forces at the molecular level, which allows the drug carriers to move freely through tissues and organs without eliciting any destructive effects or provoking immunological responses. Particles can also be designed with desired surface characteristics that prevent adsorption of proteins (reducing the formation of a protein corona), thereby enhancing their circulation time within the body. In drug delivery for cancer treatment, nanoparticles can be made hydrophilic through coating with polymers such as PEG to decreased friction and prevent identification by the immune system, hence enhancing the targeting of tumors and minimizing adverse effects [34, 35].

10.4.3 Nanotribology in tissue engineering and regenerative medicine

Nanotribology has opened up new possibilities in tissue engineering, where it aids in the development of biomimetic scaffolds that promote cell adhesion, proliferation, and differentiation while minimizing friction and wear-related damage to the newly forming tissues. When making tissue engineering scaffolds, it is necessary to replicate the mechanical characteristics of the native extracellular matrix and its ability to reduce friction. By developing nanostructured surfaces with controlled geometric characteristics, such as roughness and wettability, researchers may affect the cellular response to the scaffold. Specifically, the surface characteristics of the scaffold play an important role in determining the rate of cell migration, orientation, and differentiation. In cartilage tissue engineering, nanofibrous scaffolds of PCL and collagen have been designed for safe and effective lubrication, as well as load-bearing capabilities comparable to native cartilage. These scaffolds decrease cell damage by decreasing the amount of friction that takes place, increase cell attachment, and improve chondrogenic differentiation and aggregation to form cartilage-like tissues [36, 37].

10.4.4 Nanotribology in biosensors and medical diagnostics

Nanotribology has therefore been used to enhance the sensitivity and functionality of biosensors in medical diagnostics. Friction and adhesion at the nanoscale level are well understood in the design of more sensitive and accurate sensing elements in the detection of biomolecules. The function of numerous biosensors is based on their ability to direct the transport of biomolecules across surfaces. Nanotribology in the design of the sensor surface minimizes any interference of a frictional or adhesive nature that may hinder the sensor's ability to identify low-concentration target molecules with high specificity and negligibly low false signals from other molecules. In sensors, nanostructured surfaces are useful to increase the sensing surface for biomolecule binding and to minimize the interference interactions. Cantilever-based biosensors are based on the nanotribological concept, whereby a nanoscale cantilever is deflected by a molecular-binding event. It has also been used in the detection of proteins, DNA, and other biomarkers for diseases like cancer and diabetes [38].

10.4.5 Nanorobotics for precision surgery and targeted therapies

Nanotribology has a central position in the creation of nanorobots and microinstruments used for surgery, medication delivery, and non-invasive sustaining diagnosis. These nanorobots function normally inside biological context, where friction and adhesion control is critical for mobility and job success [39, 40]. Nanorobots not only have to move within the body's regions that are filled with fluids but also navigate across different types of tissue, whose friction coefficients may vary. The ability to investigate nanoscale friction allows for the creation of nanorobotic surfaces that minimize adhesion to undesired sites or interaction with the tissue at the microscale, ultimately reducing the friction; meanwhile, the capacity to perform higher-resolution imaging and more accurate movements. Magnetism allows for efficient drug delivery nanorobots that can be commanded to deliver treatment agents directly to the target tissue, including tumors, but with little effect on the rest of the body. These specific nanorobots depend on the concepts of nanotribology to minimize adhesion to the blood vessel as well as other biological substances during their movement [41, 42].

10.5 Conclusion

The study of nanotribology at biological interfaces is also important for manipulating the mechanics and interactions of biological interfaces with their environment at the nanoscale. This field has enriched the knowledge of how intermolecular forces, including friction, adhesion, and wear, control biological activities and how designed materials can be tailored for biomedical and biotechnological uses. Articulations such as human body joints, cells, tissues, and various biomedical implants are situated in some of the most complex dynamic environments, in which nanoscale tribological processes are critically important for optimal performance, minimized wear, and durable performance. An emerging concept from nanotribological investigations of biointerfaces is that of boundary lubricants and hydrated layers. For example, in synovial joint, it is formed from the water molecules called boundary films and the biological fluids elaborated as the lubricant like lubricin and HA minimizing the contact apposition. This understanding has been used in designing biomedical devices that function as a natural layer of lubrication to reduce wear and enhance the compatibility of the metal and its alloys with living tissues. Further, the wear and degradation of connective tissues or biological soft interfaces, mainly in load-bearing tissues such as cartilage, bone, and tendon have been seen to be governed significantly by nanoscale tribological processes. In nanotribological contact, other modes of interaction, including protein unfolding, fatigue wear, and abrasion, play a major role in initiating the tissue damage that leads to osteoarthritis and tissue degeneration. Research in nanotribology at biological interfaces in the last decade has improved the basic understanding and applications in biomedicine and biotechnology.

References

[1] Bhushan B. (2001). *Modern Tribology Handbook 1. Principles of Tribology*. CRC press.

[2] Bhushan B. (2020). Frontiers in nanotribology: Magnetic storage, bio/nanotechnology, cosmetics, and bioinspiration. *Journal of Colloid and Interface Science*, 577. 127–162.

[3] Sotres J, Arnebrant T. (2013). Experimental investigations of biological lubrication at the nanoscale: The cases of synovial joints and the oral cavity. *Lubricants*, 1. 102–131.

[4] Uzoma PC, Ding X, Wen X, Zhang L, Penkov OV, Hu H. (2022). A wear-resistant silicon nano-spherical AFM probe for robust nanotribological studies. *Physical Chemistry Chemical Physics*, 24. 23849–23857.

[5] Zhang S, Bai P, Wen X, Wen C, Cao H, Yang W, Tian Y. (2024). Effects of surface morphology on tribological properties of Arapaima gigas scales. *Friction*, 12. 1469–1482.

[6] Isaacs MA, Davies-Jones J, Davies PR, Guan S, Lee R, Morgan DJ, Palgrave R. (2021). Advanced XPS characterization: XPS-based multi-technique analyses for comprehensive understanding of functional materials. *Materials Chemistry Frontiers*, 5. 7931–7963.

[7] Das S, Devireddy R, Gartia MR. (2023). Surface plasmon resonance (SPR) sensor for cancer biomarker detection. *Biosensors*, 13. 396.

[8] Raviv U, Laurat P, Klein J. (2001). Fluidity of water confined to subnanometre films. *Nature*, 413. 51–54.

[9] Tadmor R, Chen N, Israelachvili JN. (2002). Thin film rheology and lubricity of hyaluronic acid solutions at a normal physiological concentration. *Journal of Biomedical Materials Research: An Official Journal of the Society for Biomaterials, the Japanese Society for Biomaterials, and the Australian Society for Biomaterials and the Korean Society for Biomaterials*, 61. 514–523.

[10] Tadmor R, Chen N, Israelachvili J. (2003). Normal and shear forces between mica and model membrane surfaces with adsorbed hyaluronan. *Macromolecules*, 36. 9519–9526.

[11] Benz M, Chen N, Israelachvili J. (2004). Lubrication and wear properties of grafted polyelectrolytes, hyaluronan and hylan, measured in the surface forces apparatus. *Journal of Biomedical Materials Research Part A: An Official Journal of the Society for Biomaterials, the Japanese Society for Biomaterials, and the Australian Society for Biomaterials and the Korean Society for Biomaterials*, 71. 6–15.

[12] Jay GD, Torres JR, Rhee DK, Helminen HJ, Hytinnen MM, Cha CJ, Elsaid K, Kim KS, Cui Y, Warman ML. (2007). Association between friction and wear in diarthrodial joints lacking lubricin. *Arthritis & Rheumatism: Official Journal of the American College of Rheumatology*, 56. 3662–3669.

[13] Zappone B, Ruths M, Greene GW, Jay GD, Israelachvili JN. (2007). Adsorption, lubrication, and wear of lubricin on model surfaces: Polymer brush-like behavior of a glycoprotein. *Biophysical Journal*, 92. 1693–1708.

[14] Klein J. (2013). Hydration lubrication. *Friction*, 1. 1–23.

[15] Samaroo KJ, Tan M, Eguiluz RC, Gourdon D, Putnam D, Bonassar LJ. (2017). Tunable lubricin-mimetics for boundary lubrication of cartilage. *Biotribology*, 9. 18–23.

[16] Coles JM, Chang DP, Zauscher S. (2010). Molecular mechanisms of aqueous boundary lubrication by mucinous glycoproteins. *Current Opinion in Colloid & Interface Science*, 15. 406–416.

[17] Zappone B, Ruths M, Greene GW, Jay GD, Israelachvili JN. (2007). Adsorption, lubrication, and wear of lubricin on model surfaces: Polymer brush-like behavior of a glycoprotein. *Biophysical Journal*, 92. 1693–1708.

[18] Bayer RG. editor, (2019). *Engineering Design for Wear* (Revised and Expand ed.), CRC Press.

[19] Zhang C, Chen J, Liu M, Liu Y, Liu Z, Chu H, Cheng Q, Wang J. (2022). Regulation mechanism of biomolecule interaction behaviors on the superlubricity of hydrophilic polymer coatings. *Friction*, 10. 94–109.

[20] Zheng X, Gao L, Yao Q, Li Q, Zhang M, Xie X, Qiao S, Wang G, Ma T, Di Z, Luo J. (2016). Robust ultra-low-friction state of graphene via moiré superlattice confinement. *Nature Communications*, 7. 13204.

[21] Geiger B, Bershadsky A, Pankov R, Yamada KM. (2001). Transmembrane crosstalk between the extracellular matrix and the cytoskeleton. *Nature Reviews Molecular Cell Biology*, 2. 793–805.

[22] Babaei B, Velasquez-Mao AJ, Thomopoulos S, Elson EL, Abramowitch SD, Genin GM. (2017). Discrete quasi-linear viscoelastic damping analysis of connective tissues, and the biomechanics of stretching. *Journal of the Mechanical Behavior of Biomedical Materials*, 69. 193–202.

[23] Buehler MJ. (2006). Nature designs tough collagen: Explaining the nanostructure of collagen fibrils. *Proceedings of the National Academy of Sciences*, 103. 12285–12290.

[24] Thorpe CT, Udeze CP, Birch HL, Clegg PD, Screen HR. (2012). Specialization of tendon mechanical properties results from interfascicular differences. *Journal of the Royal Society Interface*, 9. 3108–3117.

[25] Lalone EA, Giles JW, Alolabi B, Peters TM, Johnson JA, King GJ. (2013). Utility of an image-based technique to detect changes in joint congruency following simulated joint injury and repair: An in vitro study of the elbow. *Journal of Biomechanics*, 46. 677–682.

[26] Suresh S. (2007). Biomechanics and biophysics of cancer cells. *Acta Biomaterialia*, 3. 413–438.

[27] McMahon HT, Gallop JL. (2005). Membrane curvature and mechanisms of dynamic cell membrane remodelling. *Nature*, 438. 590–596.

[28] Lucas P, Constantino P, Wood B, Lawn B. (2008). Dental enamel as a dietary indicator in mammals. *BioEssays*, 30. 374–385.

[29] Addy M, Shellis RP. (2006). Interaction between attrition, abrasion and erosion in tooth wear. *Dental Erosion*, 20. 17–31.

[30] Bartlett DW. (2005). The role of erosion in tooth wear: Aetiology, prevention and management. *International Dental Journal*, 55. 277–284.

[31] Jay GD, Torres JR, Warman ML, Laderer MC, Breuer K. (2007). The role of lubricin in the mechanical behavior of synovial fluid. *Proceedings of the National Academy of Sciences*, 104. 6194–6199.

[32] Eltorai AEM, Haglin J, Antoci V. (2019). Nanotechnology applications for orthopedic implants. *Journal of Orthopaedic Surgery and Research*, 14. 1–10.

[33] Narayan R. (2009). Nanotechnology in biomedical implants. *Journal of Materials Science: Materials in Medicine*, 20. 653–656.

[34] Torchilin VP. (2009). Recent advances with liposomes as pharmaceutical carriers. *Nature Reviews Drug Discovery*, 4. 145–160.

[35] Zhang Y, Chan HF, Leong KW. (2013). Advanced materials and processing for drug delivery: The past and the future. *Advanced Drug Delivery Reviews*, 65. 104–120.

[36] Ebhodaghe SO. (2021). Natural polymeric scaffolds for tissue engineering applications. *Journal of Biomaterials Science, Polymer Edition*, 32. 2144–2194.

[37] Silver FH, Freeman JW, Seehra GP. (2003). Collagen self-assembly and the development of tendon mechanical properties. *Journal of Biomechanics*, 36. 1529–1553.

[38] Wu G, Datar RH, Hansen KM, Thundat T, Cote RJ, Majumdar A. (2001). Bioassay of prostate-specific antigen (PSA) using microcantilevers. *Nature Biotechnology*, 19. 856–860.

[39] Li J, De Ávila B E-F, Gao W, Zhang L, Wang J. (2017). Micro/nanorobots for biomedicine: Delivery, surgery, sensing, and detoxification. *Science Robotics*, 2. 6431.

[40] Aggarwal, M., & Kumar, S. (2022). The use of nanorobotics in the treatment therapy of cancer and its future aspects: a review. Cureus, 14(9). 1–8.
[41] Nelson BJ, Kaliakatsos IK, Abbott JJ. (2010). Microrobots for minimally invasive medicine. *Annual Review of Biomedical Engineering*, 12. 55–85.
[42] Banerjee H, Shen S, Ren H. (2018). Magnetically actuated minimally invasive microbots for biomedical applications. In *Electromagnetic Actuation and Sensing in Medical Robotics* (pp. 11–41).

Vartika Agarwal, Abdul Jabbar Ansari, Pavan Gangwar, Gagan Bansal, and Emad Kadum Njim

11 Clinical applications of biotribology

Abstract: Biotribology is a branch of tribology, and it concerns the investigation of friction, lubrication, and wear in the biological environment. Its relevance in clinical practice has developed very rapidly with the advancing use of medical implants, prosthetics, and tissue-engineered constructs that are subjected to various complex mechanical loads within the body. The dynamic interactions between synthetic materials and biological tissues must be analyzed with precision to limit wear and mechanical failure, which can undermine patient safety and device lifespan. This chapter reviews broadly the biotribological principles and their translational significance in clinical environments, highlighting applications in orthopedics, dentistry, ophthalmology, cardiovascular interventions, and wound care. For example, in joint replacement surgeries, knowledge of the wear behavior of biomaterials is crucial for increasing implant life and minimizing revision surgeries. In ophthalmology, tribological knowledge dictates the design of contact lenses to minimize discomfort and corneal damage. In cardiovascular devices such as artificial heart valves and stents, tribological optimization is necessary to enhance hemocompatibility and minimize the risk of thrombosis. Advances in computational modeling, surface technology, and bio-inspired materials are now making it possible to create personalized and minimally invasive treatments. By filling the gap between clinical medicines and engineering mechanics, biotribology plays a crucial role in maximizing the performance, safety, and life expectancy of biomedical devices and ultimately improving patient outcomes.

Keywords: Biotribology, biomedical implants, friction and wear, lubrication mechanisms, prosthetics and joint replacements, clinical biomechanics

11.1 Introduction

Biotribology, a genuinely interdisciplinary discipline, exists at the thrilling interface between mechanical engineering and life sciences. Fundamentally, it is the scientific examination of friction, lubrication, and wear events as they occur in biological systems [1]. This includes an enormous range of situations, from the highly complex functioning of our own natural joints to the operation of advanced synthetic medical devices implanted in the body. Biotribological principles assist in understanding how living tissues respond when subjected to dynamic loading and relative motion, and, just as crucially, the behavior of designed biomaterials within the dynamic biological environment [2]. The scope of the field is wide-ranging, extending from natural systems such as synovial joints, cardiovascular tissue, and the eye's surface to artificial

https://doi.org/10.1515/9783111563176-011

ones like orthopedic implants (hip and knee replacements), dental prostheses, contact lenses, and mechanical heart valves (shown in Fig. 11.1).

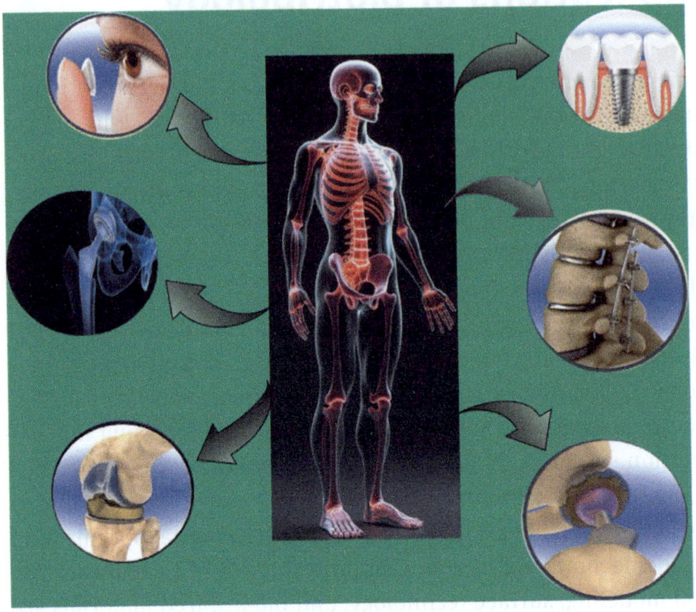

Fig. 11.1: Biomedical implants and prosthetics in the human body.

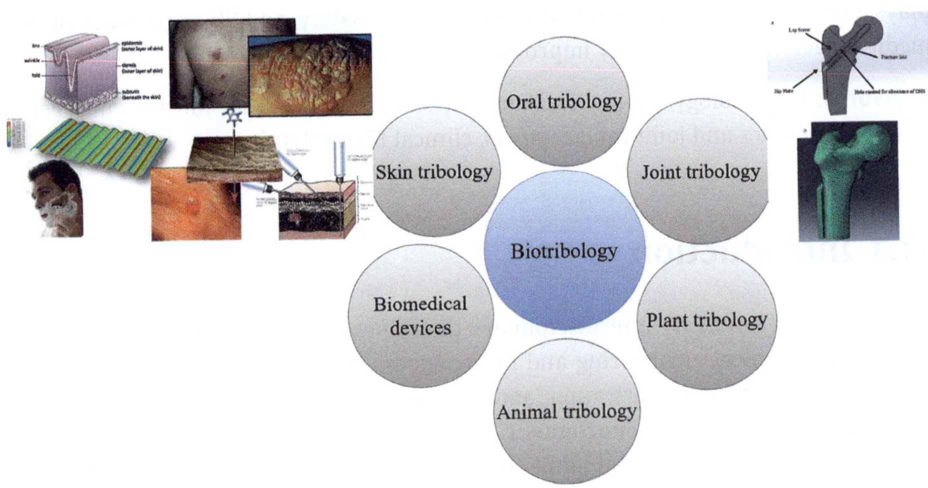

Fig. 11.2: Scope of biotribology in biological and biomedical systems.

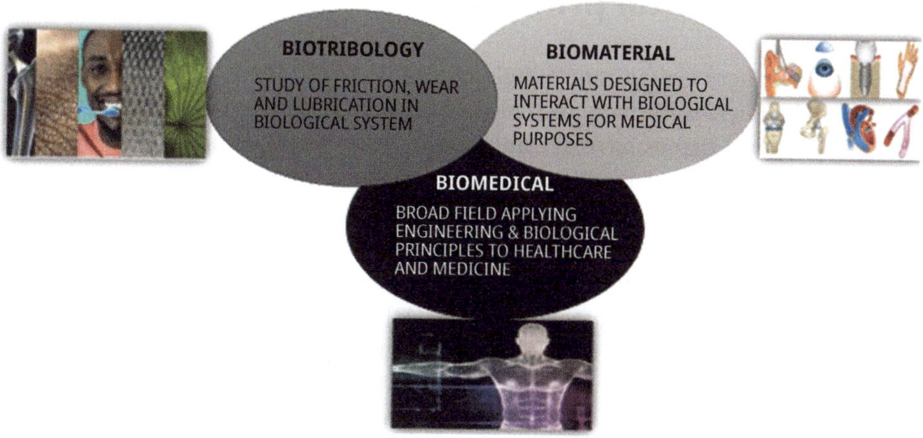

Fig. 11.3: Interdisciplinary interface of biotribology, biomaterials, and biomedical engineering.

The importance of biotribology (shown in Figs. 11.2 and 11.3) in the biomedical sciences cannot be overestimated. The mechanical functionality and, more importantly, the long-term efficacy of medical devices are inherently coupled to their tribological compatibility with the adjacent biological tissues [3]. Insufficient friction may result in mechanical inefficiency and discomfort, whereas excessive wear may create particulate debris, initiating negative biological responses such as inflammation and osteolysis, eventually causing implant failure and requiring revision procedures [4, 5]. Thus, profound knowledge of biotribological concepts is imperative for rational medical device design, the selection of proper biomaterials, and ongoing innovation in the field of biomedical engineering to develop safer, more durable, and biocompatible patient care solutions [6]. The first sparks of biotribological research can be dated back to the early twentieth century with groundbreaking studies of the extraordinary lubrication phenomena of synovial fluid in human joints [7]. However, the discipline really took off with the emerging advances in orthopedic surgery and the growing application of artificial joint replacements in the second half of the century [8]. The necessity to comprehend and minimize wear and friction in these implants became critical for longevity and patient health. As the years went by, the field of biotribology grew to include dental prosthetics, artificial organs, and other biomedical devices [9]. This growth has been largely stimulated by concurrent advancements in a multitude of allied areas, such as advanced imaging technology (e.g., advanced microscopy and MRI), high-powered computational modeling tools (e.g., finite element analysis), and novel biomaterial development with specifically tailored mechanical and surface properties [10, 11]. These advances collectively have enabled investigators to simulate, analyze, and eventually manage more effectively the sophisticated tribological interactions within the human body and between biomaterials and living tissues.

11.2 Fundamental principles

Biological interfaces are an intriguingly intricate terrain for frictional contacts. In contrast to the fairly well-characterized surfaces and conditions of many engineering contexts, biological tissues have dynamic characteristics determined by an enormous range of factors. These encompass their complex surface chemistry, typically involving complicated biomolecules; their micro- and nanoscale surface topography, which can profoundly influence contact area and adhesion; their hydration level, which is important for lubrication; and the size and character of the applied load, which can deform tissues and modify contact mechanics [12]. Biological lubrication mechanisms are also varied and, in many cases, highly efficient, especially in load-bearing joints such as cartilage, where very low coefficients of friction are necessary for pain-free and effective movement [13]. Nevertheless, even with these advanced lubrication mechanisms, wear the gradual degradation of a solid surface due to relative motion can develop over time, resulting in the formation of wear particles and possible mechanical failure of the biological structure [14].

Synovial joints, the wonders of biomechanical engineering in our body, utilize an advanced repertoire of mechanisms for lubrication to reduce friction and wear. Three basic modes of lubrication are typically known [15]. Under boundary lubrication, which is dominant in situations of low speed and high load, the articulating surfaces are brought close to each other. Lubrication is accomplished by a thin layer of large, complicated molecules, such as lubricin and phospholipids, adsorbed onto cartilage surfaces. These layers of molecules interact to minimize direct solid-solid contact and consequently decrease friction [16]. In fluid film lubrication, during high-speed movement and lower loads, synovial fluid, a thick, hyaluronic acid-rich liquid, covers the two cartilage surfaces in a continuous film, which successfully divides the two cartilage surfaces and precludes any direct contact. This mode produces extremely low friction [17]. Mixed lubrication is the most typical situation during normal joint function, with a mixture of both boundary and fluid film lubrication. Through a typical gait cycle, joints undergo changing speeds and loads, resulting in intermittent contact through the fluid film, where boundary lubrication mechanisms become operative to prevent the surfaces from contact [18]. The discipline of biotribology is highly interested in the materials that make up natural biological tissues and artificial biomedical devices. A broad range of materials is thoroughly investigated for their tribological behavior and biocompatibility [19]. Ultra-high-molecular weight polyethylene (UHMWPE) has been a horse in orthopedic implants, especially for hip and knee replacements, because it has a relatively low coefficient of friction and wear rate against metallic or ceramic counterfaces [20]. Wear of UHMWPE is still a major issue, however, resulting in the production of wear particles that can induce adverse biological reactions [21]. Titanium alloys, with good biocompatibility and mechanical strength, are common in dental implants and certain orthopedic implants. Their surface characteristics and resistance to corrosion are essential for long-term functionality [22]. Cobalt-chromium alloys, traditionally used in articulating surfaces of joint replacements, provide excellent wear resistance, but the issues of metal ion release have made other materials of interest more appealing [23]. Ceramics like alumina and zirconia have very high hardness and wear resistance, and they are of interest for bearing surfaces in joint arthroplasty and dental implants, and their biocompatibility is also usually good [24]. Hydrogels, which are networks of water-swollen polymers, have special properties such as high lubricity and biocompatibility, and they are of interest for use in applications such as contact lenses, drug delivery systems, and possibly as part of tissue-engineered constructs [25]. Additionally, understanding the inherent tribological characteristics of biological tissues such as cartilage, skin, and corneal epithelium is central to developing biomimetic materials and interfaces [26]. The elaborative representation is shown in Fig. 11.4. Considerable work is also focused on surface engineering and nanomaterial addition for improving biocompatibility, minimizing friction, and enhancing wear resistance. Technologies like hydroxyapatite coatings are designed to enhance osseointegration of implants, and nanomaterials are being investigated for their ability to form self-lubricating surfaces and to deliver therapeutic agents [27, 28].

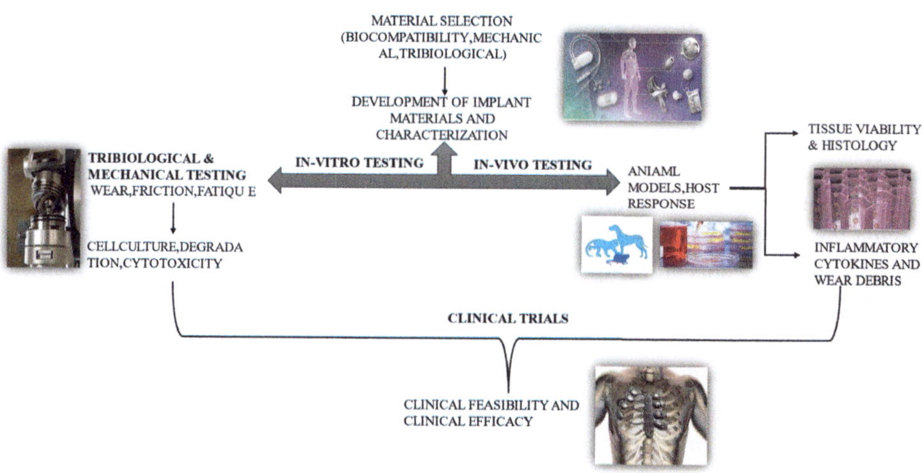

Fig. 11.4: Tribological and biological validation pathway for joint prosthetic materials.

11.3 Orthopedic applications

Within orthopedic surgery, total hip and knee arthroplasty have transformed the management of disabling diseases of the joints. The longevity and success of the prosthetic joint are crucially reliant on scrupulous tribological design [29]. One of the key features of this design is the judicious choice of material combinations for the articulating surfaces to reduce friction and wear. For instance, a typical design consists of a UHMWPE cup articulating with a cobalt-chromium, titanium alloy, or ceramic femoral head [30]. The choice of these materials depends on their wear resistance, coefficient of friction, biocompatibility, and mechanical strength. Additionally, the geometry and conformity of the articulating surfaces are important in the distribution of contact stresses and in lubrication within the joint [31]. To maximize these aspects, computational modeling, including finite element analysis, is now widely utilized preclinically to simulate and predict patterns of wear and evaluate tribological performance under simulated physiological loading conditions [32].

In spite of extensive progress in material science and prosthetic design, wear of artificial materials continues to be a main reason for late complications in joint arthroplasty [33]. Principal wear mechanisms that cause surface degradation are abrasive wear, in which hard particles such as bone cement plow through softer substrates like UHMWPE, abrading material from the surface [34]; adhesive wear, which involves microscopic welding and failure of contact points between surfaces under load [35]; and fatigue wear, in which cyclic loading induces subsurface cracks leading to eventual particle failure [36]. Most problematic are the submicron-diameter UHMWPE wear particles, which may initiate inflammatory reactions by stimulating macrophages and releasing pro-inflammatory cytokines. This results in particle-induced osteolysis – bone resorption around the prosthesis – which is a prevalent cause of implant loosening and requires intricate revision surgeries with frequently inferior outcomes [38].

Natural articular cartilage, the very specialized tissue covering our joints, attains remarkable tribological performance with low friction and high wear resistance, allowing for effortless joint movement over decades [39]. Its remarkable properties are a result of a very special composition of 70–80% water content, a well-structured extracellular matrix rich in collagen and proteoglycans, and chondrocytes – the very specialized cells that maintain the matrix [40]. This structure enables cartilage to support compressive loads, distribute stresses uniformly, and employ both fluid film and boundary lubrication mechanisms to minimize friction [41]. Recognizing the shortcomings of existing prosthetic joints, scientists are targeting cartilage tissue engineering as a viable solution. The aim is to develop cartilage substitutes that can integrate with host tissues and mimic the mechanical and tribological properties of the native tissue [42]. Typically, this involves seeding chondrocytes or stem cells onto biocompatible scaffolds composed of either natural or synthetic materials to support the development of a structured extracellular matrix mimicking that of native cartilage. However, recreating the complex architecture and composition necessary for long-term tribological success remains a key challenge in the field.

11.4 Dental tribology

Dental tribology aims at understanding the wear processes that act on natural tooth structures, that is, enamel and dentin, during mastication and exposure to various oral conditions [43]. Enamel, the surface layer of the tooth, is the hardest biological substance in the human body, consisting mostly of hydroxyapatite crystals. Its high mineral composition gives it substantial wear resistance. Dentin, the tissue that lies beneath it, is weaker and more readily abraded and eroded because of the lower mineral level and the collagen within it. The enamel and dentin wear is a multifactorial process that is affected by a variety of factors, such as the size and direction of occlusal forces during mastication, the occurrence of food and toothpaste particles with abrasive properties, the pH level of the oral cavity (acidic pH can cause erosion), and personal habits such as bruxism (teeth grinding) [44]. Understanding these wear processes is crucial for developing effective preventive strategies and for designing restorative dental materials that can withstand the harsh oral environment (Fig. 11.5).

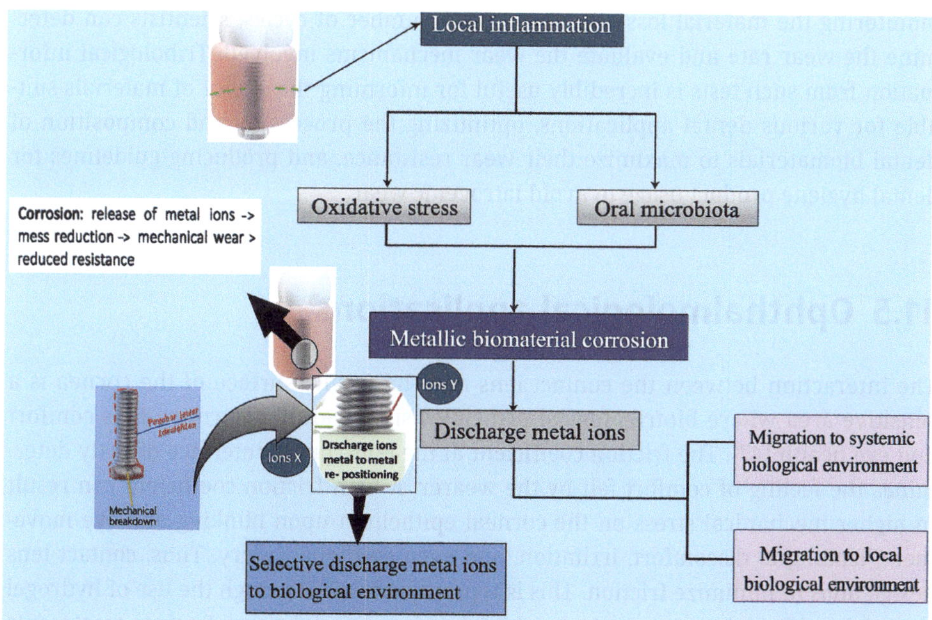

Fig. 11.5: Mechanisms and consequences of metallic biomaterial corrosion in biological environments.

The choice of restorative materials for dental treatments, including fillings, crowns, and bridges, should consider their tribological compatibility with tooth structures. Composite resins, ceramics (such as porcelain and zirconia), and dental amalgams have varying wear behaviors and may interact with opposing enamel and dentin differently during mastication. An ill-suited match in wear resistance between a restorative material and the opposing tooth can result in increased wear of the natural tooth structure, which may lead to occlusal disharmony, sensitivity, and damage [45]. Additionally, the abrasiveness of dental hygiene materials, including toothpaste and polishing agents, contributes greatly to the long-term wear of both natural teeth and restorative materials. Extremely abrasive dentifrices, though very useful for removing staining, will, on a long-term basis, lead to excessive wear in susceptible materials, especially dentin and certain resins. It follows that thoughtful appraisal of the tribological characteristics of restorative dental materials and products used for dentifrices will be crucial if the longevity and vitality of the dentition are to be assured.

In order to assess the wear resistance and tribological compatibility of dental biomaterials, several in vitro testing methods have been established, which intend to model the intricate oral environment conditions. These simulators generally expose material samples to cyclic loading and sliding against a counterface material (which may be enamel, dentin, or another restorative material) under the influence of a slurry of abrasive particles and artificial saliva to simulate the oral fluids [46]. By monitoring the material loss over a specified number of cycles, scientists can determine the wear rate and evaluate the wear mechanisms involved. Tribological information from such tests is incredibly useful for informing the choice of materials suitable for various dental applications, optimizing the processing and composition of dental biomaterials to maximize their wear resistance, and producing guidelines for dental hygiene product usage to avoid iatrogenic wear.

11.5 Ophthalmological applications

The interaction between the contact lens and the fragile surface of the cornea is a sensitive area where biotribological principles significantly determine lens comfort and eye health [47]. The friction coefficient at the lens-cornea interface directly determines the feeling of comfort felt by the wearer. A high friction coefficient can result in higher mechanical stress on the corneal epithelium upon blinking and eye movement, leading to discomfort, irritation, and even epithelial injury. Thus, contact lens design aims to minimize friction. This is typically attained through the use of hydrogel materials with high water content, which has intrinsic lubricity. Surface treatments and the addition of lubricating agents in the lens matrix are also used to increase wettability further and decrease friction, allowing for a smooth and comfortable interaction with the ocular surface [48].

The ocular surface, which includes the cornea and conjunctiva, is typically lubricated by a thin, intricate tear film. This tear film is essential for preserving the health and transparency of the cornea, creating a smooth optical surface, and preventing infection. Dry eye syndrome, a frequent condition involving inadequate tear production or excessive evaporation of tears, interferes with this delicate system of lubrication. This perturbation causes raised friction between the eyelids and the ocular surface during blinking, producing signs like dryness, irritation, burning, and blurred vision. Biotribological research is vital to comprehend mechanisms of ocular surface lubrication and the alterations happening in dry eye disease. It helps in designing artificial tear solutions that seek to restore proper lubrication and decrease friction. In addition, research in this field directs the development of new ocular surface therapies and treatment approaches to dry eye disease [49].

Intraocular lenses (IOLs) are surgically implanted artificial lenses that substitute for the natural lens of the eye, usually following cataract surgery. Although IOLs have significantly restored vision for millions, their long-term function and biocompatibility are determined by surface characteristics and behavior in the ocular environment. Intraoperative or postoperative displacement of the IOL and contact with adjacent ocular tissues can result in inflammation and other adverse consequences. Advances in tribology aim at designing IOLs with enhanced surface characteristics to reduce these unwanted effects. Low-friction coatings on the IOL surface may lower its interaction with the surrounding tissues, and hence inflammation and enhance biocompatibility. Additionally, studies are investigating the incorporation of new materials and surface treatments to increase the stability and functionality of IOLs during long-term use inside the eye [50].

11.6 Cardiovascular devices

Mechanical heart valves are artificial devices that are surgically implanted to substitute diseased or damaged native heart valves. These devices are subjected to continuous, high-frequency cycles of opening and closing, with very high mechanical stresses and tribological interactions with the flowing blood and adjacent tissues. One major design issue in the case of mechanical heart valves is wear, as wear particle generation can cause thromboembolism (blood clotting) as well as device failure. In order to reduce wear and enhance hemocompatibility, mechanical heart valves are generally manufactured using highly durable and biocompatible materials, for example, pyrolytic carbon. Carbon-based coatings are also often used for coating the valve surfaces to ensure a smooth, wear-resistant surface that reduces damage to blood cells and the possibility of thrombus formation [51]. The leaflet valve design and the dynamics of flow past the valve are also precisely engineered to reduce turbulent flow and lower shear stresses on blood constituents.

Stents are expandable mesh tubes that are implanted into narrowed or occluded arteries to re-open the arteries and restore blood flow. The vascular implants undergo cyclic mechanical stresses caused by the pulsatile blood flow and the dynamic motion of the arterial wall. Tribological fatigue, caused by continual mechanical loading, can weaken the structural integrity of the stent over time, leading to fracture and device failure. Relevant research aims at creating superelastic alloys, including nitinol (nickel-titanium), that have a high fatigue resistance and can sustain huge deformations and return to their initial [52]. In addition, it is essential to optimize the surface properties of the stents to decrease friction between the stent struts and the arterial wall, avoiding vascular damage and enhancing endothelialization (development of a healthy inner lining of the blood vessel).

Hemocompatibility, the property of a material to be compatible with blood and not induce any adverse effect, is an overriding concern in the development of all cardiovascular devices in contact with blood. Excessive shear stress, the tangential force applied by the moving blood to the surfaces of the device, can cause hemolysis (red blood cell rupture) and platelet activation, with an associated increase in thrombosis risk. Surface engineering techniques are utilized to maximize the flow profiles in the vicinity of cardiovascular devices and reduce areas of high shear stress. This involves the use of smooth, biocompatible materials and the deposition of surface coatings which enhance endothelial cell adhesion and proliferation, forming a more natural and blood-compatible interface [53]. Investigations are ongoing to investigate new surface modifications and materials with increased hemocompatibility in order to enhance the safety and long-term durability of cardiovascular implants.

11.7 Tribology in skin and wound healing

The largest organ of the human body, skin, is continuously under frictional force due to contact with clothing, medical devices, and prosthetic limbs [54]. It is important to understand the tribological properties of the skin in order to design skin-friendly interfaces for minimizing blistering and irritation. The coefficient of friction of skin depends heavily on skin hydration, surface roughness, and the applied normal load. In prosthetics, high friction at the skin-socket interface can result in skin breakdown and discomfort in amputees. Thus, attention is being paid to creating prosthetic socket materials and interface liners with low friction coefficients and moisture management capabilities to enhance comfort and avoid skin damage [55].

Pressure ulcers, or bedsores, constitute a major clinical issue, especially in immobilized patients. Constant pressure and shear forces at the skin-support surface interface lead to tissue injury and ulceration. Principles of biotribology find relevance in applying the role played by friction in the causation of pressure ulcers. Low-friction support surfaces and bedclothes can decrease shear forces on the skin, reducing the risk of ulcer formation. The same is true for the structure of bandage materials. Low-friction dressings can minimize trauma to the underlying skin upon application and removal, facilitate healing, and reduce patient discomfort [56].

Wound care is in a constant development phase, wherein advanced wound dressing is being studied to not just cover the wound but to drive healing actively too. Emerging wound dressing biomaterials like hydrogels, silicone films, and bioactive polymers are being developed for optimal tribology [57]. These products seek to create a moist wound environment that is favorable for healing and has low friction to avoid trauma on dressing changes and decrease attachment to the wound bed. Additionally, some of these new dressings contain bioactive agents that may be delivered in a controlled way to induce tissue regeneration and prevent infection [58]. Their tribological behavior is an important aspect of their clinical effectiveness and acceptance by patients.

11.8 Computational and in vitro models

Finite element simulation (FEM) and multibody dynamics analyses are now essential tools for the long-term prediction of wear behavior in prosthetic joints. The mathematical models make it possible to mimic the intricate contact mechanics and motion patterns within the joint during physiological loading conditions. By including material behavior and wear laws, FEM is able to forecast the location and extent of wear on articulating surfaces and gain useful insights into the long-term functionality and potential failure mechanisms of various implant designs. Such simulations can be used to improve implant geometry, material choice, and surgical methods for reducing wear and prolonging the life of joint replacements.

Microfluidic devices, or lab-on-a-chip systems, provide novel platforms for performing small-scale tribological testing of biomaterials under controlled conditions. Microfluidic devices enable accurate control of fluid flow, contact forces, and environmental conditions, facilitating high-throughput screening of various materials and surface modifications. Microfluidic tribometers are capable of replicating the lubricated conditions present within biological systems, including synovial joints and the ocular surface, generating useful information regarding friction and wear at a diminished scale while using less material [59]. These platforms are well-suited to the investigation of the tribological response of new biomaterials and the determination of the impact of biological fluids and molecules on surface interactions [60].

The blending of machine learning (ML) and artificial intelligence (AI) methods is revolutionizing many fields of biomedical engineering, such as biotribology [61]. Large databases of clinical and experimental tribology data can be used to train AI algorithms to generate predictive models of wear behavior in implants. These models may consider parameters such as patient activity levels, implant design parameters, and material properties in order to forecast the risk of wear-related complications and maximize material selection for specific patients. AI can also aid in real-time decision-making during implant design and preclinical evaluation, accelerating the development of longer-lasting and more biocompatible medical devices [62].

11.9 Challenges and future perspectives

Even with dramatic improvements in computational and in vitro modeling, existing models continue to suffer from limitations in being able to completely mimic the complex biological detail of in vivo systems [63]. The dynamic and variable character of physiological loading, the impact of biochemical variables (e.g., enzymes and proteins), and the active biological response of the surrounding tissues to the implant and wear debris are frequently challenging to completely integrate into reduced models. Future studies must concentrate on the generation of more advanced and physiologically realistic models that have the ability to accurately capture the complex interactions between the bio-material interfaces. This will involve the merging of advanced computational methods with greater knowledge of the biological environment and how the body responds to implanted devices.

The growing trend toward personalized medicine offers exciting opportunities for advancing the field of biotribology [64]. Recognizing the significant variability in patient anatomy, activity levels, and physiological conditions, the development of custom implants tailored to individual patient needs holds the promise of improved clinical outcomes and implant longevity. Biotribological principles will be central to the design and material choice of these patient-specific implants, promoting maximum mechanical performance and reducing wear and failure risk based on the individual's unique biomechanics. This will presumably entail the application of advanced imaging technologies, computational modeling, and additive manufacturing methodologies to produce implants with geometries and material properties that are finely tailored.

Nanotribology, that is, friction, adhesion, lubrication, and nanoscale wear, is well on its way to becoming evermore important to the creation of next-generation biomedical devices. Control and comprehension of tribology at this length scale enable the design of surfaces with unprecedented properties, including ultra-low-friction coatings that can improve biocompatibility and minimize wear on implants [65]. Nanotribological concepts are also important in the development of new drug delivery systems, where friction and adhesion, under control at the nanoscale, can facilitate targeted and sustained release of drug agents. Nanotribology is also helping to design smart biomaterials with responsive surfaces that have the ability to change according to their biological setting, with the potential to enhance their integration and long-term function.

11.10 Conclusion

Biotribology has solidly positioned itself as a pillar in the design and clinical application of biomedical devices. Its principles offer the essential perspective by which we comprehend the complex interactions between biological tissues and the materials we place within the body. From the macro world of speaking articulating joint replacements to the microscopic contact lens interfaces and the dynamic milieu of cardiovascular devices, control of friction, lubrication, and wear is of utmost importance to guarantee device life, biocompatibility, and ultimately patient welfare.

The history of biotribology has been one of relentless progression, propelled by advances in engineering, materials science, and our increasing comprehension of biological systems. The lessons learned have translated directly into the development of safer, more effective medical solutions in a range of clinical applications, such as orthopedics, dentistry, ophthalmology, and cardiovascular treatment.

In the future, the discipline sits at the threshold of thrilling new horizons. Overcoming the shortcomings of the existing in vitro and computational models to more closely mimic the complicated in vivo situation is a key challenge. The emerging age of personalized medicine brings with it the possibility of using patient-specific need and activity-based design and materials selection for tailoring implants to improve clinical success. Additionally, the new frontier of nanotribology presents great potential for next-generation biomedical devices with revolutionary precision in surface engineering, drug delivery, and smart biomaterials. Ultimately, the future of biotribology depends on ongoing multidisciplinary collaborations that integrate engineers, biologists, clinicians, and data scientists' expertise seamlessly. Together, we can reveal novel understanding of the fundamental principles of biological tribology and deliver on discoveries in innovative solutions addressing unmet clinical needs and improving patients' quality of life globally.

References

[1] Persson BNT. (2014). *History of Tribology* (pp. 791–797). *Springer eBooks*, https://doi.org/10.1007/978-3-642-22647-2_80.

[2] Sahoo P, Das SK, Davim JP. (2019). *Tribology of Materials for Biomedical Applications* (pp. 1–45). *Elsevier eBooks*, https://doi.org/10.1016/b978-0-08-102174-3.00001-2.

[3] Zhou Z, Jin Z. (2015). Biotribology: Recent progresses and future perspectives. *Biosurface and Biotribology*, 1(1), 3–24. https://doi.org/10.1016/j.bsbt.2015.03.001.

[4] Hu JCY, Athanasiou KA. (2003). *Structure and Function of Articular Cartilage* (pp. 73–95). *Humana Press eBooks*, https://doi.org/10.1007/978-1-59259-417-7_4.

[5] Walker PS, Dowson D, Longfield MD, Wright V. (1969). Lubrication of human joints. *Annals of the Rheumatic Diseases*, 28(2), 194. https://doi.org/10.1136/ard.28.2.194-a.

[6] Lancaster JG, Dowson D, Isaac GH, Fisher J. (1997). The wear of ultra-high molecular weight polyethylene sliding on metallic and ceramic counterfaces representative of current femoral

surfaces in joint replacement. *Proceedings of the Institution of Mechanical Engineers Part H Journal of Engineering in Medicine*, 211(1), 17–24. https://doi.org/10.1243/0954411971534647.

[7] MacConaill MA. (1948). THE MOVEMENTS OF BONES AND JOINTS. *Journal of Bone and Joint Surgery – British Volume*, 30-B(2), 322–326. https://doi.org/10.1302/0301-620x.30b2.322.

[8] Charnley J. (1961). ARTHROPLASTY OF THE HIP a new operation. *The Lancet*, 277(7187), 1129–1132. https://doi.org/10.1016/s0140-6736(61)92063-3.

[9] Williams DF. (2009). On the nature of biomaterials. *Biomaterials*, 30(30), 5897–5909. https://doi.org/10.1016/j.biomaterials.2009.07.027.

[10] Jin Z. (2020). *Fundamentals of Computational Modelling of Biomechanics in the Musculoskeletal System* (pp. 5–12). *Elsevier eBooks*, https://doi.org/10.1016/b978-0-12-819531-4.00002-x.

[11] Jagadeesh P, Rangappa SM, Siengchin S. (2023). Advanced characterization techniques for nanostructured materials in biomedical applications. *Advanced Industrial and Engineering Polymer Research*, https://doi.org/10.1016/j.aiepr.2023.03.002.

[12] Bhushan B. (2008). Nanotribology and nanomechanics in nano/biotechnology. *Philosophical Transactions of the Royal Society a Mathematical Physical and Engineering Sciences*, 366(1870), 1499–1537. https://doi.org/10.1098/rsta.2007.2170.

[13] Hills BA. (2000). Boundary lubrication in vivo. *Proceedings of the Institution of Mechanical Engineers Part H Journal of Engineering in Medicine*, 214(1), 83–94. https://doi.org/10.1243/0954411001535264.

[14] Dowling NE. (1993). *Mechanical behavior of materials: engineering methods for deformation, fracture, and fatigue*. http://ci.nii.ac.jp/ncid/BA1978131X

[15] Abreu E. (2005). Mow VC, Huiskes R: Basic Orthopaedic Biomechanics and Mechano-Biology. *BioMedical Engineering OnLine*, 4(1). https://doi.org/10.1186/1475-925x-4-28.

[16] Bishop A. (2021). Wound assessment and dressing selection: An overview. *British Journal of Nursing*, 30(5), S12–S20. https://doi.org/10.12968/bjon.2021.30.5.s12.

[17] Hamrock BJ, Schmid SR, Jacobson BO. (2004). *Fundamentals of Fluid Film Lubrication. CRC Press eBooks*, https://doi.org/10.1201/9780203021187.

[18] Morris N, Johns-Rahnejat P, Rahnejat H. (2020). Tribology and Dowson. *Lubricants*, 8(6), 63. https://doi.org/10.3390/lubricants8060063.

[19] Black J, Black J. (2005). *Biological Performance of Materials. CRC Press eBooks*, https://doi.org/10.1201/9781420057843.

[20] McKellop H, Shen F, Lu B, Campbell P, Salovey R. (1999). Development of an extremely wear-resistant ultra high molecular weight polythylene for total hip replacements. *Journal of Orthopaedic Research®*, 17(2), 157–167. https://doi.org/10.1002/jor.1100170203.

[21] Shanbhag A, Jacobs J, Glant T, Gilbert J, Black J, Galante J. (1994). Composition and morphology of wear debris in failed uncemented total hip replacement. *Journal of Bone and Joint Surgery – British Volume*, 76-B(1), 60–67. https://doi.org/10.1302/0301-620x.76b1.8300684.

[22] Brunette DM, Tengvall P, Textor M, Thomsen P. (2001). *Titanium in Medicine. Engineering materials*, https://doi.org/10.1007/978-3-642-56486-4.

[23] Jacobs JJ, Skipor AK, Patterson LM, Hallab NJ, Paprosky WG, Black J, Galante JO. (1998). Metal release in patients who have had a primary total hip arthroplasty. a prospective, controlled, longitudinal study*. *Journal of Bone and Joint Surgery*, 80(10), 1447–1458. https://doi.org/10.2106/00004623-199810000-00006.

[24] Bierbaum BE, Nairus J, Kuesis D, Morrison JC, Ward D. (2002). Ceramic-on-Ceramic bearings in total hip arthroplasty. *Clinical Orthopaedics and Related Research*, 405, 158–163. https://doi.org/10.1097/00003086-200212000-00019.

[25] Hoffman AS. (2012). Hydrogels for biomedical applications. *Advanced Drug Delivery Reviews*, 64, 18–23. https://doi.org/10.1016/j.addr.2012.09.010.

[26] Derler S, Gerhardt L. (2011). Tribology of Skin: Review and analysis of experimental results for the friction coefficient of human skin. *Tribology Letters*, 45(1), 1–27. https://doi.org/10.1007/s11249-011-9854-y.

[27] Geetha M, Singh A, Asokamani R, Gogia A. (2008). Ti based biomaterials, the ultimate choice for orthopaedic implants – A review. *Progress in Materials Science*, 54(3), 397–425. https://doi.org/10.1016/j.pmatsci.2008.06.004.

[28] Webster TJ. (2001). *Nanophase Ceramics: The Future Orthopedic and Dental Implant Material* (pp. 125–166). *Advances in chemical engineering*, https://doi.org/10.1016/s0065-2377(01)27005-7.

[29] Hc A. (1970). Complications of total hip replacement. *PubMed*, 72, 123–137. https://pubmed.ncbi.nlm.nih.gov/4990007.

[30] Bedard NA, Callaghan JJ, Stefl MD, Willman TJ, Liu SS, Goetz DD. (2014). Fixation and wear with a contemporary acetabular component and Cross-Linked polyethylene at minimum 10-Year Follow-Up. *The Journal of Arthroplasty*, 29(10), 1961–1969. https://doi.org/10.1016/j.arth.2014.05.008.

[31] Banaszkiewicz PA. (2013). *The Effect of Conformity, Thickness, and Material on Stresses in Ultra-High Molecular Weight Components for Total Joint Replacement* (pp. 93–96). *Springer eBooks*, https://doi.org/10.1007/978-1-4471-5451-8_22.

[32] Huiskes R, Chao E. (1983). A survey of finite element analysis in orthopedic biomechanics: The first decade. *Journal of Biomechanics*, 16(6), 385–409. https://doi.org/10.1016/0021-9290(83)90072-6.

[33] Billi F, Benya P, Kavanaugh A, Adams J, Ebramzadeh E, McKellop H. (2011). The John Charnley Award: An accurate and sensitive method to separate, display, and characterize wear debris: Part 1: Polyethylene particles. *Clinical Orthopaedics and Related Research*, 470(2), 329–338. https://doi.org/10.1007/s11999-011-2057-x.

[34] Evans D, Lancaster J. (1979). *The Wear of Polymers* (pp. 85–139). *Treatise on materials science and technology*, https://doi.org/10.1016/s0161-9160(13)70066-8.

[35] Rabinowicz E, Tanner RI. (1965). Friction and wear of materials. *Wear*, 8(6), 491. https://doi.org/10.1016/0043-1648(65)90145-6.

[36] Suh NP. (1973). The delamination theory of wear. *Wear*, 25(1), 111–124. https://doi.org/10.1016/0043-1648(73)90125-7.

[37] Drees P, Eckardt A, Gay RE, Gay S, Huber LC. (2007). Mechanisms of Disease: Molecular insights into aseptic loosening of orthopedic implants. *Nature Clinical Practice Rheumatology*, 3(3), 165–171. https://doi.org/10.1038/ncprheum0428.

[38] Harris WH. (2001). Wear and periprosthetic osteolysis. *Clinical Orthopaedics and Related Research*, 393, 66–70. https://doi.org/10.1097/00003086-200112000-00007.

[39] Hall AJ. (1996). Operative Arthroscopy. *Journal of Bone and Joint Surgery – British Volume*, 78-B(6), 999–1000. https://doi.org/10.1302/0301-620x.78b6.0780999a.

[40] Buckwalter JA, Mankin HJ. (1998). Articular cartilage: Tissue design and chondrocyte-matrix interactions. *PubMed*, 47, 477–486. https://pubmed.ncbi.nlm.nih.gov/9571449.

[41] Daniel M. (2012). *Role of Surface-Active Lipids in Cartilage Lubrication* (pp. 225–243). *Deleted Journal*, https://doi.org/10.1016/b978-0-12-396533-2.00007-0.

[42] Langer R, Vacanti JP. (1993). Tissue engineering. *Science*, 260(5110), 920–926. https://doi.org/10.1126/science.8493529.

[43] Grippo JO. (1991). Abfractions: A new classification of hard tissue lesions of teeth. *Journal of Esthetic and Restorative Dentistry*, 3(1), 14–19. https://doi.org/10.1111/j.1708-8240.1991.tb00799.x.

[44] Buzalaf MAR, Magalhães AC, Rios D. (2018). Prevention of erosive tooth wear: Targeting nutritional and patient-related risks factors. *BDJ*, 224(5), 371–378. https://doi.org/10.1038/sj.bdj.2018.173.

[45] Tsujimoto A, Barkmeier WW, Fischer NG, Nojiri K, Nagura Y, Takamizawa T, Latta MA, Miazaki M. (2017). Wear of resin composites: Current insights into underlying mechanisms, evaluation

methods and influential factors. *Japanese Dental Science Review*, 54(2), 76–87. https://doi.org/10.1016/j.jdsr.2017.11.002.

[46] Zadeh PN, Lümkemann N, Sener B, Eichberger M, Stawarczyk B. (2018). Flexural strength, fracture toughness, and translucency of cubic/tetragonal zirconia materials. *Journal of Prosthetic Dentistry*, 120(6), 948–954. https://doi.org/10.1016/j.prosdent.2017.12.021.

[47] Mann A, Tighe B. (2013). Contact lens interactions with the tear film. *Experimental Eye Research*, 117, 88–98. https://doi.org/10.1016/j.exer.2013.07.013.

[48] Jones L, Downie LE, Korb D, Benitez-Del-Castillo JM, Dana R, Deng SX, Dong PN, Geerling G, Hida RY, Liu Y, Seo KY, Tauber J, Wakamatsu TH, Xu J, Wolffsohn JS, Craig JP. (2017). TFOS DEWS II Management and Therapy Report. *The Ocular Surface*, 15(3), 575–628. https://doi.org/10.1016/j.jtos.2017.05.006.

[49] Craig JP, Nichols KK, Akpek EK, Caffery B, Dua HS, Joo C, Liu Z, Nelson JD, Nichols JJ, Tsubota K, Stapleton F. (2017). TFOS DEWS II Definition and Classification Report. *The Ocular Surface*, 15(3), 276–283. https://doi.org/10.1016/j.jtos.2017.05.008.

[50] Huang Q, Cheng GP, Chiu K, Wang G. (2016). Surface modification of intraocular lenses. *Chinese Medical Journal*, 129(2), 206–214. https://doi.org/10.4103/0366-6999.173496.

[51] Yacoub MH, Cohn LH. (2004). Novel approaches to cardiac valve repair. *Circulation*, 109(9), 1064–1072. https://doi.org/10.1161/01.cir.0000115634.66549.4d.

[52] Machado L, Savi M. (2003). Medical applications of shape memory alloys. *Brazilian Journal of Medical and Biological Research*, 36(6), 683–691. https://doi.org/10.1590/s0100-879x2003000600001.

[53] Baier RE. (1975). *Blood Compatibility of Synthetic Polymers: Perspective and Problems* (pp. 139–159). *Springer eBooks*, https://doi.org/10.1007/978-1-4684-7744-3_10.

[54] Guan Z, Zhou L, Li L, Wei H, Chen W, Wang S, Li C. (2020). Friction properties of in vivo human skin from visualized friction testing. *Journal of the Mechanical Behavior of Biomedical Materials/Journal of Mechanical Behavior of Biomedical Materials*, 104, 103692. https://doi.org/10.1016/j.jmbbm.2020.103692.

[55] Klute GK, Glaister BC, Berge JS. (2010). Prosthetic liners for lower limb amputees. *Prosthetics and Orthotics International*, 34(2), 146–153. https://doi.org/10.3109/03093641003645528.

[56] Kharola A, Sharma K, Choudhary V. (2024). Novel multi-step-based process modelling and prediction of thermodynamic properties of steam using machine learning. *International Journal of Manufacturing Research*, 19(3), 239–265.

[57] Bansal G, Gautam RK, Misra JP, Mishra A. (2024). Tribological behavior of silver-doped eggshell-derived hydroxyapatite reinforcement in PMMA-based composite. *Proceedings of the Institution of Mechanical Engineers, Part L: Journal of Materials: Design and Applications*, 238(11), 2142–2157.

[58] Huang G, Huang H. (2018). Hyaluronic acid-based biopharmaceutical delivery and tumor-targeted drug delivery system. *Journal of Controlled Release*, 278, 122–126. https://doi.org/10.1016/j.jconrel.2018.04.015.

[59] Jayamohan H, Romanov V, Li H, Son J, Samuel R, Nelson J, Gale B. (2016). *Advances in Microfluidics and Lab-on-a-Chip Technologies* (pp. 197–217). *Elsevier eBooks*, https://doi.org/10.1016/b978-0-12-802971-8.00011-0.

[60] Zhang F, Sun Y, Luo C. (2021). Microfluidic approaches for synthetic gene circuits' construction and analysis. *Quantitative Biology*, 9(1), 47–60. https://doi.org/10.15302/j-qb-021-0235.

[61] Nahavandi D, Alizadehsani R, Khosravi A, Acharya UR. (2021). Application of artificial intelligence in wearable devices: Opportunities and challenges. *Computer Methods and Programs in Biomedicine*, 213, 106541. https://doi.org/10.1016/j.cmpb.2021.106541.

[62] Revilla-León M, Gómez-Polo M, Vyas S, Barmak BA, Galluci GO, Att W, Krishnamurthy VR. (2021). Artificial intelligence applications in implant dentistry: A systematic review. *Journal of Prosthetic Dentistry*, 129(2), 293–300. https://doi.org/10.1016/j.prosdent.2021.05.008.
[63] Rieker CB. (2016). Tribology of total hip arthroplasty prostheses. *EFORT Open Reviews*, 1(2), 52–57. https://doi.org/10.1302/2058-5241.1.000004.
[64] Chandra S, Verma S, Lim WM, Kumar S, Donthu N. (2022). Personalization in personalized marketing: Trends and ways forward. *Psychology and Marketing*, 39(8), 1529–1562. https://doi.org/10.1002/mar.21670.
[65] Bhushan B. (2017). *Nanotribology and Nanomechanics. Springer eBooks*, https://doi.org/10.1007/978-3-319-51433-8.

Venkata Satya Prasad Somayajula and Priyadarsini Morampudi

12 Experimental techniques in biotribology

Abstract: Biotribology, one of the rapidly evolving fields, integrates the concepts pertaining to tribology and biology to tackle biomedical engineering as well as materials science issues. It involves studying and evaluating the friction, wear, and lubrication characteristics in biological systems. In a biological environment, it is of utmost necessity to learn about the interactions of biological tissues as well as synthetic materials to design, develop, and optimize various medical devices such as implants and prosthetics to safeguard their biocompatibility and warrant longevity. A diverse range of experimental techniques is employed in Biotribology to evaluate the complexities in biological systems. The current book chapter focuses on an elaborated discussion on these experimental techniques, their working principles, and the outcomes of these techniques. Description related to the application of these experimental techniques in understanding material characteristic and their interpretation is also included in this chapter.

Keywords: Biotribology, biomaterials, nanotribology, wear, surface roughness

12.1 Introduction

Biotribology is significant for understanding friction, wear, and lubrication within biological systems, helps to improve the performance of joints, organs, and implants. Experimental methods are central in this field, for providing valuable insights on tissue-implant interactions [1]. This discipline has wide applications in orthopedics, dentistry, as well as tissue engineering, to interpret friction and wear in biological environments [2]. Such research is extremely needed for the design of enduring implants and prosthetics. The primary goals include enhancement in patient quality of life, minimal wear debris, assurance of effective lubrication, and improved durability. Researchers are aiming to assess wear mechanisms, optimize surface properties, and develop biolubricants. This review discusses experimental techniques for biotribological phenomena-related study, stressing upon their potential in advancement of this field [3, 4]. Biotribology is relevant in both natural and artificial biological systems. Numerous tribological interfaces exist within the human body, such as synovial joints, cartilage, and soft tissues, where it is essential to maintain low friction and high wear resistance for proper functioning. The synovial joints are lubricated structures, which exhibit exceptional lubrication mechanisms, thereby reducing friction and wear in decades of use. In medicine, biotribology is essential to create implants, prostheses, and surgical tools that sustain biological loads without causing undue wear [3]. It also contributes to the improvement of contact lenses, dental fillings, and biolubricants,

https://doi.org/10.1515/9783111563176-012

ensuring better comfort and activity. Moreover, the progress in biotribology reflects on the appearance of new disciplines, such as bioimitated robotics and smart clothes, where the imitation of natural lubrication and friction factors can lead to the optimization of functional principles [5, 6].

12.2 Key areas of study in biotribology

Joint tribology could be exemplified by human joints, such as the knee and the hip, which are typical biological tribosystems. Particularly, the presence of synovial fluid that acts as a natural lubricant, allows for smooth movement since it reduces the friction between articulating cartilage surfaces. Consequently, research in the field of joint tribology is aimed at improving the design of artificial implants, namely hip and knee replacements, by creating biomaterials that closely resemble the natural cartilage properties [7].

Dental tribology is a new branch of biotribology that studies friction, wear, and lubrication of teeth for their optimal functioning in eating different types of food. The teeth interact with the food by compression, which generates complex tribological processes at the interface of the enamel and dentin. In the process of eating, the enamel and dentin should not only be durable under mechanical stresses but also wear-resistant [8].

Skin and soft tissue tribology: The frictional behavior of skin is crucial in various applications, such as prosthetics, wearable medical devices, and cosmetics. Understanding skin tribology helps to reduce irritation, optimize adhesive patches, and develop better protective materials for individuals with sensitive skin [9].

Ocular tribology, which is the study of tribology in the eye, mainly looks into the lubrication mechanism of the tear film. This mechanism protects the cornea and ensures smooth blinking. Studies in this field lead to progress in artificial tears, contact lens materials, and treatments for dry eye syndrome [10].

Biomaterial tribology: Implants, prosthetics, and medical devices interact with biological tissues and fluids, making their tribological properties crucial for long-term functionality. Biocompatible materials with low friction and high wear resistance, such as ultra-high-molecular-weight polyethylene (UHMWPE) and ceramic coatings, are widely researched for improving implant longevity [11, 12]. The other key areas pertaining to biotribology are represented in Fig. 12.1.

Despite significant advancements, biotribology faces challenges in replicating the complexity of biological lubrication and wear mechanisms. The dynamic nature of biological tissues, variations in environmental conditions, and interactions with biochemical processes make experimental analysis difficult [13]. Computational modeling and advanced simulation techniques are being integrated to predict tribological behaviors in biological systems more accurately. Future research is expected to focus on bioinspired materials, smart lubrication systems, and regenerative approaches to en-

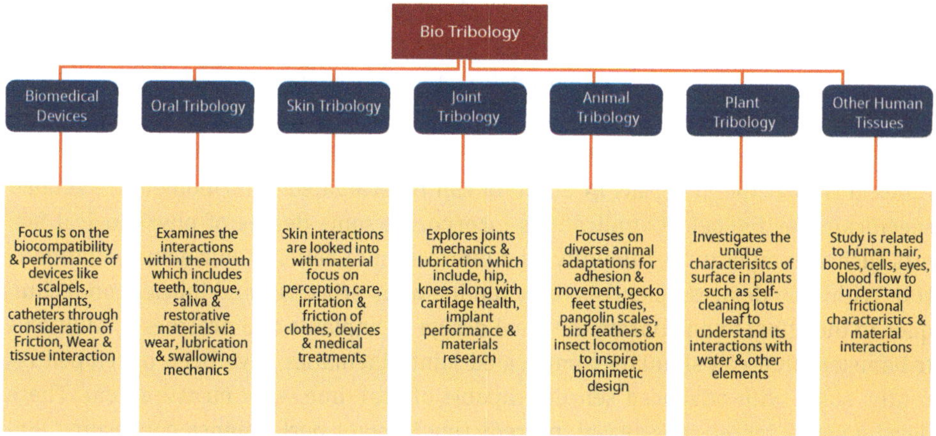

Fig. 12.1: Key areas of biotribology.

hance tissue durability and repair. As biotribology continues to evolve, its impact on medical technology, healthcare, and bioengineering will become increasingly profound.

12.3 Experimental techniques in biotribology

Experimental techniques in biotribology are crucial for studying how biological surfaces interact under mechanical stress, including friction, wear, and lubrication. The joints, skin, teeth, and implanted prosthetics have their functioning in environments that are dynamic and complex. Hence, the standard testing methods of tribology are not enough to capture their behavior in real world. Confrontation of these challenges by researchers is via specialized experimental techniques developed that imitate the physiological environments, which enable profound understanding of material characteristics and biomaterial optimization needed for medical applications [14, 15]. It is of utmost importance that materials created have close resemblance of the natural biological surfaces with refined biolubrication strategies. This is only possible through accurate experimental testing. Hence, these advancements benefit applications related to artificial joints and contact lenses, where smooth movement and longevity are absolutely critical.

12.4 Tribometers for biotribology studies

Tribometers are vital for studying friction and wear in biological interfaces since they aid obliquely in evaluating biomaterials under controlled conditions. The more common setups of pin-on-disk and ball-on-flat tribometers are configured to mimic joint rotation so that the materials in question can be evaluated. Biotribometers intended specifically for biological applications incorporate some degree of physiological factors such as temperature, humidity, and lubrication to better mimic actual conditions [16]. With such devices, scientists can examine wear processes and lubrication techniques in the effort to ameliorate the ideal design of implants and prostheses. Multiaxis tribometers that can simulate complicated joint kinematics have been developed recently, along with advanced sensors capable of real-time wear measurement. These developments increase accuracy in predicting implant performance concerning longevity and reliability under physiological conditions.

12.4.1 Pin-on-disk tribometers

The pin-on-disk tribometer is a common equipment used in the study of friction and wear characteristics of materials. It consists of a stationary disk and a pin-shaped specimen, which may be held against its flat surface or the periphery of a rotating disk, called the pin-on-drum configuration. The geometry and shape of the pin and disk are different for each particular test (Fig. 12.2a) [17, 18]. This type of tribometer is more widely accepted than other tests of this type because of its high level of control over the conditions or parameters of an experiment, which provides reliable and repeatable results. Unlike some other methods, it permits experimentation in a steady state, which helps to reduce the variability in the contact area and the sliding velocity.

However, keeping a constant normal load on the pin can be most difficult. These vibrations or instability can occur because of the buildup of transfer films, surface roughening with wear, or even changes in material property, especially in ceramics where grain orientation determines the direction of wear. To improve reliability, it has been suggested that, instead of conventional hanging weights, springs or hydraulic systems with pressure stabilizers should be used to apply the load. The essential parameters of friction coefficient, wear rate, and surface temperature are recorded by tribometer records to give treasured material performance insights under diverse operating environments. ASTM G99 and ISO 7148 standards ensure accuracy and repeatability on test setup, sample preparation, and data analysis.

12.4.2 Ball-on-disk tribometer

The study of friction, wear, and lubrication interactions between materials in artificial sliding contact is done using the ball-on-disk tribometer depicted in Fig. 12.2(b). The equipment consists of a disk or plate, circular in shape, that anchors a spherical tool or ball made of steel or ceramic and permits axial motion while sliding in rolling motion [18]. It is very common in engineering and scientific research in development to analyze the behavior of materials in differing performance conditions, such as changes in loading, speed, and the effects of the environment [19, 20].

This experiment has the spherical tool rotating against a stationary disk set within a rigid frame, the so-called counterface, with a controlled applied load. As relative motion takes place between the two interacting surfaces, it is possible to assess friction and wear characteristics under varying conditions. Adding lubricants can help mimic realistic operating conditions more closely. Scientists can evaluate the effect of materials performance using loads, sliding speeds, and temperatures as control parameter modifications.

Friction force, the rate of wear, material loss, and surface loss are essential results from this analysis, further analyzed using surface analysis methods such as profilometry and microscopy techniques. Parts and materials to be tested, optimized, and improved, coatings and lubricants, as well as materials used in engineering, all call for tribology studies, which is what this ball-on-disk tribometer is engineered for in materials science, engineering, and industry [21, 22]. Its simple yet efficient design allows for controlled and repeatable experiments to gain insights about how the material behaves when sliding. While this approach provides an accurate representation of the purely tribological aspects of real-world interactions, the application of this test does not consider the complexities encountered in practical applications. This is why, when interpreting laboratory results, we must do so carefully, so that we make sure the results have real-world implications [23, 24].

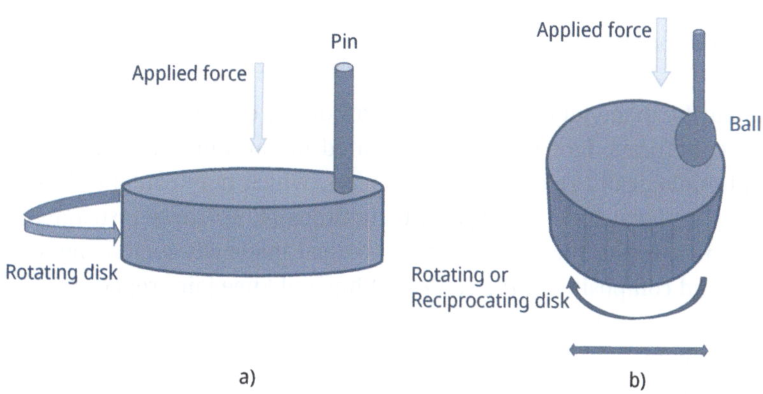

Fig. 12.2: Illustrations of **(a)** pin on disk (POD) and **(b)** ball on disk.

12.4.3 Biotribometers

Biotribometers, depicted in Fig. 12.3, are specialized devices engineered to mimic the biological conditions present in artificial joint implants. These machines are used to test the wear and friction of biomaterials within an accelerated time period, which is controlled, making them essential [18]. Other researchers usually test materials that include UHMWPE, aluminum oxide (Al_2O_3), polycarbonate urethanes (PCU), and titanium alloy (Ti–6Al–4V) on biotribometers. These materials substitute critical joint parts like acetabular cup, femur head, meniscus, and tibial epiphysis saucers, respectively. For injectable artificial lubricants used in knee and hip joints, animal-based phospholipid-based hydrogels and hyaluronic acid are commonly used to enhance lubrication and resemble the conditions of natural joints. Biotribometers exceed normal conditions, which helps them to better understand diagnosing and improving the longevity and physiological function of biomedical implants.

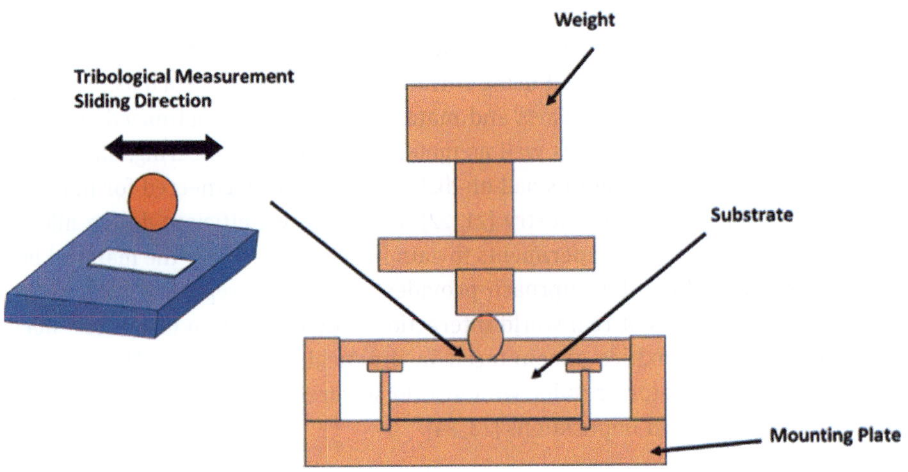

Fig. 12.3: Diagram of the biotribometer.

As opposed to standard tribometers, which were traditionally designed and operated under on-complex conditions, biotribometers operated under simulated biological conditions using physiologically relevant environments, which may contain physiological fluids such as synovial fluid or artificial body fluids [4, 5]. These instruments play a pivotal role in assessing the efficacy of biomedical materials such as metals, polymers, ceramics, and composites in the context of hip and knee joint replacements, dental implants, and various other medical devices.

Like traditional tribometers, biotribometers enable accurate manipulation of key parameters, including applied load, sliding speed, and temperature, which allows simulation of relevant physiological scenarios during testing. They quantify important parameters like friction, wear rate, and material degradation in biological or simulated biological environments [25]. To further understand material performance, they can utilize wear particle analysis, surface characterization, and biological response assessments based on analytical techniques.

With valuable data on durability, biocompatibility, and friction resistance, biotribometers serve as key contributors to the progress of biomedical materials and implant design. Their use is critical in demonstrating that medical devices meet these equally demanding regulatory requirements for long-term safety and effectiveness. Topics such as biotribology help accelerate the development of novel materials, coatings, and devices that improve patient outcomes in various medical applications, including implants and surgical instruments.

The in situ biotribometer is a specialized form designed to measure dynamic interactions between biological interfaces in real time at contact and sliding. Trained on data until October 2023, this sophisticated instrument allows for simultaneous optical analysis in situ during indentation and friction experiments on living biological materials. It also provides statistically significant gene expression analysis, linking the worlds of tribology, rheology, and biology. The use of in situ biotribology instruments enables the creation of more dependable and successful medical implants by facilitating a better understanding of complex biomedical interactions.

12.5 Atomic force microscopy (AFM) in biotribology

Atomic force microscopy (AFM) is a commonly used technique for surface interaction analysis at the nanoscale, which is very useful for investigating soft biological tissues and biofilms. With the ability to characterize friction forces, adhesion, and surface roughness at the nanoscale level, AFM allows for an improved understanding of cell interactions at diverse surfaces and biomaterials' performance at physiological conditions. It is a key technique for studying polymeric coatings, hydrogel interfaces, and natural joint lubrication. New techniques, like force spectroscopy and high-speed AFM imaging, have allowed scientists to observe molecular structures in real time, providing unprecedented information about the operations of lubricating films and wear particles. These advances have led to the development of biomimetic materials with superior tribological properties that can improve the performance of medical device coatings and synthetic cartilage [26–28].

AFM creates detailed topographical images by scanning the surface of a sample with an exceedingly fine probe (Fig. 12.4). A typical AFM consists of a cantilever with tips at the nanoscale, a laser, a four-quadrant photodiode, and a scanner. The geometry allows for the depth of surface detail, so that structures only a few nanometers across can be resolved [29], which leads to a wide range of resolution (about 100 μm up to less than 1 μm can be achieved).

The imaging is done by moving a sharp tip (usually made of silicon or Si_3N_4) at a distance of a few angstroms (about 10 Å) to the sample. The annular tip is typically 3–15 μm long, and the tip is attached to a 100–500 μm long, flexible cantilever. This cantilever is literally like a spring, which makes it very sensitive to the respective surface changes. When the tip scans the sample, it senses even the minutest variation in topography, thus creating a map of the surface structure. Due to its ability to obtain nanometer-scale images, AFM has become a valuable technology used for both basic research and biomedical applications [29].

As the AFM tip, which is fixed to the free end of the cantilever, approaches the sample surface, the cantilever bends up or down due to the interplay of attractive and repulsive forces acting on it. The amount of bending is precisely monitored by bouncing a laser beam off the back of the cantilever onto a detector. Since the cantilever essentially behaves like a spring, the force acting between the tip and the sample is related to its stiffness (spring constant) and the distance between them. This interaction adheres to Hooke's law, a mathematical model explaining the relationship between force, stiffness, and deflection [29, 30]:

$$F = -k \cdot x \tag{12.1}$$

where F is the force, k is the spring constant (stiffness of the cantilever), and x is the cantilever deflection.

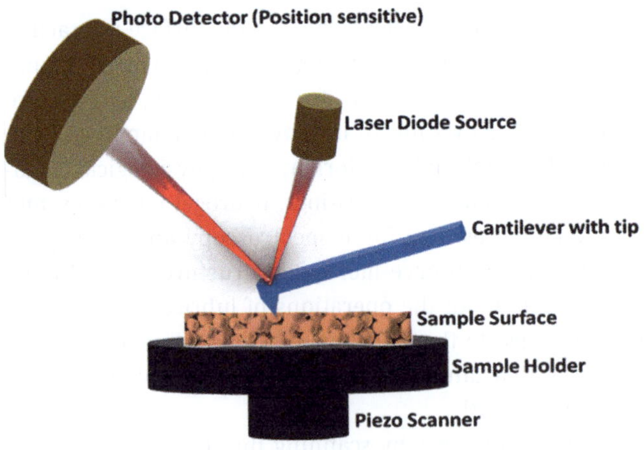

Fig. 12.4: Principle of AFM.

If the surface interacts with a more flexible cantilever, it bends in response to the forces acting on it, and this bending is carefully monitored. As the tip moves across the sample, the AFM adjusts its position to trace the contours of the surface. These movements are guided by a host of forces, such as electrostatic, magnetic, capillary, and Van der Waals forces. The schematic diagram of AFM is shown in fig. 12.5. By carefully measuring how the tip moves, the scientists can obtain an extremely detailed map of the driving surface.

Instead of changing the AFM probe, the sample is often moved with a piezoelectric stage, which facilitates nanometer-level precision in the *x*, *y*, and *z* axes. Piezoelectric materials, which are ceramics that expand or contract when an electric voltage is applied, enable highly accurate positioning [31]. A laser beam is aimed at the back of the cantilever and reflects onto a four-quadrant photodiode detector, which continuously monitors changes in the cantilever's position. Since the cantilever bends in response to extremely small forces between the tip and the sample, the detector captures these subtle shifts, allowing researchers to construct a highly detailed topographical image of the surface.

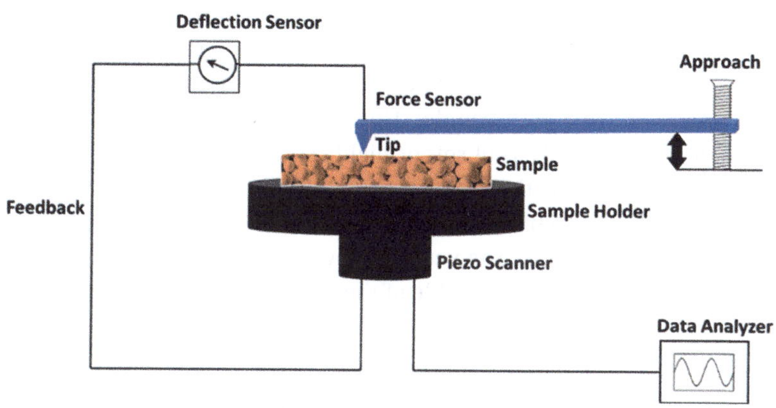

Fig. 12.5: Schematics of AFM operation.

12.6 Operation modes of atomic force microscopy (AFM)

– **Contact mode:** In contact mode, the AFM tip remains in continuous, gentle contact with the sample as it scans the surface. Depending on the settings, the tip can either maintain a fixed height above the sample or operate under a constant applied force. However, friction and adhesion between the tip and the surface can influence the measurement and, in some cases, cause damage to delicate samples. If the cantilever is more flexible than the sample, it bends upon contact, generating a repulsive force on the tip. To ensure consistent force interaction, a feedback system continuously adjusts the cantilever's deflection, maintaining a steady force between the tip and the sample. This controlled movement allows for the creation of high-resolution surface images [29].

- **Dynamic mode (tapping):** Also known as intermittent contact mode, tapping mode helps reduce the effects of friction and adhesion by allowing the tip to make brief, periodic contact with the surface while oscillating at controlled amplitude. This method is less aggressive compared to contact mode because the amplitude is finely tuned to avoid the adhesive forces keeping the tip on the surface. In this mode, a cantilever vibrates near its natural resonance frequency, with an electronic feedback circuit keeping the oscillation amplitude constant. In this method, the tip and the sample remain in a dipole that never leads to any direct bonding, thus allowing both delicate materials and high surface resolution to be preserved [29].
- **Non-contact mode:** The AFM tip does not physically touch the sample but rather hovers over it during scanning. Thus, this technique bases its functionality on weak attractive forces, such as the force of attraction between the tip and the surface (van der Waals), to create an image. The system continuously monitors the variation in the oscillation amplitude, which allows scientists to study surface morphology without altering the sample. The technique becomes especially handy when it comes to the analysis of soft or highly sensitive samples, as it eliminates direct contact and lessens the risk of surface alteration. The topography is elaborately mapped out by calculating variations in oscillation amplitude, resonance frequency, and phase shifts. The vertical movement of the piezoscanner is then used to build an accurate height profile of the sample in question [30].

12.7 Microtribological and nanotribological techniques

By understanding friction, wear, and lubrication at very small scales, micro- and nanotribology provide information on how certain biological materials respond to friction, wear, and lubrication. Nanoindentation, in conjunction with scanning probe microscopy, analyzes the properties of biological tissues, such as hardness and elasticity. Such studies are crucial in improving the biomaterial used in medical applications so that they work efficiently inside the human body [32, 33]. Among these techniques, SPM stands out in biotribology because it allows scientists to see how biomaterials, tissues, and cells operate together under physiological conditions. Studying these nanoscale interactions provides the knowledge necessary for producing materials with better durability and resistance to wear. The threat from friction and degradation of biomaterials is, however, more intensified, as this process is more prevalent in biomedical applications such as joint prostheses, interfacial prostheses, dental restorations, and contact lenses, whereby the lifetime is determined on minimal friction

[32]. Developments in nanoscale tribology have seen the new generation of biomolecular force sensors and real-time imaging technology that allow a closer look at protein-lubricant interactions. These developments promise a more in-depth glimpse into the molecular mechanisms that control biotribological performance. A corollary of the new insights provided by these innovations will be their integration in computational models for designing biomaterials with increased similarity to their natural biological surfaces, resulting in dependable and long-lasting medical implements that enhance patient outcomes [33].

12.7.1 Nanoindentation

This depth-sensing indentation (DSI) approach relies on the controlled application of a certain load or depth on the surface to cause localized surface deformation (Fig. 12.6). The main components of a nanoindentation experiment usually comprise the test material, the sensors and actuators used to apply and measure the mechanical load and corresponding indenter displacement, and the indenter tip [34]. The latter piece is conventionally diamond-tipped, shaped into a sharp symmetric structure, such as the three-sided Berkovich pyramid. The pyramidal shape was selected in part for its nominally geometrically self-similar state, thus allowing a more easily interpretable analysis using continuum mechanics techniques. The standard nanoindentation test measures force and displacement as the indenter tip is driven into the surface of the test material following a certain loading and unloading profile. The general shape of the *P-h* curve differs from one material to another and usually indicates different mechanical properties [35].

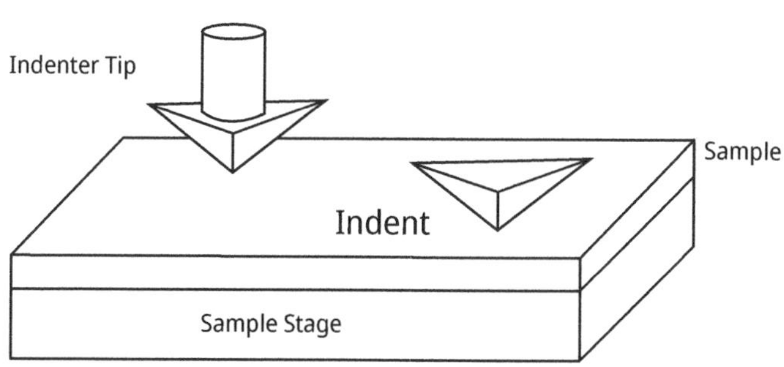

Fig. 12.6: Nanoindentation principle.

12.8 Surface force apparatus (SFA) in biotribology

The surface force apparatus (SFA) is an extremely sensitive tool in biotribology, studying intermolecular and surface interactions with nanometer resolution (Fig. 12.7) [36].

It allows the measurement of adhesion, friction, lubrication, and surface forces between biological materials and is, therefore, applicable in the study of synovial joint lubrication, tear film stability, and biocompatible coatings for prosthetics [37]. The SFA provides biological and synthetic lubricants with critical insights into their molecular-level performance by mimicking the physiological conditions, therefore allowing new biomedical understanding to develop.

The SFA works by bringing two extremely smooth surfaces close together (usually mica sheets) in very controlled ways and manipulating them, thus mimicking natural lubrication environments by coating them with biological molecules like lubricin, hyaluronic acid, or phospholipids [38]. The distance between the surfaces is measured through multiple-beam interferometry, enabling nanometer-scale accuracy. The SFA analyzes the critical interaction forces, including adhesion (whether surfaces attract or repel each other) and lubrication (the ability of biological fluids to minimize friction and deformation) - how biological surfaces respond under compressive forces - by applying both normal and lateral forces. Such observations would offer insights into understanding biological materials' tribological behavior. Furthermore, specialized SFA can work in a liquid environment, which can simulate biological conditions such as pH, ionic strength, temperature, and protein concentration on the lubrication mechanism [38, 39].

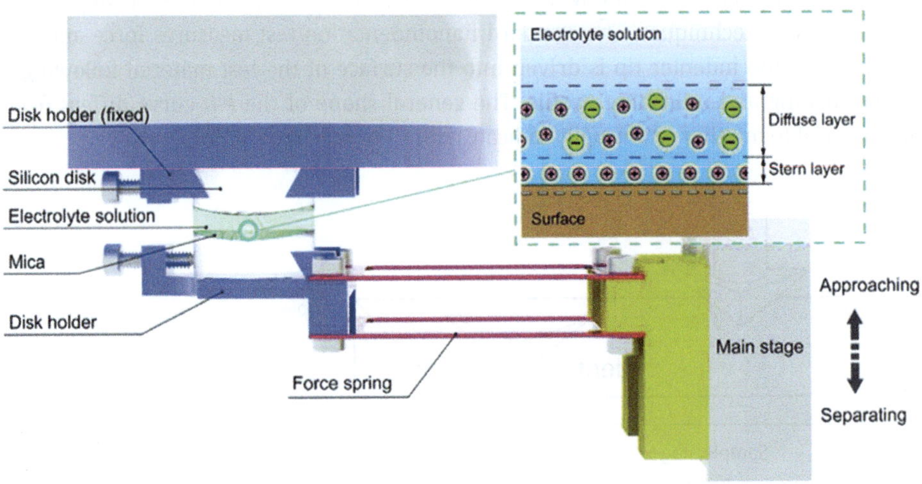

Fig. 12.7: Schematics of the surface force apparatus (SFA) system (licensed under CC by MDPI).

12.9 Applications of SFA in biotribology

a) **Synovial joint lubrication**

SFA has been instrumental in studying the lubrication mechanisms of synovial fluid, which contains hyaluronic acid and lubricin, two key molecules that reduce friction in joints [37]. By measuring how these molecules interact with cartilage-like surfaces, researchers can evaluate:

- The role of hydration lubrication in maintaining smooth joint movement
- How aging or diseases like osteoarthritis alter the lubricating properties of synovial fluid
- The effectiveness of synthetic biolubricants designed to enhance joint lubrication and protect against cartilage wear

b) **Tear film and ocular biotribology**

In ophthalmology, SFA has been used to analyze the lubrication properties of the tear film, which protects the corneal surface [37]. By studying the interactions of mucin, lipids, and aqueous layers, researchers can develop:

- Better treatments for dry eye syndrome, which is caused by poor lubrication
- Advanced contact lens coatings that improve comfort and reduce friction

c) **Oral and mucosal lubrication**

SFA studies have provided insights into saliva lubrication, which is essential for oral health and food processing [37]. By examining how mucin proteins reduce friction between oral surfaces, researchers can:

- Improve artificial saliva formulations for patients with dry mouth conditions
- Develop biomaterials that enhance dentures and oral prosthetics by reducing friction-related discomfort

d) **Soft tissue and biomedical coatings**

SFA has been applied to study the frictional behavior of soft tissues, such as skin and mucous membranes, under different physiological conditions [37]. This research has led to advancements in:

- Biomedical coatings that reduce wear on implants, stents, and catheters
- Bioinspired lubricants that mimic the adaptive properties of natural lubrication systems, improving long-term performance of medical devices

12.10 In vitro and in vivo biotribology testing

In vitro testing is a vital tool for studying biological tribosystems, allowing researchers to replicate real-world conditions in a controlled laboratory setting. This approach helps evaluate the performance of artificial joints, dental materials, and tissue substitutes by subjecting them to precisely controlled environments. To closely mimic natural physiological conditions, these experiments often use artificial synovial fluids and specialized wear simulators. Current research is focused on refining in vitro models to better understand how frictional shear affects cell behavior and biological processes [40–42].

To recreate the interactions between biomedical implants and biological surfaces, researchers conduct experiments where soft materials slide against cultured mammalian cells. These cells are maintained in conditions that support their natural functions and can be grown as monolayers or multilayers. In fundamental tribological studies, simplified models using one or two cell types are often employed, allowing for precise control over factors such as material selection, applied force, contact pressure, sliding speed, and test duration [43–45]. However, testing multiple cells in contact poses challenges due to the low elastic modulus and the delicate nature of cell monolayers [45]. Fluorescence microscopy is an invaluable tool in this field, commonly used by cell biologists to analyze cell layers before and after testing, and ideally in real time during experiments to gain deeper insights into their behavior.

In contrast, in vivo studies involve testing directly within living organisms to assess the long-term performance of biomedical implants and the biological response to wear debris [46–48]. These studies are critical for evaluating how implants function under actual physiological loading conditions, offering essential data for regulatory approvals and clinical applications. By integrating both in vitro and in vivo techniques, researchers acquire a more thorough knowledge of biotribological behavior, eventually leading to the creation of safer and more effective medical implants and therapies.

12.11 Advancements in biotribological testing

Recent progress in biotribology has thrust new technologies like high-resolution imaging and real-time wear monitoring, and multi-scale modeling, which provide a deeper understanding of material behavior and surface interactions. The appearance of 3D-printed biomaterials and bioengineered surfaces is another change that contexts this field by providing novel solutions to long-standing experiences in biomedical applications [49, 50]. In addition, the integration of machine learning and predictive modeling has allowed data analysis to reform, further providing researchers with the ability to explicitly optimize materials. Thus, the new computational biotribology allows simulating contact stresses and lubrication mechanisms in different scales, helping refine the designs of the implants before they go to clinical trials. Moreover, bioreactor-based wear testing systems that seek to mimic the dynamic nature of biological tribosystems have also augmented the accuracy and trustworthiness of experimental studies.

12.12 Conclusion

Friction, wear, and lubrication in natural biological systems are well understood through the influence of biotribology in biomedical engineering. In biological tribosystems, research has advanced greatly owing to novel experimental approaches, tribometers, AFM, SFA, and in vitro/in vivo testing. These techniques have provided an avenue to optimize biomaterials for implants, prostheses, contact lenses, and biomedical coatings with maximum biocompatibility, durability, and performance. The mimicry of biological lubrication is complex, owing to dynamic interactions at cellular and molecular levels. High-resolution imaging, machine learning, and computational modeling enhancements are improving material characterization and implant optimization. Advancements in biotribology will provide improved patient outcomes, dependability of medical devices, and quality of life for patients with biomedical implants and prostheses. The future will include research on bioinspired materials and self-healing lubricants to imitate natural lubrication systems; nanostructured coatings for improved wear resistance; bioreactor wear testing; multi-scale simulations for precision implant design; and predictive modeling for increased implant life and smaller failure rates.

References

[1] Zhou ZR, Jin ZM. (2015 Mar 1). Biotribology recent progresses and future perspectives. *Biosurface and Biotribology*, 1(1), 3–24.

[2] Wagner RM, Maiti R, Carré MJ, Perrault CM, Evans PC, Lewis R. (2021 Mar 1). Bio-tribology of vascular devices: A review of tissue/device friction research. *Biotribology*, 25, 100169.

[3] Singh H, Kaur K. (2024 Mar 28). Biotribology: Recent advancements, Applications, Challenges and Future Directions. In *Metal Matrix Composites: A Modern Approach to Manufacturing* (pp. 147–175).

[4] Freschi M, Paniz A, Cerqueni E, Colella G, Dotelli G. (2022 Jun 17). The twelve principles of green tribology: Studies, research, and case studies – A brief anthology. *Lubricants*, 10(6), 129.

[5] Sinha SK. (2021 Nov 23). Advancements in Biotribology. In *Progress in Lubrication and Nano-and Biotribology* (pp. 1–27). CRC Press.

[6] Hammood AS, Brittain R. (2021 Aug 26). Biomedical Tribology. In *Tribology and Sustainability* (pp. 295–322). CRC Press.

[7] Gebeshuber IC. (2013 Aug 28). Biomimetic inspiration regarding nano-tribology and materials issues in MEMS. In *Nano-tribology and Materials in MEMS* (pp. 53–79). Berlin, Heidelberg: Springer Berlin Heidelberg.

[8] Lanza A, Ruggiero A, Sbordone L. (2019 Jun 20). Tribology and dentistry: A commentary. *Lubricants*, 7(6), 52.

[9] Pal A, Gour R, Kumar A, Kumar A, Kumar A, Chowdhury N. (2024 Jun 30). Tribological measurements of human skin. In *Applications of Biotribology in Biomedical Systems* (pp. 149–165). Cham: Springer Nature Switzerland.

[10] Lievens CW, Rayborn E. (2022 Mar). Tribology and the ocular surface. *Clinical Ophthalmology*, 30, 973–980.

[11] Moghadasi K, Isa MS, Ariffin MA, Raja S, Wu B, Yamani M, Muhamad MR, Yusof F, Jamaludin MF, Bin Ab Karim Ms, Bin Yusoff N. (2022 Mar 1). A review on biomedical implant materials and the effect of friction stir based techniques on their mechanical and tribological properties. *Journal of Materials Research and Technology*, 17, 1054–1121.

[12] Faisal S, Chowdhury N, Kumar A, Kumar A, Kumar A, Choudhari A, Gupta AK. (2024 Nov 14). Lubricant: Soft materials (hydrogel). In *Performance Characterization of Lubricants* (pp. 191–212). CRC Press.

[13] Kovarik ML, Gach PC, Ornoff DM, Wang Y, Balowski J, Farrag L, Allbritton NL. (Jan 17 2012). Micro total analysis systems for cell biology and biochemical assays. *Analytical Chemistry*, 84(2), 516–540.

[14] Choudhari A, Gupta AK, Kumar A, Kumar A, Gupta A, Chowdhury N, Kumar A. (Jun 2024). Wear and friction mechanism study in knee and hip rehabilitation: A comprehensive review. *Applications of Biotribology in Biomedical Systems*, 30, 345–432.

[15] Simson D, Subbu SK. (Sep 4 2024). Investigating the tribological performance of bioimplants. In *Bioimplants Manufacturing* (pp. 258–283). CRC Press.

[16] Wollina U. (Jan 1 2011). Mechanical skin irritations due to textiles. In *Handbook of Medical Textiles* (pp. 248–268e). Woodhead Publishing.

[17] Wahlström J, Lyu Y, Matjeka V, Söderberg A. (Aug 15 2017). A pin-on-disc tribometer study of disc brake contact pairs with respect to wear and airborne particle emissions. *Wear*, 384, 124–130.

[18] Muralidharan K, Vignesh V, RV V, Govindaraju M, Baghad A, Narassima MS, Suyambulingam I, Siengchin S. (Sep 2024). Comprehensive overview of nano, micro, and macro tribometers in practice. *Journal of Bio-and Tribo-Corrosion*, 10(3), 44.

[19] Sukumaran AK, Zhang C, Nisar A, Rengifo S, Agarwal A. (Oct 15 2023). Tribological behavior of Al 6061 and Ti6Al4V alloys against lunar regolith simulants at extreme temperatures. *Wear*, 530, 205028.

[20] Thrush SJ, Comfort AS, Dusenbury JS, Nautiyal P, Elinski MB, Carpick RW, Demas NG, Gould BJ, Han X, Wang X, Qu H. (Nov 1 2022). Growth and morphology of thermally assisted sinterable zirconia nanoparticle tribofilm. *Tribology International*, 175, 107820.

[21] Coors T, Faqiri Y, Saure F, Pape F, Hassel T, Poll G. (Jul 2023). Wear of tailored forming steels. *Advanced Engineering Materials*, 25(13), 2201740.

[22] Poshtahani AG, Roostaie S, Azadi M. (May 1 2023). Plasma nitriding effect on tribological and corrosion properties of Stellite 6 and 12 PTA weld clad hardfaced on stainless steel 410. *Results in Surfaces and Interfaces*, 11, 100108.

[23] Lorenzo-Bonet E, Shaji S, Negrin-Gonzalez J, Perez-Acosta O, Ortega-Saenz JA, Hernandez-Rodriguez MA. (Jun 15 2023). Tribological analysis of ultra-high molecular weight polyethylene composites with boron carbide micro and nanoparticles. *Wear*, 523, 204861.

[24] Moon HH, Choi EJ, Yun SH, Kim YC, Premkumar T, Song C. (2022). Aqueous lubrication and wear properties of nonionic bottle-brush polymers. *RSC Advances*, 12(28), 17740–17746.

[25] Bhushan B. (2017). *MEMS/NEMS and BioMEMS/BioNEMS: Tribology, Mechanics, Materials and Devices* (pp. 1331–1416). Springer Handbook of Nanotechnology.

[26] Li M, Xun K, Zhu X, Liu D, Liu X, Jin X, Wu M. (Dec 12 2023). Research on AFM tip-related nanofabrication of two-dimensional materials. *Nanotechnology Reviews*, 12(1), 20230153.

[27] Thakur A, Kumar A, Kaya S, Marzouki R, Zhang F, Guo L. (Oct 3 2022). Recent advancements in surface modification, characterization and functionalization for enhancing the biocompatibility and corrosion resistance of biomedical implants. *Coatings*, 12(10), 1459.

[28] Beake BD, Harris AJ, Liskiewicz TW. (Jan 2016). Advanced nanomechanical test techniques. In *Materials Characterization: Modern Methods and Applications* (Vol. 5, pp. 17–18).

[29] Voigtländer B. (May 23 2019). *Atomic Force Microscopy*. Berlin, Germany: Springer.

[30] Modigunta JK, Vadivel S, Murali G, In I, Sawangphruk M. (Jun 28 2022). Atomic force microscopy: An advanced imaging technique – from molecules to morphologies. In *Microscopic Techniques for the Non-Expert* (pp. 115–136). Cham: Springer International Publishing.
[31] Voigtländer B, Voigtländer B. (2015). Technical aspects of atomic force microscopy (AFM). In *Scanning Probe Microscopy: Atomic Force Microscopy and Scanning Tunneling Microscopy* (pp. 157–175).
[32] Tsukruk VV, Singamaneni S. (Jan 9 2012). *Scanning Probe Microscopy of Soft Matter: Fundamentals and Practices*. John Wiley & Sons.
[33] Bhushan B, Marti O. (2017). Scanning probe microscopy – Principle of operation, instrumentation, and probes. In *Nanotribology and Nanomechanics: An Introduction* (pp. 33–93).
[34] Qian L, Zhao H. (Dec 11 2018). Nanoindentation of soft biological materials. *Micromachines*, 9(12), 654.
[35] Oyen M. (Jan 2013). Nanoindentation of biological and biomimetic materials. *Experimental Techniques*, 37(1), 73–87.
[36] Kan Y, Yang Q, Zhang X, Zhang Y. (Sep 9 2020). Effect of hofmeister anions on interfacial properties of mica surface in concentrated aqueous solution. *Coatings*, 10(9), 872.
[37] Drummond C, Richetti P. (Nov 6 2014). Surface Forces Apparatus in Nanotribology. In *Fundamentals of Friction and Wear on the Nanoscale* (pp. 17–34). Cham: Springer International Publishing.
[38] Andresen Eguiluz RC, Cook SG, Tan M, Brown CN, Pacifici NJ, Samak MS, Bonassar LJ, Putnam D, Gourdon D. (Jun 28 2017). Synergistic interactions of a synthetic lubricin-mimetic with fibronectin for enhanced wear protection. *Frontiers in Bioengineering and Biotechnology*, 5, 36.
[39] Jin R, Skedung L, Cazeneuve C, Chang JC, Rutland MW, Ruths M, Luengo GS. (Dec 1 2020). Bioinspired self-assembled 3D patterned polymer textures as skin coatings models: Tribology and tactile behavior. *Biotribology*, 24, 100151.
[40] Gordon E, Schimmel L, Frye M. (Jun 18 2020). The importance of mechanical forces for in vitro endothelial cell biology. *Frontiers in Physiology*, 11, 684.
[41] Jackson ML, Bond AR, George SJ. (Oct 2023). Mechanobiology of the endothelium in vascular health and disease: In vitro shear stress models. *Cardiovascular Drugs and Therapy*, 37(5), 997–1010.
[42] Wagner RM, Maiti R, Carré MJ, Perrault CM, Evans PC, Lewis R. (Mar 1 2021). Bio-tribology of vascular devices: A review of tissue/device friction research. *Biotribology*, 25, 100169.
[43] Sose AT, Joshi SY, Kunche LK, Wang F, Deshmukh SA. (2023). A review of recent advances and applications of machine learning in tribology. *Physical Chemistry Chemical Physics*, 25(6), 4408–4443.
[44] Findik F. (May 1 2014). Latest progress on tribological properties of industrial materials. *Materials & Design*, 57, 218–244.
[45] Wagner RM, Maiti R, Carré MJ, Perrault CM, Evans PC, Lewis R. (Mar 1 2021). Bio-tribology of vascular devices: A review of tissue/device friction research. *Biotribology*, 25, 100169.
[46] Thrivikraman G, Madras G, Basu B. (2014). In vitro/in vivo assessment and mechanisms of toxicity of bioceramic materials and its wear particulates. *Rsc Advances*, 4(25), 12763–12781.
[47] Li C, Guo C, Fitzpatrick V, Ibrahim A, Zwierstra MJ, Hanna P, Lechtig A, Nazarian A, Lin SJ, Kaplan DL. (Jan 2020). Design of biodegradable, implantable devices towards clinical translation. *Nature Reviews Materials*, 5(1), 61–81.
[48] Al-Zyoud W, Haddadin D, Hasan SA, Jaradat H, Kanoun O. (Oct 26 2023). Biocompatibility testing for implants: A novel tool for selection and characterization. *Materials*, 16(21), 6881.
[49] Sheng X, Wang A, Wang Z, Liu H, Wang J, Li C. (Mar 1 2022). Advanced surface modification for 3D-printed titanium alloy implant interface functionalization. *Frontiers in Bioengineering and Biotechnology*, 10, 850110.
[50] Li Y, Li H, Kong N, Lu S, Sun C, Tian R, Wang K, Li D, Yang P. (Jul 1 2024). Biotribological properties of 3D printed high-oriented short carbon fibre reinforced polymer composites for artificial joints. *Biomaterials Advances*, 161, 213888.

M. Karthik

13 Exploring the nanotribology: characterization, friction, and wear mechanisms at biological interfaces with implications for biomedicine and biotechnology

Abstract: Nanotribology, the study of friction, wear, and lubrication at the nanoscale, has become a vital discipline in advancing our understanding and application of interactions at biological interfaces. This field integrates fundamental principles of tribology with the complex requirements of biological and biomedical systems, leading to notable progress in several areas. By investigating nanoscale interactions between biological surfaces and materials, nanotribology has significantly improved the design and functionality of medical implants, drug delivery systems, and diagnostic tools. In the domain of medical implants, nanotribological research has led to the development of advanced wear-resistant coatings and surface modifications. These innovations enhance the durability and performance of implants by reducing friction and wear, thereby increasing their lifespan and reliability. For drug delivery systems, insights from nanotribology have facilitated the creation of nanoparticles designed to navigate biological barriers more effectively and deliver therapeutic agents with greater precision. This has resulted in improved targeted treatments and reduced side effects, representing a significant advancement in personalized medicine. Additionally, the field has contributed to the development of highly sensitive diagnostic tools, including nanoscale biosensors and advanced imaging techniques, which provide unprecedented resolution and accuracy for disease detection and biomolecular analysis.

Despite these achievements, several challenges persist. The inherent complexity and variability of biological systems make it difficult to replicate and study their tribological properties accurately. Furthermore, integrating nanomaterials with biological environments raises concerns about biocompatibility, stability, and scalability. Addressing these issues requires overcoming significant manufacturing and regulatory hurdles.

Future research in nanotribology must focus on developing more refined experimental techniques and models to better understand and predict nanoscale interactions in biological systems. Improving the integration of nanomaterials with biological interfaces will involve innovative approaches to enhance their biocompatibility and performance. Additionally, addressing scalability and cost-effectiveness will be crucial for the widespread adoption of these technologies. Standardizing methodologies and ensuring regulatory compliance will also play a critical role in advancing the field. By overcoming these challenges, nanotribology promises to drive further innovations in

https://doi.org/10.1515/9783111563176-013

healthcare and biotechnology, offering transformative solutions that enhance medical treatments, diagnostic capabilities, and our understanding of nanoscale interactions in biological systems.

Keywords: Biotribology, nanotribology, engineering, multidisciplinary, biomedical

13.1 Introduction

Nanotribology, a specialized field within tribology, focuses on the study of friction, wear, and lubrication at the nanoscale – an area where traditional macroscopic theories often fail to fully capture the complexities of material interactions. As the size of the system under investigation approaches the nanometer scale, quantum and molecular effects become increasingly significant, altering the fundamental mechanics of contact and movement [1]. The term "nanotribology" encompasses a broad range of phenomena, from the fundamental forces that govern interactions at this scale to the practical implications of these interactions in various applications. At the nanoscale, the primary forces influencing tribological behavior include van der Waals forces, which arise from transient dipole interactions between molecules, and electrostatic forces, which play a critical role when dealing with charged surfaces or particles. Additionally, mechanical forces at this scale can exhibit behaviors not seen at larger scales such as size-dependent friction and wear. These unique properties necessitate advanced experimental techniques to accurately measure and understand nanoscale interactions [2]. Atomic force microscopy (AFM), for instance, has become a cornerstone of nanotribological research, allowing scientists to probe surface properties with unprecedented resolution. Scanning tunneling microscopy (STM) and surface force apparatus (SFA) are other critical tools that provide insights into the nanoscale phenomena governing material behavior. Nanotribology has profound implications across various fields, particularly in the development of advanced materials and technologies. In electronics, for example, the increasing miniaturization of components requires an understanding of nanoscale friction and wear to ensure the reliability and longevity of devices. In biotechnology, nanotribology contributes to the design of more effective drug delivery systems and improved biomaterials by revealing how nanoscale interactions influence biological processes [3]. Additionally, advancements in nanotribology offer potential solutions to longstanding challenges in fields such as materials science and mechanical engineering by enabling the design of materials with tailored frictional properties and enhanced durability. Despite its advancements, the field faces significant challenges, including the inherent complexity of biological systems, the difficulty of replicating and standardizing nanoscale measurements, and the need for interdisciplinary approaches to address multifaceted problems [4]. As research progresses, continued development of sophisticated measurement techniques and innovative materials is expected to drive further breakthroughs, making nanotribology a vital area of study for both fundamental science and practical applications. Table 13.1 provides a comparison of the standard operating parameters for SFA, STM, and AFM/FFM in the context of micro/nanotribological research.

Tab. 13.1: Overview of the typical operating parameters for SFA, STM, and AFM/FFM in micro/nanotribological studies.

Operating Parameter	SFA	STM[a]	AFM/FFM
Radius of mating surface/tip	~10 mm	5–100 nm	5–100 nm
Radius of contact area	10–40 μm	n.a.	0.05–0.5 nm
Normal load	10–100 mN	n.a.	<0.1–500 nN
Sliding velocity	0.001–100 μm s^{-1}	0.02–200 μm s^{-1} (scan size ~1 × 1 nm^2 to 125 × 125 μm^2; scan rate <1–122 Hz)	0.02–200 μm s^{-1} (scan size ~1 × 1 nm^2 to 125 × 125 μm^2; scan rate <1–122 Hz)
Sample limitations	Typically atomically smooth, optically transparent mica; opaque ceramic, smooth surfaces can also be used	Electrically conducting samples	None of the above

13.2 Fundamental principles of nanotribology

Nanotribology is a specialized area within tribology that investigates friction, wear, and lubrication at the nanoscale, where classical mechanical theories often fall short and molecular or quantum effects become significant. This field focuses on how forces at the nanoscale differ from those at larger scales, impacting frictional behavior, wear mechanisms, and lubrication strategies [5].

At the nanoscale, the interactions between surfaces are influenced by a combination of van der Waals forces, electrostatic forces, and mechanical forces. Van der Waals forces, which result from transient dipole interactions, play a crucial role in determining the adhesive and frictional properties of materials. These forces become particularly important when surfaces are separated by just a few nanometers, leading to unique frictional and adhesive characteristics not observed at macroscopic scales.

Electrostatic forces are also significant, especially with charged surfaces or particles. The interactions between these charged entities can create attractive or repulsive forces, greatly affecting friction and wear [6]. For instance, electrical double layers in liquid environments can alter the stability and behavior of nanoscale particles and surfaces.

Mechanical forces at this scale have distinct features due to the smaller contact area and the effects of surface roughness and elasticity. Size-dependent friction phenomena can emerge, indicating that frictional forces may vary with the size of the contact area. Traditional friction laws, like Coulomb's law, often do not adequately describe these behaviors, prompting the need for new models that address the unique aspects of nanoscale interactions.

Wear mechanisms at the nanoscale can differ from those at larger scales as well. At this level, wear can result from processes such as atomic-scale plowing or material transfer due to adhesion, potentially altering the properties and performance of nanoscale devices.

Measuring friction, wear, and lubrication at this scale presents unique challenges that require advanced techniques. AFM is a key tool in this realm, providing high-resolution imaging and precise force measurements essential for studying nanoscale interactions [7]. AFM allows researchers to measure forces with nanometer accuracy, facilitating a deeper understanding of nanotribology. Other techniques, such as STM and SFA, offer additional insights. STM enables atomic-level imaging and manipulation, while SFA measures forces between closely spaced surfaces.

Investigating lubrication at the nanoscale involves understanding how lubricants behave in confined spaces and interact with surfaces at the molecular level. Traditional lubrication theories may not apply as lubricant behavior can be affected by molecular size, surface interactions, and external fields. Therefore, innovative models and experimental methods are necessary to accurately describe and optimize lubrication processes in nanoscale systems.

The principles of nanotribology have important implications for various applications, including the development of advanced materials, nanoscale devices, and biomedical technologies. By gaining insights into the fundamental interactions at the nanoscale, researchers can design materials with customized frictional properties, enhance the performance of nanoscale components, and create innovative solutions for technological challenges [8].

Despite the advancements in nanotribology, ongoing research is vital to overcome existing challenges and deepen our understanding of nanoscale interactions. As new technologies and methodologies emerge, the field is positioned to reveal further insights into the principles governing nanoscale friction, wear, and lubrication, ultimately leading to innovations that benefit multiple industries.

13.3 Biological interfaces in nanotribology

Biological interfaces, the points of contact between biological systems and external materials or structures, are critical to understanding how nanoscale interactions influence biological functions and processes. Nanotribology at these interfaces explores how friction, wear, and lubrication phenomena at the nanoscale affect are affected by biological components. These interactions have profound implications for a range of fields, from medicine to biotechnology, and require an understanding of both biological and nanoscale tribological principles.

At the core of biological interfaces are cell membranes and complex structures composed of lipids, proteins, and carbohydrates. These membranes act as barriers and gatekeepers, regulating the movement of substances into and out of cells and mediating interactions with the external environment [9]. The friction and wear properties of these membranes can significantly impact cellular processes such as adhesion, migration, and signal transduction. Nanotribological studies of cell membranes often involve probing their mechanical properties and interactions with nanoparticles or other surfaces. Techniques like AFM are employed to measure the forces involved in membrane deformation and adhesion, providing insights into how nanoscale particles or devices might interact with or affect cellular function.

Protein-protein and protein-surface interactions are also crucial components of biological interfaces. Proteins often interact with each other and with surfaces in highly specific ways, governed by their molecular structures and the properties of the surfaces they encounter. At the nanoscale, surface modifications can dramatically alter these interactions, influencing everything from protein adsorption to the formation of biofilms [10, 11]. Understanding these interactions is essential for designing surfaces that can promote or inhibit specific biological responses such as enhancing the biocompatibility of implants or developing sensors for detecting specific biomolecules.

Biological lubricants, such as synovial fluid in joints, play a critical role in reducing friction and wear in biological systems. Synovial fluid, a viscous liquid found in the cavities of synovial joints, contains a complex mixture of proteins, glycoproteins, and hyaluronic acid. This fluid acts as a natural lubricant, reducing friction between cartilage surfaces and facilitating smooth joint movement. Nanotribological studies of biological lubricants focus on understanding how these fluids behave at the nanoscale, including how their molecular components interact with surfaces and how they affect frictional properties in confined geometries. Insights gained from these studies can lead to improvements in artificial joint design and the development of new materials for orthopedic applications.

Interactions between nanoscale materials and biological systems are the key focus of nanotribology in biological contexts. Nanoscale materials, such as nanoparticles or nanostructured surfaces, can have unique interactions with biological molecules and cells. These interactions can influence drug delivery, cellular uptake, and the overall effectiveness of nanomedicines [12]. For instance, nanoparticles designed for drug delivery must be engineered to interact with cellular membranes in a controlled manner to ensure that drugs are delivered effectively without causing unintended damage. Nanotribological studies can help optimize the design of these particles by providing detailed information on how they interact with biological surfaces at the nanoscale.

Surface modifications and coatings play a significant role in enhancing the biocompatibility of implants and other medical devices. By altering the nanoscale properties of surfaces, researchers can improve the performance and longevity of these devices, reducing the risk of wear and failure. For example, coating materials with specific functional groups can promote favorable interactions with biological tissues, leading to better integration and reduced risk of rejection. Nanotribology helps in developing these coatings by providing insights into how surface properties affect friction, wear, and adhesion at the nanoscale.

In addition to medical applications, nanotribology at biological interfaces has implications for diagnostics and imaging. Nanoscale probes and sensors can be used to detect specific biological molecules with high sensitivity. By understanding how these probes interact with biological surfaces, researchers can design more effective diagnostic tools that provide accurate and reliable results. For example, nanostructured probes can be engineered to bind selectively to specific biomarkers, enabling early detection of diseases or monitoring of biological processes.

Despite the advancements in understanding biological interfaces through nanotribology, several challenges remain. The complexity and variability of biological systems pose difficulties in replicating and standardizing nanoscale measurements. Biological interfaces are often dynamic and responsive to environmental changes, making it challenging to predict and control interactions accurately. Additionally, interdisciplinary approaches are required to address the multifaceted nature of biological systems, combining insights from nanotechnology, biology, and materials science to develop effective solutions.

13.4 Nanoscale characterization of biological surfaces and interfaces

The nanoscale characterization of biological surfaces and interfaces is essential for understanding the interactions between biological systems and nanoscale materials. This area of research is crucial for applications in drug delivery, biomaterials development, and diagnostic technologies [13]. The ability to probe biological surfaces and interfaces at the nanoscale provides insights into their mechanical, chemical, and biological properties, which are critical for designing effective and biocompatible materials. Various advanced techniques are employed to achieve detailed characterization of biological surfaces and interfaces at the nanoscale.

13.4.1 Atomic force microscopy (AFM)

AFM is an effective method for examining biological surfaces at the nanoscale, as illustrated in Fig. 13.1. It employs a sharp tip attached to a cantilever to probe the sample's surface. During this process, the tip interacts with the surface, detecting various forces like van der Waals, electrostatic, and mechanical forces. This information is then utilized to create high-resolution images and assess surface properties [14].

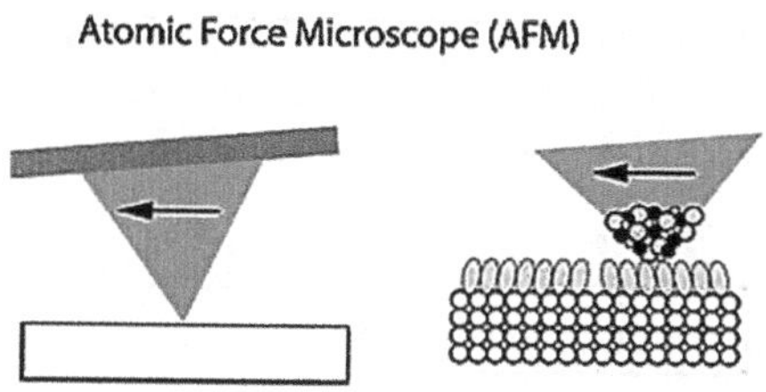

Fig. 13.1: Atomic force microscopy.

- **Principles and operation:** AFM functions by monitoring the deflection of a cantilever as the tip moves across the surface. This deflection is measured by directing a laser beam onto the cantilever, which then reflects onto a photodetector. AFM can be operated in several modes, including contact mode, where the tip makes direct contact with the surface, and non-contact mode, where the tip remains close to the surface without touching it.
- **Applications in biological systems:** AFM is used to study the topography of biological surfaces such as cell membranes and proteins. It can measure mechanical properties like stiffness and elasticity, which are important for understanding cell behavior and interactions with biomaterials. AFM is also used to investigate the binding interactions between biomolecules and surfaces, providing insights into processes like cell adhesion and protein adsorption.

13.4.2 Scanning electron microscopy (SEM)

Scanning electron microscopy (SEM) provides high-resolution images of biological surfaces by scanning them with a focused electron beam. SEM can reveal detailed information about surface morphology, structure, and texture at the nanoscale [15]:
- **Principles and operation:** SEM involves scanning a specimen with a focused electron beam and detecting the secondary electrons emitted from the surface. The detected electrons are used to create an image that reflects the surface topography and morphology. SEM can achieve high magnifications and resolutions, allowing for the observation of nanoscale features.
- **Applications in biological systems:** SEM is used to examine the surface structure of biological samples such as tissues, cells, and microorganisms. It is valuable

for studying the morphology of cell membranes, the distribution of cellular organelles, and the interaction of cells with nanomaterials. SEM can also be used to investigate the effects of surface modifications on biological interfaces.

13.4.3 Scanning tunneling microscopy (STM)

- **Principles and operation:** STM is operated by scanning a sharp metallic tip very close to a conductive surface [16]. The tunneling current between the tip and the surface is measured and used to generate images. The resolution of STM can reach atomic levels, making it suitable for studying surfaces with atomic precision.
- **Applications in biological systems:** Although STM is less commonly used for biological samples due to the requirement for conductive surfaces, it has been employed to study the surface properties of conductive biological components and the interactions between biological molecules and conductive substrates.

13.4.4 Surface force apparatus (SFA)

The SFA measures the forces between two surfaces as they approach and retract from each other as shown in Fig. 13.2. It provides detailed information about intermolecular forces and surface interactions:
- **Principles and operation:** SFA is involved in placing two surfaces together and measuring the forces between them as they are moved toward and away from each other. The instrument measures forces such as van der Waals, electrostatic, and hydrodynamic forces. It can provide insights into the interactions at very short distances and the effects of surface modifications.
- **Applications in biological systems:** SFA is used to study the interactions between biological molecules and surfaces such as protein adsorption and cell adhesion. It is also employed to investigate the effects of surface coatings and modifications on biological interactions, providing valuable information for designing biocompatible materials.

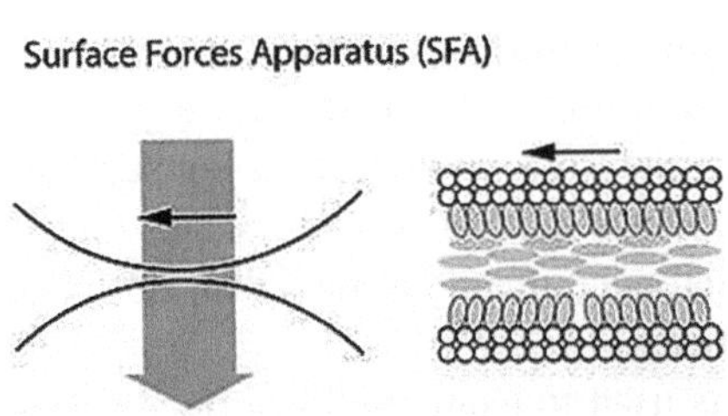

Fig. 13.2: Surface force apparatus.

13.4.5 X-ray photoelectron spectroscopy (XPS)

- **Principles and operation:** XPS involves in irradiating a sample with X-rays and detecting the emitted photoelectrons. The binding energies of the photoelectrons are used to identify the elements present on the surface and their chemical states [17]. XPS provides information on surface chemistry with high sensitivity.
- **Applications in biological systems:** XPS is used to analyze the chemical composition of biological surfaces and interfaces including the identification of surface-bound biomolecules and the effects of surface modifications. It is valuable for understanding how chemical changes at the surface influence biological interactions.

13.4.6 Contact angle measurement

Contact angle measurement assesses the wettability of surfaces by measuring the angle formed between a liquid droplet and the surface. This technique provides information about surface energy and interactions with liquids [18]:

- **Principles and operation:** Contact angle measurement involves placing a liquid droplet on a surface and measuring the angle between the liquid and the surface. The contact angle reflects the surface's hydrophobicity or hydrophilicity and provides insights into surface interactions with biological fluids.
- **Applications in biological systems:** Contact angle measurements are used to study the wettability of biological surfaces and biomaterials. They provide information on how biological fluids interact with surfaces, which is important for applications such as tissue engineering and the development of medical devices.

13.4.7 Atomic force microscopy-based techniques

In addition to standard AFM imaging, several specialized AFM-based techniques are used to probe biological surfaces and interfaces:

- **Force spectroscopy:** Measures the force between the AFM tip and the surface as a function of distance. It is used to study binding interactions between biomolecules and surfaces as well as the mechanical properties of biological samples.
- **AFM-based nanoindentation:** Provides information on the hardness and elastic modulus of surfaces by applying a controlled force with the AFM tip and measuring the resulting indentation. It is useful for characterizing the mechanical properties of biological tissues and cells.

13.4.8 Fluorescence microscopy

Fluorescence microscopy, while not a direct nanoscale technique, is often used in conjunction with nanoscale characterization methods to study biological systems [19]. It involves tagging specific molecules with fluorescent dyes and imaging their distribution and interactions:

- **Principles and operation:** Fluorescence microscopy uses fluorescent dyes or proteins that emit light upon excitation with a specific wavelength. The emitted light is detected to produce images of the tagged molecules within the biological sample.
- **Applications in biological systems:** Fluorescence microscopy is used to visualize and track specific biomolecules, study cellular processes, and investigate interactions at biological interfaces. It provides complementary information to nanoscale characterization techniques.

The nanoscale characterization of biological surfaces and interfaces is critical for advancing our understanding of biological systems and developing innovative materials and technologies. Techniques such as AFM, SEM, STM, and SFA provide detailed insights into surface properties, interactions, and mechanical characteristics at the nanoscale. XPS and contact angle measurements offer information on surface chemistry and wettability, while AFM-based techniques and fluorescence microscopy provide additional details on mechanical properties and molecular interactions [20]. Together, these techniques enable researchers to explore the complex interactions between biological systems and nanoscale materials, leading to advancements in areas such as drug delivery, biomaterials development, and diagnostics.

13.5 Molecular mechanisms of friction and wear in biological systems

The molecular mechanisms of friction and wear in biological systems are integral to understanding how biological surfaces interact under various conditions and how these interactions affect physiological and pathological processes. At the nanoscale, friction and wear phenomena are governed by complex interactions between molecules, proteins, and cellular components [21]. These mechanisms have significant implications for the design of medical devices, the understanding of joint diseases, and the development of biomaterials. Here, we explore the molecular mechanisms underlying friction and wear in biological systems, focusing on cellular and molecular interactions, lubricating mechanisms, and the effects of surface properties.

13.5.1 Molecular interactions and surface mechanics

- **Van der Waals forces:** These are weak, short-range forces arising from transient dipole interactions between molecules. In biological systems, van der Waals forces contribute to the adhesive properties of surfaces such as those of cell membranes or protein-coated implants. These forces are crucial in understanding cell adhesion and the interaction between cells and biomaterials.
- **Electrostatic forces:** Electrostatic interactions occur between charged molecules or surfaces. In biological systems, these forces can influence protein adsorption, cell adhesion, and the formation of biofilms. For instance, electrostatic forces play a role in the interaction between negatively charged cell membranes and positively charged nanoparticles or proteins.
- **Hydrogen bonding:** Hydrogen bonds are strong dipole-dipole interactions between molecules containing hydrogen and electronegative atoms like oxygen or nitrogen. In biological systems, hydrogen bonds are important for protein folding, enzyme-substrate interactions, and the formation of cellular structures like the extracellular matrix. They also contribute to the frictional properties of surfaces in contact such as those in synovial joints.

13.5.2 Lubrication mechanisms

Lubrication is a critical mechanism for reducing friction and wear in biological systems [22]. Biological fluids and surface coatings provide lubrication by creating a thin film between surfaces, minimizing direct contact and reducing friction:

- **Synovial fluid:** In joints, synovial fluid acts as a natural lubricant. It consists of hyaluronic acid, lubricin, and other glycoproteins that form a viscoelastic film between the articular surfaces. This fluid reduces friction and wear by preventing direct contact between cartilage surfaces, which helps in smooth joint movement. The molecular structure of hyaluronic acid and lubricin is crucial for maintaining the lubricating properties of synovial fluid.
- **Glycosaminoglycans (GAGs) and proteoglycans:** GAGs and proteoglycans are components of the extracellular matrix in tissues like cartilage and tendons. They contribute to lubrication by attracting and holding water, forming a gel-like matrix that reduces friction. The interaction between GAGs and surface proteins affects the wear resistance of these tissues.
- **Mucin proteins:** Mucin proteins present in mucus and other bodily fluids act as lubricants by forming thick, gel-like layers that reduce friction. These proteins are involved in protecting epithelial surfaces and reducing wear in areas such as the gastrointestinal tract and the eyes.

13.5.3 Surface properties and frictional behavior

The frictional behavior of biological surfaces is influenced by their surface properties including roughness, elasticity, and chemical composition:

- **Surface roughness:** The roughness of biological surfaces affects friction and wear. For instance, the roughness of cartilage surfaces can influence the frictional resistance in joints. Microscopic and nanoscale roughness can lead to increased contact points and higher friction, while smoother surfaces tend to reduce friction and wear.
- **Elasticity and stiffness:** The mechanical properties of biological surfaces, such as their elasticity and stiffness, play a role in friction and wear. For example, the elasticity of cartilage affects its ability to absorb and distribute mechanical loads, impacting the frictional resistance in joints.
- **Chemical composition:** The chemical composition of biological surfaces, including the presence of proteins, lipids, and carbohydrates, influences friction and wear. Surface modifications, such as the adsorption of proteins or the formation of biofilms, can alter the frictional properties of surfaces.

13.5.4 Wear mechanisms in biological systems

Wear mechanisms in biological systems are complex and can involve several processes including abrasion, adhesion, and fatigue:

- **Abrasion:** Abrasive wear occurs when hard particles or rough surfaces slide across softer biological tissues, causing material removal. In joints, abrasive wear can result from the interaction between roughened cartilage surfaces or the presence of particulate debris. This type of wear can lead to the degradation of cartilage and contribute to conditions such as osteoarthritis.
- **Adhesion:** Adhesive wear involves the transfer of material from one surface to another due to adhesive interactions. In biological systems, adhesive wear can occur when proteins or cells adhere to surfaces and are displaced during sliding contact. This process can lead to changes in surface properties and contribute to wear in medical implants or prosthetic devices.
- **Fatigue:** Fatigue wear results from repeated loading and unloading cycles, leading to the gradual failure of materials. In biological systems, fatigue wear can affect tissues and implants subjected to cyclic mechanical stress. For example, the repetitive loading of cartilage in joints can lead to fatigue-induced damage and wear.

13.5.5 Implications for medical devices and biomaterials

- **Implant design:** The design of implants and prosthetic devices must consider the frictional and wear properties of materials to ensure long-term functionality and biocompatibility. Surface modifications and coatings can be used to improve lubrication and reduce wear, enhancing the performance and longevity of implants.
- **Biomaterial development:** The development of biomaterials for tissue engineering and regenerative medicine requires an understanding of friction and wear mechanisms. Biomaterials must be designed to mimic the mechanical properties and lubrication characteristics of natural tissues to ensure successful integration and function.
- **Joint disease management:** Insights into friction and wear mechanisms can aid in the management and treatment of joint diseases such as osteoarthritis. Strategies to reduce friction and wear in affected joints, such as the use of lubricating agents or joint resurfacing techniques, can help alleviate symptoms and improve joint function.

The molecular mechanisms of friction and wear in biological systems involve complex interactions between molecules, proteins, and cellular components. Van der Waals forces, electrostatic interactions, and hydrogen bonding play crucial roles in determining frictional behavior [23]. Lubrication mechanisms, such as the presence of synovial fluid and GAGs, are essential for reducing friction and wear. Surface properties, including roughness and elasticity, influence frictional behavior, while wear mechanisms such as abrasion, adhesion, and fatigue contribute to material degradation. Understanding these mechanisms is vital for designing effective medical devices, developing advanced biomaterials, and managing joint diseases, ultimately improving health outcomes and enhancing the performance of medical technologies.

13.6 Applications of nanotribology in biomedicine and biotechnology

Nanotribology, the study of friction, wear, and lubrication at the nanoscale, has transformative implications for biomedicine and biotechnology [24]. By understanding and manipulating nanoscale interactions, researchers and engineers can develop advanced materials and technologies with enhanced performance and functionality. The applications of nanotribology in these fields are diverse, spanning from improving medical implants to advancing drug delivery systems and developing novel diagnostic tools.

13.6.1 Medical implants and prosthetics

13.6.1.1 Wear-resistant coatings

Nanotribology contributes significantly to the development of wear-resistant coatings for medical implants and prosthetics. For example, hip and knee implants experience significant mechanical stress and friction, leading to wear over time. Nanotribological studies help design coatings that reduce friction and wear, enhancing the durability and longevity of these implants. Coatings made from nanostructured materials, such as titanium nitride or diamond-like carbon, offer improved wear resistance and biocompatibility.

13.6.1.2 Surface modification

Surface modifications at the nanoscale can improve the performance and integration of implants. Techniques such as plasma spraying, chemical vapor deposition, and atomic layer deposition are used to apply nanoscale coatings that enhance the interaction between the implant and surrounding biological tissues. For instance, nanoscale roughness can promote better osseointegration of bone implants, improving their stability and function.

13.6.1.3 Lubrication in joint implants

The study of synovial fluid and other natural lubricants at the nanoscale helps in designing better lubrication systems for joint implants. Nanotribological research can optimize the composition and delivery of lubricants to reduce friction and wear in artificial joints, mimicking the natural lubrication mechanisms found in healthy joints.

13.6.2 Drug delivery systems

13.6.2.1 Nanoparticle-based delivery

Nanotribology is essential for advancing nanoparticle-based drug delivery systems. By studying the frictional and adhesive interactions between nanoparticles and cellular membranes, researchers can design particles that effectively penetrate cells and release therapeutic agents. For instance, liposomes and polymeric nanoparticles are engineered to optimize drug uptake and ensure controlled release, thereby enhancing therapeutic effectiveness.

13.6.2.2 Targeted delivery

Nanotribological studies enable precise control over how nanoparticles interact with specific biological targets. By altering the surface properties of nanoparticles – such as incorporating targeting ligands or adjusting surface charge – researchers can enhance the specificity of drug delivery to diseased tissues or cells. This approach helps minimize side effects and improves treatment outcomes.

13.6.2.3 Overcoming biological barriers

Nanotribology is instrumental in overcoming challenges related to drug delivery across biological barriers including the blood-brain barrier. By analyzing and optimizing nanoscale interactions between drug carriers and these barriers, researchers can create strategies that enhance drug penetration and effectiveness for treating central nervous system disorders.

13.6.3 Diagnostic tools

13.6.3.1 Nanoscale sensors

Nanotribological principles are used to develop highly sensitive nanoscale sensors for diagnostic applications. For instance, surface plasmon resonance sensors and cantilever-based biosensors rely on nanoscale interactions to detect specific biomolecules with high sensitivity. These sensors are used for early disease detection, monitoring biomolecular interactions, and analyzing complex biological samples.

13.6.3.2 Imaging techniques

Advanced imaging techniques, such as AFM and STM, provide detailed information about biological surfaces and interfaces. These techniques are used to visualize and analyze cellular structures, protein interactions, and molecular conformations, aiding in the diagnosis and understanding of various diseases.

13.6.4 Tissue engineering

13.6.4.1 Scaffold design

Nanotribology contributes to the design of scaffolds for tissue engineering by optimizing their mechanical properties and surface interactions. Scaffolds with nanoscale features can better mimic the extracellular matrix, providing appropriate mechanical support and promoting cell adhesion, proliferation, and differentiation. Techniques such as electrospinning and 3D printing are used to create scaffolds with precise nanoscale structures.

13.6.4.2 Cellular interactions

Understanding the frictional and adhesive interactions between cells and scaffold materials at the nanoscale is crucial for tissue engineering. Nanotribological studies help design scaffolds that enhance cell attachment and growth, improving the functionality and integration of engineered tissues.

13.6.5 Biosensors and bioelectronics

13.6.5.1 Nanostructured electrodes

Nanotribology is used to develop nanostructured electrodes for biosensors and bioelectronics. These electrodes are designed to have high-surface area and optimal frictional properties to enhance the performance of biosensors. By optimizing the nanoscale surface features and material properties, researchers can improve the sensitivity and reliability of biosensors used for detecting biomolecules and monitoring physiological processes.

13.6.5.2 Wearable biosensors

Wearable biosensors that monitor physiological parameters, such as glucose levels or heart rate, benefit from nanotribological research. Nanostructured materials and coatings improve the performance, durability, and comfort of wearable sensors, making them more effective and user-friendly.

13.6.6 Antimicrobial and antifouling surfaces

13.6.6.1 Antimicrobial coatings

Nanotribological studies contribute to the development of antimicrobial coatings for medical devices and surfaces. Nanostructured coatings with antimicrobial properties can reduce the risk of infections and biofilm formation on medical implants and surfaces. These coatings are designed to inhibit the growth of microorganisms while maintaining the desired frictional and mechanical properties.

13.6.6.2 Antifouling coatings

In addition to antimicrobial coatings, nanotribology helps design antifouling surfaces that prevent the adhesion of biological molecules and cells. Such coatings are important for preventing unwanted biofouling on medical devices and diagnostic tools, ensuring their performance and longevity.

13.6.7 Understanding biological friction and wear

13.6.7.1 Joint health research

Nanotribology provides insights into the mechanisms of friction and wear in natural joints, helping to understand conditions such as osteoarthritis. By studying the nanoscale interactions between cartilage surfaces and synovial fluid, researchers can develop better treatments and preventive measures for joint diseases.

13.6.7.2 Cellular and molecular dynamics

The study of friction and wear at the cellular and molecular levels helps in understanding cellular processes and interactions with materials. This knowledge is valuable for designing biomaterials that interact with biological systems in a controlled and predictable manner, improving the outcomes of medical treatments and interventions.

Nanotribology has a profound impact on biomedicine and biotechnology by providing insights into the nanoscale interactions that govern the performance and functionality of medical devices, drug delivery systems, diagnostic tools, and tissue engineering applications [25]. By optimizing frictional, wear, and lubrication properties at the nanoscale, researchers and engineers can develop advanced materials and technologies that enhance the effectiveness and safety of medical treatments, improve patient outcomes, and advance the field of biotechnology. As our understanding of nanoscale interactions continues to grow, the applications of nanotribology in these fields are expected to expand, leading to innovative solutions and breakthroughs in healthcare and technology.

13.7 Challenges and future directions in nanotribology for biomedicine and biotechnology

As the field of nanotribology advances, particularly in its applications to biomedicine and biotechnology, several challenges arise, alongside promising future directions that could enhance the effectiveness and impact of this research [26]. Addressing these challenges and exploring new avenues will be crucial for the continued development of innovative solutions in healthcare and technology.

13.7.1 Challenges: complexity of biological environments

Biological systems are highly complex and dynamic, presenting significant challenges in replicating and studying their tribological properties accurately at the nanoscale. Key challenges include:

- **Heterogeneity of biological surfaces:** Biological surfaces, such as cell membranes and tissues, are not uniform. They exhibit varying degrees of roughness, elasticity, and chemical composition, complicating the study and prediction of their frictional and wear properties.
- **Dynamic nature:** Biological systems are in constant flux due to processes like cell movement, tissue remodeling, and fluid flow. Capturing and understanding these dynamic interactions at the nanoscale require advanced techniques and methodologies.

13.7.2 Integration of nanomaterials with biological systems

Integrating nanomaterials with biological systems poses several challenges:

- **Biocompatibility:** Ensuring that nanomaterials are biocompatible and do not elicit adverse reactions is a critical challenge [27]. Nanomaterials must be designed to avoid toxicity and immunogenic responses while achieving their intended function.

- **Stability and degradation:** Nanomaterials must maintain their performance over time in the biological environment, which often involves complex interactions with fluids, enzymes, and cells that can affect their stability and degradation.

13.7.3 Scale-up and manufacturing

Translating nanotribological research from the laboratory to practical applications involves several obstacles:

- **Manufacturing precision:** Achieving the precision required for nanostructured materials and coatings on a large scale is technically demanding and costly. Scaling up fabrication processes while maintaining nanoscale accuracy remains a significant challenge.
- **Cost-effectiveness:** The production and integration of nanomaterials can be expensive. Developing cost-effective methods for manufacturing and incorporating these materials into biomedical devices and technologies is crucial for widespread adoption.

13.7.4 Regulatory and safety concerns

The use of nanomaterials in biomedicine and biotechnology raises regulatory and safety issues:

- **Regulation:** Nanomaterials are subject to regulatory scrutiny, which can vary by region and application. Ensuring compliance with regulations and standards for safety, efficacy, and environmental impact is essential.
- **Safety assessment:** Comprehensive safety assessments, including long-term studies on potential toxicity and environmental impact, are required to ensure that nanomaterials do not pose risks to patients or ecosystems.

13.7.5 Standardization of measurement techniques

Standardizing measurement techniques for nanoscale friction and wear is crucial for ensuring reliable and reproducible results:

- **Consistency in methodology:** Variations in experimental conditions, measurement techniques, and data interpretation can lead to inconsistencies in results. Developing standardized methodologies and protocols is necessary for advancing research and comparing findings across studies.
- **Resolution and sensitivity:** Improving the resolution and sensitivity of measurement techniques to accurately capture nanoscale interactions is an ongoing chal-

lenge. Advances in instrumentation and data analysis are needed to address these limitations.

13.7.6 Future directions: advanced nanomaterial design

Future research should focus on designing advanced nanomaterials with tailored properties for specific biomedical and biotechnological applications:
- **Smart nanomaterials:** Developing smart nanomaterials that respond to environmental changes, such as pH or temperature, could lead to more effective drug delivery systems and responsive medical implants.
- **Bio-inspired materials:** Drawing inspiration from natural systems to design nanomaterials with optimized frictional and wear properties can lead to innovations in biomimetic materials and coatings.

13.7.7 Integration of nanotribology with other technologies

Combining nanotribology with other advanced technologies holds promise for enhancing biomedical and biotechnological applications:
- **Nanorobotics and precision medicine:** Integrating nanotribological principles with nanorobotics could lead to precise, targeted interventions at the cellular or molecular level. This integration could revolutionize surgical techniques, drug delivery, and diagnostics.
- **Advanced imaging and sensing:** Combining nanotribology with advanced imaging and sensing technologies, such as high-resolution AFM and optical microscopy, can provide deeper insights into nanoscale interactions and improve the development of diagnostic tools and biomaterials.

13.7.8 Personalized medicine

Nanotribology can contribute to personalized medicine by enabling the development of customized materials and devices tailored to individual patient needs:
- **Tailored implants and prosthetics:** Using nanotribological data to design implants and prosthetics that match the mechanical and biological properties of individual patients could enhance the fit, function, and longevity of these devices.
- **Customized drug delivery systems:** Developing drug delivery systems that are personalized based on patient-specific factors, such as genetic profile or disease state, can improve therapeutic outcomes and minimize side effects.

13.7.9 Enhanced understanding of biological tribology

Gaining a deeper understanding of biological tribology can lead to improved solutions for various medical and biotechnological challenges:
- **Mechanistic insights:** Further research into the molecular mechanisms of friction and wear in biological systems can inform the development of better therapeutic and diagnostic technologies [28]. Understanding how biological surfaces interact at the nanoscale can lead to innovations in tissue engineering, joint health, and disease treatment.
- **Long-term studies:** Conducting long-term studies to assess the durability and performance of nanomaterials and devices in biological environments will provide valuable data for optimizing their design and application.

13.8 Sustainable and eco-friendly technologies

Developing sustainable and eco-friendly technologies is an important future direction:
- **Green nanotechnology:** Focusing on environmentally friendly methods for synthesizing and disposing of nanomaterials can reduce the environmental impact and promote the responsible use of nanotechnology in biomedicine and biotechnology.
- **Recycling and reuse:** Implementing strategies for recycling and reusing nanomaterials and devices can contribute to sustainability and reduce waste associated with their use.

The field of nanotribology holds great promise for advancing biomedicine and biotechnology, but it faces several challenges that need to be addressed [29]. Overcoming these challenges and exploring future directions will be crucial for developing innovative solutions that enhance medical treatments, improve diagnostic technologies, and advance our understanding of nanoscale interactions in biological systems [30]. By focusing on advanced material design, integration with other technologies, personalized medicine, and sustainable practices, the field of nanotribology can continue to make significant contributions to healthcare and technology.

13.9 Conclusion

Nanotribology at biological interfaces has emerged as a pivotal field, bridging the gap between fundamental tribological principles and their application in biomedical and biotechnological contexts. By examining friction, wear, and lubrication at the nanoscale, this field has led to significant advancements in medical implants, drug delivery systems, and diagnostic tools. The detailed understanding of nanoscale interactions between biological surfaces and materials has enabled the development of innovative technologies that enhance the performance, durability, and effectiveness of various medical devices and treatments.

In the realm of medical implants, nanotribology has revolutionized the design and optimization of wear-resistant coatings and surface modifications. These advancements have improved the longevity and functionality of implants by reducing friction and wear, thereby enhancing patient outcomes and device reliability. Similarly, in drug delivery systems, nanotribological insights have facilitated the development of nanoparticles that can precisely navigate biological barriers and deliver therapeutic agents more effectively. This has led to more targeted treatments with minimized side effects, marking a significant leap forward in personalized medicine. The field has also contributed to the creation of highly sensitive diagnostic tools, such as nanoscale biosensors and advanced imaging techniques, which offer unprecedented resolution and accuracy in disease detection and biomolecular analysis. Despite these notable achievements, several challenges remain in the field of nanotribology at biological interfaces. The complexity of biological systems poses a significant obstacle, as these systems are dynamic and heterogeneous, with varying surface properties and interactions that are difficult to replicate and study. Additionally, integrating nanomaterials with biological systems requires overcoming issues related to biocompatibility, stability, and scalability. Ensuring that nanomaterials perform effectively within the complex biological environment while maintaining safety and minimizing adverse effects is crucial. Furthermore, translating nanoscale insights into practical applications involves addressing challenges in manufacturing precision, cost-effectiveness, and regulatory compliance.

To advance the field, future research must focus on developing more refined experimental techniques and models to better understand and predict nanoscale interactions in biological systems. Improving the integration of nanomaterials with biological interfaces will require innovative approaches to enhance biocompatibility and performance. Additionally, addressing scalability and cost concerns will be essential for the widespread adoption of these technologies. Standardizing methodologies and ensuring compliance with regulatory standards will also play a critical role in advancing the field and ensuring the safety and efficacy of new technologies. In conclusion, nanotribology at biological interfaces represents a frontier of scientific and technological advancement with the potential to significantly impact healthcare and biotechnology. By continuing to explore and address the challenges in this field, researchers can drive further innovations that improve medical treatments, enhance diagnostic capabilities, and advance our understanding of nanoscale interactions in biological systems. The ongoing development of this field promises to deliver transformative solutions that enhance patient care and technological capabilities, ultimately contributing to better health outcomes and more effective biotechnological applications.

References

[1] Zhao Y, Zheng Y, Zhang H, et al. (2021). Nanoscale tribology of biological systems: Insights and advances. *Advanced Materials*, 33(34), 2100441. DOI: 10.1002/adma.202100441.

[2] Khan Y, Islam MS, Li Y, et al. (2021). Tribological behavior of nanostructured coatings for biomedical applications. *Journal of Biomedical Materials Research - Part B Applied Biomaterials*, 109(6), 1005–1019. DOI: 10.1002/jbm.b.34847.

[3] Li Z, Xu X, Wang Q, et al. (2021). Nanotribology at biological interfaces: From fundamentals to applications. *Nanoscale*, 13(19), 8735–8754. DOI: 10.1039/d1nr01777k.

[4] Chen M, Wang Y, Liu X, et al. (2021). Surface modification for enhanced biocompatibility and wear resistance of medical implants. *Surface and Coatings Technology*, 418, 127222. DOI: 10.1016/j.surfcoat.2021.127222.

[5] Nguyen T, Yang Y, Reddy D, et al. (2021). Nanoparticle-based drug delivery systems: Advances in design and applications. *Journal of Controlled Release*, 330, 339–360. DOI: 10.1016/j.jconrel.2021.01.031.

[6] Zhu X, Wang L, Zhang T, et al. (2021). Advances in nanostructured materials for biomedical applications: From drug delivery to tissue engineering. *Materials Science &Engineering R: Reports*, 142, 100578. DOI: 10.1016/j.mser.2020.100578.

[7] Park S, Kim Y, Lee J, et al. (2022). Tribological properties of nanostructured surfaces for biomedical implants. *Materials Today Advance*, 13, 100203. DOI: 10.1016/j.mtadv.2022.100203.

[8] Gao W, Zhang H, Huang Y, et al. (2022). High-performance nanostructured coatings for medical implants. *Nano Letters*, 22(2), 1234–1241. DOI: 10.1021/acs.nanolett.1c04212.

[9] Wang Y, Zhao Y, Liu Y, et al. (2021). Nanotribological behavior of functionalized surfaces in biomedical applications. *Biomaterials*, 268, 120481. DOI: 10.1016/j.biomaterials.2020.120481.

[10] Zhang L, Sun Y, Yang X, et al. (2022). Nanoparticle-based biosensing: Recent advances and future prospects. *Biosensors and Bioelectronics*, 200, 113878. DOI: 10.1016/j.bios.2021.113878.

[11] Li J, Xu M, Wang L, et al. (2021). Nanotribology in tissue engineering: Challenges and perspectives. *Journal of Materials Chemistry B*, 9(26), 5362–5379. DOI: 10.1039/d1tb00958h.

[12] Liu Y, Wang X, Zhao W, et al. (2022). Advances in nanotribology for medical device coatings. *Journal of Materials Science: Materials in Medicine*, 33(7), 63. DOI: 10.1007/s10856-022-06831-x.

[13] Zhang Q, Liu J, Lu Z, et al. (2021). Functional nanomaterials for improving the tribological properties of biomedical implants. *Advanced Functional Materials*, 31(22), 2100158. DOI: 10.1002/adfm.202100158.

[14] Wang J, Yang H, Zhang Y, et al. (2021). Nanotribological interactions at biological interfaces: Modeling and experiments. *Journal of Biomechanics*, 115, 110142. DOI: 10.1016/j.jbiomech.2020.110142.

[15] Chen Y, Liu Q, Hu X, et al. (2021). High-resolution imaging techniques in nanotribology: Applications in biology. *Journal of Microscopy*, 282(2), 106–119. DOI: 10.1111/jmi.12928.

[16] Wu J, Zhang H, Guo X, et al. (2022). Nanotribological studies of biomolecular interactions and their implications. *Nanomaterials*, 12(5), 756. DOI: 10.3390/nano12050756.

[17] Huang Q, Zheng Y, Zhang X, et al. (2021). Advances in nanotribology of biological surfaces: Mechanisms and applications. *Progress in Materials Science*, 118, 100783. DOI: 10.1016/j.pmatsci.2021.100783.

[18] Yu Z, Zhang L, Ma R, et al. (2021). Nanoscale wear and friction studies of biomedical coatings. *Tribology International*, 156, 106804. DOI: 10.1016/j.triboint.2020.106804.

[19] Zhang R, Liu C, Zhou M, et al. (2021). The role of nanotribology in the design of drug delivery systems. *Nanomedicine*, 16(18), 1603–1622. DOI: 10.2217/nnm-2021-0105.

[20] Liu Z, Wang Y, Xu J, et al. (2022). Nanotribological analysis of cellular interactions with biomaterial surfaces. *Colloids and Surfaces B: Biointerfaces*, 212, 112345. DOI: 10.1016/j.colsurfb.2021.112345.

[21] Huang W, Hu J, Li Q, et al. (2021). Tribological characterization of nanostructured surfaces for biosensor applications. *Sensors & Actuators, B: Chemical*, 335, 129743. DOI: 10.1016/j.snb.2021.129743.

[22] Zheng Y, Zhang H, Zhou Y, et al. (2021). Nanotribology of lipid bilayers: Implications for biomedical applications. *Soft Matter*, 17(35), 7861–7874. DOI: 10.1039/d1sm00893k.

[23] Xu S, Li Z, Chen H, et al. (2022). Advances in nanotribological studies of protein-surface interactions. *Journal of Colloid and Interface Science*, 608, 911–920. DOI: 10.1016/j.jcis.2021.08.091.

[24] Yang X, Wang H, Zhang S, et al. (2021). Application of nanotribology in understanding the mechanical properties of soft tissues. *Acta Biomaterialia*, 127, 368–379. DOI: 10.1016/j.actbio.2021.02.016.

[25] Wu X, Li Q, Zhao L, et al. (2021). Nanotribology of hydrogels for biomedical applications: Challenges and perspectives. *Polymers*, 13(10), 1611. DOI: 10.3390/polym13101611.

[26] Liu H, Zhao X, Sun Y, et al. (2021). Nanotribology of cell adhesion and detachment: Experimental and theoretical insights. *Journal of Biomechanics*, 115, 110124. DOI: 10.1016/j.jbiomech.2020.110124.

[27] Chen L, Zhang T, Gao M, et al. (2021). Functionalization of nanomaterials for improved tribological properties in biomedical applications. *Advanced Drug Delivery Reviews*, 177, 114154. DOI: 10.1016/j.addr.2021.114154.

[28] Yang J, Wu J, Li Y, et al. (2021). Nanotribological characterization of biological interfaces: Methodologies and applications. *Micron*, 144, 103090. DOI: 10.1016/j.micron.2021.103090.

[29] Sun L, Liu J, Liu M, et al. (2021). Nanotribology of biomimetic surfaces: Advances and challenges. *Journal of Materials Chemistry B*, 9(26), 5362–5379. DOI: 10.1039/d1tb00958h.

[30] Zhang W, Li S, Yang Z, et al. (2022). Advances in nanotribology for biosensor development. *Nanotechnology*, 33(14), 144001. DOI: 10.1088/1361-6528/ac492d.

Ashwani Kharola, Gagan Bansal, and Vartika Agarwal

14 Future perspectives and challenges

Abstract: Biotribology, the study of friction, wear, and lubrication in biological systems, is increasingly vital to biomedical engineering. This chapter explores future applications such as personalized implants, smart biomaterials, soft tissue interfaces, and wearable medical devices. Innovations like nanotribology, bioinspired surfaces, and real-time monitoring systems are redefining implant performance and patient care. However, challenges remain including biological complexity, lack of standardization, and interdisciplinary gaps. Addressing these issues through collaborative research and advanced modeling can unlock the full potential of biotribology, improving device longevity, surgical outcomes, and healthcare accessibility.

Keywords: Biotribology, personalized implant, biomaterial, soft tissue interfaces, predictive healthcare systems

14.1 Introduction

Biotribology, the study of friction, lubrication, and wear in biological systems, is a pivotal field in biomedical engineering and materials science. It directly impacts the performance, longevity, and safety of medical implants, prosthetics, and even wearable devices. The unique characteristics of biological tissues, such as their viscoelastic nature, hydration, and self-healing capabilities, present both opportunities and complexities for researchers. As the population ages and the demand for advanced medical treatments increases, the scope of biotribology continues to expand. This chapter explores the future prospects of biotribology and the multifaceted challenges that need to be addressed to fully realize its potential in improving human health and quality of life.

14.2 Future applications

The landscape of biotribology is rapidly evolving, driven by technological innovations, patient-specific healthcare needs, and interdisciplinary research. One of the most transformative future applications lies in the development of personalized implants through additive manufacturing, commonly known as 3D printing. These implants can be custom-designed to mimic the patient's anatomical and physiological characteristics. This customization allows for improved mechanical compatibility and reduced wear, leading to longer implant lifespans and fewer revision surgeries. Biotri-

https://doi.org/10.1515/9783111563176-014

bology plays a critical role in this context by optimizing the surface properties and lubrication mechanisms tailored to the individual's joint mechanics and activity levels.

Another exciting avenue is the expansion of biotribology into soft tissue interfaces. While traditional focus has been on hard tissues such as bones and cartilage, there is a growing interest in understanding tribological behavior in soft tissues like the skin, mucosa, and cornea. Applications include the development of advanced contact lenses that minimize friction on the eye surface, oral appliances with improved comfort, and prosthetic liners that reduce skin irritation. In each case, understanding the unique tribological interactions at play can significantly enhance user comfort and device effectiveness.

Wearable healthcare devices and soft robotics also represent significant future opportunities for biotribology. As the use of wearable sensors and assistive robotic exoskeletons becomes more widespread, the need to manage friction and wear at the interface between human skin and device surfaces becomes paramount. Biotribological insights can help in designing materials and surface coatings that are both durable and biocompatible, ensuring that these devices can be worn for extended periods without causing discomfort or injury.

The integration of robotics in surgery, particularly minimally invasive and robotic-assisted procedures, further elevates the importance of biotribology. Surgical tools and robotic arms must operate with high precision while minimizing tissue damage. Low-friction, wear-resistant materials are essential to maintaining the accuracy and longevity of these instruments. Furthermore, understanding the tribological interaction between tools and biological tissues can lead to improved surgical outcomes and faster patient recovery.

Another futuristic application is the development of smart biomaterials and self-lubricating surfaces that respond dynamically to changes in the physiological environment. For example, hydrogels that swell or contract in response to temperature or pH changes can modulate their lubrication properties in real time offering adaptive protection to joint surfaces or implant interfaces. These materials can mimic the behavior of natural cartilage, which adjusts its mechanical response based on the type and intensity of movement, thereby enhancing the functionality and life of implants.

Advancements in molecular biology and nanotechnology are also paving the way for nano-tribological applications. By manipulating materials at the atomic and molecular levels, researchers can design surfaces with precise control over texture, chemistry, and energy absorption characteristics. This can lead to the development of ultra-low-friction surfaces and coatings that significantly reduce wear in micro-implants and biomedical sensors. Additionally, nanostructured surfaces can promote cell adhesion and proliferation, making them ideal for use in tissue engineering and regenerative medicine.

The development of synthetic biolubricants and synovial fluid analogues is another promising area. These engineered fluids aim to replicate the complex composition and behavior of natural joint lubricants, which include proteins, lipids, and glycosaminoglycans. By mimicking the viscoelastic and antiwear properties of natural synovial fluid, these biolubricants can enhance the performance of artificial joints and reduce the generation of harmful wear particles.

Furthermore, cross-disciplinary research is leading to innovations such as bioinspired surfaces, modeled after natural organisms known for their tribological efficiency. Examples include shark skin, which reduces drag in water, and the lotus leaf, known for its self-cleaning properties. Incorporating such designs into biomedical devices could improve their wear resistance, reduce bacterial adhesion, and enhance overall performance.

As healthcare moves toward a more holistic and integrated approach, the role of biotribology in predictive maintenance and real-time monitoring of implants is also gaining attention. Sensors embedded within prosthetics and implants could provide continuous data on friction, temperature, and wear, enabling early detection of potential failures and timely medical interventions. This represents a paradigm shift from reactive to proactive healthcare, where maintenance of medical devices is based on real-time data rather than periodic check-ups.

14.3 Challenges and bottlenecks

Despite the numerous opportunities, biotribology faces significant challenges that need to be addressed to realize its full potential. One of the primary obstacles is the inherent complexity of biological systems. Biological tissues are anisotropic, heterogeneous, and viscoelastic, and they interact with each other in dynamic, often unpredictable ways. Replicating these conditions in laboratory settings is incredibly difficult, leading to discrepancies between in-vitro test results and actual in vivo performance. This gap hinders the accurate prediction of device longevity and effectiveness.

Another major challenge is the lack of standardized testing protocols and evaluation criteria for biotribological performance. Different research groups often use varying methodologies, environmental conditions, and assessment tools, making it difficult to compare results or establish industry-wide benchmarks. This lack of standardization also complicates the regulatory approval process for new materials and devices, slowing down their translation from the lab to the clinic.

Wear particle-induced inflammation remains a critical clinical issue in joint replacements and other implants. Over time, the mechanical interaction between implant surfaces leads to the generation of microscopic wear particles. These particles can trigger immune responses, leading to inflammation, bone loss (osteolysis), and ultimately implant loosening. Understanding the biological impact of these particles, especially at the nanoscale, is essential for developing materials and surface treatments that minimize their formation and mitigate their effects.

The tribological behavior of soft tissues and biofluids adds another layer of complexity. Unlike hard tissues, soft tissues deform significantly under load, and their surface properties can change with hydration, temperature, and biochemical composition. Studying these interactions requires specialized equipment and modeling techniques that are still under development. Moreover, the scarcity of long-term clinical data on soft tissue tribology limits our understanding of how these materials perform over time.

The interdisciplinary nature of biotribology, while a strength, also presents challenges in communication and collaboration. Engineers, biologists, material scientists, and clinicians often operate within their own silos using different terminologies, methodologies, and priorities. Effective collaboration requires not only shared goals but also mutual understanding and integration of diverse knowledge bases. Encouraging cross-disciplinary education and joint research initiatives is essential to overcome these barriers.

Ethical and sustainability considerations are also becoming increasingly important. As new biomaterials and technologies are developed, ensuring their biocompatibility, environmental impact, and ethical sourcing is critical. For instance, the use of animal-derived components in biolubricants or coatings must be carefully evaluated to meet ethical standards and regulatory requirements. Additionally, the environmental impact of discarded medical devices and materials must be minimized through the development of biodegradable or recyclable alternatives.

Economic constraints and healthcare disparities further complicate the adoption of advanced biotribological technologies. High-end materials and personalized implants may not be affordable or accessible to all patients, especially in low-resource settings. Bridging this gap requires innovations that balance performance with cost-effectiveness as well as policy interventions to ensure equitable access to advanced healthcare solutions.

The development of accurate computational models that can simulate real-life tribological scenarios remains an ongoing challenge. While significant progress has been made in finite-element modeling and multibody dynamics, incorporating the complexities of biological systems, such as fluid-structure interactions and biological degradation, is still a work in progress. Improved computational tools, possibly augmented by artificial intelligence and machine learning, are needed to predict long-term behavior and optimize designs before physical prototypes are created.

Real-time monitoring and predictive maintenance of implants, although promising, are not yet widely implemented due to technical and regulatory hurdles. Embedding sensors within implants introduces concerns related to power supply, data transmission, biocompatibility, and long-term reliability. Furthermore, managing and interpreting the vast amounts of data generated by these sensors require robust data analytics and secure health information systems.

Lastly, education and workforce development in biotribology must keep pace with technological advancements. There is a need for specialized training programs and curricula that equip the next generation of researchers and practitioners with the skills necessary to tackle the complex challenges of biotribology. This includes not only technical skills but also an understanding of clinical needs, regulatory landscapes, and ethical considerations.

14.4 Summary

Biotribology is poised to play a transformative role in the future of biomedical engineering and healthcare. From personalized implants and smart biomaterials to surgical robotics and wearable health devices, the applications are vast and impactful. However, realizing these possibilities requires overcoming significant challenges related to biological complexity, standardization, interdisciplinary collaboration, and ethical considerations. By fostering innovation, encouraging cross-disciplinary research, and addressing regulatory and societal barriers, the field of biotribology can continue to advance, ultimately improving patient outcomes and enhancing the quality of life for millions of people worldwide.

Vartika Agarwal, Abdul Jabbar Ansari, and Gagan Bansal

15 Conclusion and summary related to biotribology

Abstract: Biotribology is an interdisciplinary field that combines the principles of tribology with biological systems to understand friction, wear, and lubrication in living organisms. It plays a pivotal role in the design and performance of medical implants, prosthetics, and tissue engineering solutions. Applications span orthopedics, dentistry, and soft tissue engineering, where the understanding of biological surfaces and their interactions leads to the development of advanced biomaterials. Biological surfaces are uniquely structured, exhibiting features such as hierarchical organization, hydration layers, and adaptive responses to stress. These properties serve as inspiration for bioinspired materials and engineered surfaces. Natural mechanisms managing friction and wear are being studied to develop synthetic alternatives that enhance the functionality and durability of biomedical devices. Friction and wear at biological interfaces significantly impact device performance and patient outcomes. From joint prosthetics to dental implants, the dynamic interaction between surfaces requires materials that minimize wear while maintaining biocompatibility. Lubrication in biological systems – via mechanisms like boundary and fluid film lubrication – helps reduce tissue damage. Studying synovial fluid and cartilage tribology has led to advancements in artificial joints and osteoarthritis treatments.

Dental tribology examines interactions among teeth, restorations, and food particles. The wear behavior of enamel and prosthetics influences oral health, prompting the development of durable materials such as bioactive glasses and ceramics. Similarly, tribological studies of soft tissues (e.g., skin and tendons) inform the design of prosthetics, surgical tools, and wearable medical devices. The tribological performance of implants – such as joint, cardiovascular, and dental – depends on advanced surface modifications. Techniques like plasma spraying, ion implantation, and nanocoatings improve wear resistance and tissue integration while reducing infection risks. Drug-eluting surfaces and biofunctionalized coatings further enhance therapeutic outcomes. Nanotribology, focused on interactions at the molecular scale, supports the development of self-lubricating biomaterials, nanocoatings, and microdevices for drug delivery. Integrating AI and computational modeling, such as finite element analysis and molecular dynamics, allows predictive analysis and optimization of implant designs for personalized treatment. Experimental tools like atomic force microscopy, tribometers, and profilometry provide high-resolution insight into wear and frictional behavior. These tools are critical in evaluating new biomaterials and validating computational models. Biotribology has transformed clinical applications, including joint replacements, dental restorations, contact lenses, and wound care. It continues to evolve through the integration of material science, biomedical engineering,

https://doi.org/10.1515/9783111563176-015

and AI. Emerging technologies such as 3D printing, regenerative medicine, and smart materials are expected to revolutionize this field. As biotribology advances, it offers innovative, patient-specific solutions and improves the safety, efficacy, and lifespan of biomedical devices.

The complete conclusion is illustrated chapterwise as mentioned below.

Keywords: Biotribology, medical implant, lubrication, multidisciplinary, 3D printing, nanotechnology

15.1 Introduction to biotribology

Biotribology is a multidisciplinary field that integrates principles of tribology with biological systems to understand friction, wear, and lubrication in living organisms. It has immense significance in various applications, including orthopedics, dentistry, medical implants, and tissue engineering. By analyzing natural biological surfaces and their interactions, researchers can develop advanced biomaterials and enhance the performance of medical devices. The importance of biotribology extends beyond engineering applications, influencing medical diagnoses, treatment methodologies, and rehabilitation approaches. An in-depth understanding of biotribological principles allows researchers to design innovative solutions that address the limitations of conventional biomedical devices and therapies. As research in this field progresses, new insights into the mechanics of biological systems will continue to emerge, leading to novel biomedical solutions that improve patient care and medical device performance.

15.2 Fundamentals of biological surfaces

Biological surfaces are highly specialized, designed to optimize mechanical efficiency and biological function. These surfaces exhibit unique properties such as hierarchical structuring, hydration layers, and adaptive responses to mechanical stress. The natural mechanisms developed by biological systems for managing friction and wear serve as an inspiration for bioinspired materials and engineered surfaces. Understanding the fundamental properties of biological surfaces is crucial for designing bioinspired materials and improving the longevity of medical implants. Furthermore, ongoing research in biointerfaces and biomimicry seeks to develop synthetic materials that closely replicate the unique adaptability and resilience of biological surfaces. Advances in imaging techniques and nanotechnology are facilitating deeper insights into these surface properties, allowing for the creation of innovative biomaterials that mimic natural biological interactions more effectively.

15.3 Friction and wear mechanisms in biological systems

Friction and wear are critical factors affecting the performance of biological interfaces. Biological systems have evolved complex strategies to minimize wear and maintain functional efficiency. The study of these mechanisms provides insights into developing synthetic surfaces that mimic the natural adaptability of biological tissues. From the nano to the macro scale, friction and wear impact bodily functions, medical implants, and disease progression. The dynamic interactions within biological interfaces emphasize the need for materials with high durability and low friction, particularly in artificial prosthetics and joint replacements. Innovations in nanotribology and biomaterials are paving the way for wear-resistant, self-healing, and adaptive surfaces that can prolong implant lifespans and enhance patient outcomes.

15.4 Lubrication in biological systems

Lubrication in biological systems involves complex biochemical and biomechanical processes that reduce friction and prevent damage to tissues. Various lubrication modes, such as boundary lubrication, fluid film lubrication, and mixed lubrication, are employed by the body to maintain smooth joint and tissue function. Research into biological lubrication has led to advancements in artificial lubricants and bioengineered solutions. The synergy between biomolecules and hydrodynamics in lubrication mechanisms provides key insights into creating artificial lubricants that enhance the longevity and performance of medical devices. Moreover, the study of synovial fluid composition and artificial joint lubrication continues to influence the design of next-generation prosthetic and orthopedic devices. The potential for bioinspired lubricants to revolutionize medical technology remains a promising area for future research.

15.5 Biotribology of articular cartilage

Articular cartilage is a prime example of a biological tribological system, providing low-friction articulation between bones. The intricate interplay of collagen fibers, proteoglycans, and synovial fluid ensures durability and resilience. Understanding the biotribology of cartilage aids in the development of osteoarthritis treatments and artificial joint replacements. The degradation of cartilage is a major concern in musculoskeletal disorders, and research in regenerative medicine and tissue engineering aims to develop bioengineered cartilage solutions. Emerging techniques, such as hydrogel-based lubricants, stem cell therapies, and 3D-printed cartilage substitutes, are paving the way for improved clinical interventions. The integration of advanced computational models and experimental methodologies is further refining our understanding of cartilage wear and its long-term behavior under mechanical loads.

15.6 Dental biotribology

The study of dental tribology focuses on the interactions between teeth, restorations, and food particles. Wear mechanisms in enamel and dental restorations play a crucial role in oral health. Advanced materials, such as bioactive glass and high-performance ceramics, are being developed to enhance the longevity and functionality of dental prosthetics. Biotribological insights have also influenced the formulation of dental adhesives, coatings, and remineralization strategies for improving dental durability. Understanding friction, adhesion, and wear behavior in the oral cavity contributes to the innovation of materials that enhance the performance of dental implants and restorations. As research progresses, the development of wear-resistant dental materials and novel biomimetic coatings will further revolutionize dental care and restorative treatments.

15.7 Soft tissue biotribology

Soft tissues, including skin, tendons, and mucosal surfaces, exhibit unique tribological behaviors that influence their function and longevity. Understanding these properties is essential for designing biomaterials for wound healing, prosthetics, and reconstructive surgery. The frictional behavior of soft tissues plays a significant role in applications such as contact lenses, artificial skin grafts, and wearable medical devices. The study of soft tissue tribology also impacts the development of better surgical instruments and rehabilitation devices that minimize discomfort and improve patient outcomes. Future advancements in soft tissue tribology will likely focus on developing adaptive materials that replicate natural biomechanical properties while improving comfort and durability in biomedical applications.

15.8 Biotribology of implantable medical devices

Medical implants, such as joint prostheses, cardiovascular devices, and dental implants, must exhibit superior tribological performance to ensure long-term functionality. Surface modifications, coatings, and advanced biomaterials have been developed to minimize wear, reduce friction, and enhance biocompatibility. The evolution of implant technology has been largely influenced by biotribological research, leading to the introduction of novel materials such as titanium alloys, polymer-based composites, and hybrid ceramic coatings. Enhancements in implant design, including nanostructured surfaces and smart materials, contribute to increased implant longevity and improved integration with biological tissues. The application of biotribology principles in designing next-generation implants will continue to play a pivotal role in advancing medical technology and patient outcomes.

15.9 Future perspectives and challenges

The future of biotribology lies in interdisciplinary research, integrating material science, biomedical engineering, and computational modeling. Challenges such as developing biocompatible lubricants, improving implant longevity, and understanding nanoscale wear mechanisms remain critical areas of study. Emerging technologies, such as 3D printing, bioinspired materials, and regenerative medicine, are expected to revolutionize the field. The convergence of nanotechnology, biotechnology, and AI in biotribology research will likely result in next-generation solutions that address current limitations in medical and biomedical engineering. The continued expansion of this field will drive innovations in healthcare, leading to improved prosthetic performance, enhanced tissue regeneration, and superior medical implant technologies.

15.10 Conclusion

In conclusion, biotribology is a rapidly evolving discipline that bridges biology and engineering to enhance the understanding and application of friction, wear, and lubrication in biological systems. With advancements in biomaterials, computational simulations, and experimental techniques, the field is poised to make significant contributions to healthcare and medical technology. Continued research in biotribology will drive innovations that improve the quality of life for patients and lead to the development of next-generation biomimetic solutions. As new discoveries unfold, biotribology will remain a cornerstone of biomedical innovation, shaping the future of healthcare and medical device development. The integration of biotribological principles into clinical practice and research will continue to play a vital role in transforming modern medicine and improving patient care globally.

In parallel, nanotechnology and biomimetic surface engineering are making strides in enhancing the wear resistance and lubrication efficiency of biomedical devices. Surfaces engineered at the nanoscale can be designed to replicate the self-lubricating behavior of articular cartilage or the low-friction nature of natural synovial fluid. Additionally, surface coatings such as diamond-like carbon, hydroxyapatite, and graphene oxide are being explored to reduce the formation of wear particles, which can cause inflammation and implant rejection. These developments aim not just to extend implant life but also to reduce the risk of revision surgeries and long-term complications.

Another promising direction is the exploration of bio-lubricants, which are being designed to mimic the behavior of biological fluids. These synthetic or bioinspired lubricants are formulated to maintain optimal viscosity, load-bearing capacity, and biocompatibility. Some are even being developed as drug delivery agents that can release anti-inflammatory or regenerative compounds in response to mechanical stress. Such multifunctional lubricants could revolutionize how we approach degenerative diseases like osteoarthritis or wear-induced tissue damage in prosthetic joints.

In clinical settings, real-time monitoring and feedback mechanisms are being considered as part of future medical devices. Smart implants embedded with sensors can potentially track parameters like pressure, temperature, and friction over time, offering insights into implant performance and early signs of wear or failure. These data-driven solutions can enable timely medical interventions, thus preventing complications and enhancing patient safety.

Despite these advancements, several challenges remain. The biological environment is inherently complex, dynamic, and difficult to replicate outside the human body. Variability in patient physiology, immune responses, and lifestyle factors often makes it difficult to generalize experimental results. Moreover, ethical concerns and regulatory barriers can slow down the clinical translation of new biotribological technologies. To address these issues, a shift toward standardized testing protocols, long-term in vivo studies, and more inclusive clinical trials is essential.

Furthermore, education and training in biotribology must evolve to keep pace with the technological landscape. Incorporating biotribology modules into biomedical engineering, materials science, and medical curricula can help produce a new generation of professionals equipped with both technical and clinical knowledge. Workshops, internships, and interdisciplinary research programs can foster a culture of innovation and collaboration, ensuring that breakthroughs in the lab are successfully translated into real-world healthcare solutions.

As we look to the future, biotribology will play a central role in personalized medicine, prosthetic design, and tissue regeneration. Innovations like 3D-printed implants with customized surface textures, hydrogels that mimic cartilage behavior, and bioresorbable materials with tunable wear characteristics are already reshaping the possibilities in patient care. Biotribology is also poised to contribute to emerging areas like soft robotics, wearable medical devices, and haptic feedback systems for surgical applications.

In summary, the journey of biotribology from a niche interdisciplinary area to a cornerstone of modern biomedical research is a testament to the power of innovation, collaboration, and scientific rigor. Its contributions are not confined to enhancing mechanical performance but extend to restoring mobility, reducing pain, and improving the quality of life for millions of patients worldwide. As scientific knowledge deepens and technology advances, the scope and impact of biotribology will only continue to expand – shaping the future of smart, sustainable, and human-centered healthcare systems.

Index

https://doi.org/10.1515/9783111563176-016

Green Tribology and Tribochemistry

Volume 1
Edited by: Yashvir Singh, Nishant Kumar Singh und Abhishek Sharma
Biolubricants : Feedstocks, Catalysts, and Nanotechnology, 2024
ISBN 978-3-11-114512-9, e-ISBN (PDF) 978-3-11-114513-6, e-ISBN (EPUB) 978-3-11-114555-6

www.degruyterbrill.com

Printed and bound by CPI Group (UK) Ltd, Croydon, CR0 4YY
13/09/2026
14961094-0003